Advances in Spatial Science

For further volumes:
http://www.springer.com/series/3302

Karima Kourtit • Peter Nijkamp •
Roger R. Stough
Editors

Drivers of Innovation, Entrepreneurship and Regional Dynamics

Springer

Editors
Dr. Karima Kourtit
Prof. Dr. Peter Nijkamp
VU University Amsterdam
Department of Spatial Economics
De Boelelaan 1105
1081 HV Amsterdam, The Netherlands
k.kourtit@vu.nl
p.nijkamp@vu.nl

Prof. Dr. Roger R. Stough
George Mason University
School of Public Policy
University Drive 4400
Fairfax, VA 22030, USA
rstough@gmu.edu

Advances in Spatial Science ISSN 1430-9602
ISBN 978-3-642-17939-6 e-ISBN 978-3-642-17940-2
DOI 10.1007/978-3-642-17940-2
Springer Heidelberg Dordrecht London New York

Library of Congress Control Number: 2011930132

Cover design: eStudio Calamar S.L.

Printed on acid-free paper

Springer is part of Springer Science+Business Media (www.springer.com)

Editorial Preface

The annual Tinbergen Workshop held over the last several years has brought together regional scientists from Europe, North America and Australia – and sometimes beyond – to address issues relating to the general field of innovation, entrepreneurship and regional development. The theme of the June 2009 Workshop was *Creative, Intellectual and Entrepreneurial Resources for Regional Development: Analysis and Policy*. In addressing these issues it is inevitable that we will be focusing our attention on forces and processes in regional development that are largely *endogenous* to a city or region and how policy may play a role to enhance regional growth performance through the roles that institutions and leadership might play in the context of regional development policy to help cultivate creativity, human capital development and innovation and entrepreneurial activity as drivers of economic development to give a city or region its competitive edge.

The present volume is based on a selection of papers presented at the above workshop. It aims to provide an overview of thinking about endogenous forces and processes that may enhance the economic performance of a city or region and the type of empirical evidence that supports the notion that creativity, intellectual and entrepreneurial resources, along with leadership and institutions, are crucial drivers of the regional development process and consequently are key factors differentiating between high and low performing cities and regions.

The various contributions in this volume have been carefully reviewed and may be seen as novel contributions to the emerging field of creative, intellectual and entrepreneurial resources for regional development. The editors wish to thank Elfie Bonke and Ellen Woudstra for their assistance in composing this volume.

Amsterdam, The Netherlands
Fairfax, VA

Karima Kourtit, Peter Nijkamp
Roger R. Stough

Contents

Contributors

George Anastasiou Department of Planning and Regional Development, School of Engineering, University of Thessaly, Pedion Areos, 38 334 Volos, Greece

Jane Andrew PhD Candidate, Australian Institute for Social Research, University of Adelaide Lecturer, Director matchstudio, Art, Architecture and Design School, University of South Australia

Roberto Basile ISAE (Institute for Studies and Economic Analyses), Piazza dell'Indipendenza, 4, 00185 Rome, Italy

Scott Baum Urban Research Program, Griffith University, Nathan, Brisbane, QLD 4111, Australia

William B. Beyers Department of Geography, University of Washington, 426 Smith Hall, Seattle, WA 98195-3550, USA

Ron Boschma Faculty of Geosciences, Department of Economic Geography, University of Utrecht, PO Box 80 115 3508 TC Utrecht, The Netherlands

Martijn J. Burger Department of Applied Economics, Erasmus University Rotterdam, Room H13-27, P.O. Box 17383000 DR Rotterdam, The Netherlands

Roberta Capello Dipartimento BEST, Politecnico di Milano, Piazza Leonardo da Vinci 32, 20133 Milano, Italy

Andrea Caragliu Dipartimento BEST, Politecnico di Milano, Piazza Leonardo da Vinci 32, 20133 Milano, Italy

Shaoming Cheng Professor, Department of Public Administration, Florida International University, 11200 SW 8th Street, Miami, FL 33199, USA

Jenny Grek Jönköping International Business School, P.O. Box 1026, SE-551 11 Jönköping, Sweden

Charlie Karlsson Jönköping International Business School, Jönköping University, P.O. Box 1026, SE-551 11 Jönköping, Sweden

Johan Klaesson Jönköping International Business School, Jönköping University, P.O. Box 1026, SE-551 11 Jönköping, Sweden

Karima Kourtit Department of Spatial Economics, VU University Amsterdam, De Boelelaan 1105, 1081 HV Amsterdam, The Netherlands

William Latham Department of Economics, University of Delaware, Newark, DE 19716, USA

John Martin Centre for Sustainable Regional Communities, Law and Management, La Trobe University, PO Box 199, Bendigo, VIC 3552, Australia

Mark Middleton Division of Resource Management, Regional Research Institute, West Virginia University, P.O. Box 6108, Morgantown, WV 26506-6108, USA

William Mitchell Centre of Full Employment and Equity, University of Newcastle, Callaghan, NSW 2308, Australia

Marcella Nicolini Fondazione ENI Enrico Mattei, Corso Magenta 63, 20123 Milan, Italy

Peter Nijkamp Department of Spatial Economics, VU University Amsterdam, De Boelelaan 1105, 1081 HV Amsterdam, The Netherlands

Apostolos Papadoulis Department of Planning and Regional Development, School of Engineering, University of Thessaly, Pedion Areos, 38 334 Volos, Greece

George Petrakos Department of Planning and Regional Development, School of Engineering, University of Thessaly, Pedion Areos, 38 334 Volos, Greece

Otto Raspe Urban and Regional research centre Utrecht (URU), Utrecht University, P.O. Box 8011 53508 TC, Utrecht, The Netherlands

Laura Resmini Faculty of Political Science and International Relations, Università della Valle d'Aosta, Loc. Grand Chemin 73/75, 11020 Saint Christophe (AO), Italy

Alistair Robson The Institute for Social Science Research, The University of Queensland, Building 31B, St Lucia, QLD 4072, Australia

Peter Schaeffer Assistant Professor, Division of Resource Management and Regional Research Institute, West Virginia University, P.O. Box 6108, Morgantown, WV 26506-6108, USA

Tung-Kai Shyy School of Geography, Planning and Environmental Management, The University of Queensland, Brisbane, QLD 4072, Australia

Pantoleon Skayannis Department of Planning and Regional Development, School of Engineering, University of Thessaly, Pedion Areos, 38 334 Volos, Greece

John Spoehr Australian Institute for Social Research, The University of Adelaide

Robert J. Stimson School of Geography, Planning and Environmental Management, The University of Queensland, Brisbane, QLD, 4072, Australia

Roger R. Stough School of Public Policy, George Mason University, Fairfax, VA 22030-4444, USA

Ryan C. Sutter School of Public Policy, George Mason University, Fairfax, VA 22030-4444, USA

Frank G. van Oort Department of Economic Geography, Utrecht University, P.O. Box 8011 53508 TC, Utrecht, The Netherlands

Hong Yin Department of Economics, University of Delaware, Newark, DE 19716, USA

Junbo Yu Regional Research Institute, 886 Chestnut Ridge Road, PO Box 6825, Morgantown, WV 26505-6825, USA

Junyang Yuan School of Public Policy, George Mason University, PMB 3704, 4450 Rivanna Ln, Fairfax, VA 22030, USA

Part I
Concepts and Models

An Endogenous Perspective on Regional Development and Growth

Roger R. Stough, Robert J. Stimson, and Peter Nijkamp

1 The Evolution of Regional Economic Development Theory

Over the past two decades or so the emphasis in regional economic development theory has shifted from a focus primarily on *exogenous* factors to an increasing focus on *endogenous* factors. Traditional regional economic development approaches were erected on neo-classical economic growth theory, based largely on the Solow (1956, 2000) growth model. The new approach – while recognizing that development is framed by exogenous factors – attributes a much more significant role for endogenous forces. In this context, a suite of models and arguments that broadly convey the *new growth theory* have been directed towards endogenous factors and processes (see, e.g., Johansson et al. 2001).

These developments are of great interest to regional economic development analysts and practitioners for several reasons, including the recognition of the importance of cities and regions in the development process and also because they introduce an explicit *spatial* variable into economic growth theory, which was a mostly ignored element in neo-classical thinking. This evolutionary development is particularly significant as the importance of regions in national economies – and in particular the role of many of the world's mega city regions – has changed considerably since the 1970s as a result of globalization, deregulation, and structural change and adjustment. Understanding these newly recognized processes of change is crucial for analysing and understanding different patterns of regional economic performance and in formulating and implementing regional economic development planning strategy.

R.R. Stough (✉)
School of Public Policy, George Mason University, Fairfax, VA 22030-4444, USA
e-mail: rstough@gmu.ed

R.J. Stimson
School of Geography, Planning and Environmental Management, The University of Queensland, Brisbane, QLD 4072, Australia

P. Nijkamp
Department of Spatial Economics, VU University Amsterdam, De Boelelaan 1105, 1081 HV Amsterdam, The Netherlands

K. Kourtit et al. (eds.), *Drivers of Innovation, Entrepreneurship and Regional Dynamics*, Advances in Spatial Science, DOI 10.1007/978-3-642-17940-2_1,

1.1 The Nature of Regional Development

Stimson et al. (2006: p. 4) have observed that it is often difficult in regional economic development planning strategy formulation and implementation to match desired *outcomes* of regional economic development with the *processes* that create them. That gap in understanding the relationship between the apparent causes and effects of development poses a dilemma for those responsible for managing regional economic development in the making of policies and strategies, and their implementation of plans. The dilemma they face is how to achieve some form of congruence between desired outcomes and appropriate and acceptable economic development tools and processes. The dilemma is further compounded by the frequently unstable and changing nature of economic environments, where "externalities" or exogenous factors (such as exchange rates, new technologies, and foreign competition) increasingly impact the decision-making processes that influence economic policy and strategy in cities and regions.

Blakely (1994) has emphasized how regional economic development needs to be viewed as both a *product* and a *process* but often not by the same groups or actors in the development milieu. For example, economic agents that live, work and invest in regions are those most concerned with economic development outputs or products such as job and wealth creation, investment, quality of life or standards of living and conditions of the work environment. Contrary to this view is the more process orientation of regional scientists, development planners and practitioners where concern focuses on the creation of infrastructure, labor force preparation, human capital and market development. So it is important when considering regional economic development to maintain an awareness of its *product* and *process* aspects.

Regional economic development also is known in terms of *quantitative* and *qualitative* attributes. In that context, and with respect to the benefits it creates, our concern has typically been with the quantitative measurement of such factors as increasing/decreasing wealth and income levels, job creation or employment levels, the availability of goods and services, and improving financial security. At the same time – and especially in recent times – our concern has also been with such qualitative considerations as generating creative capital, creating greater social and financial equity, achieving sustainable development, creating a spread in the range of employment, and gaining improvements in the quality of life. Thus the regional economic development process needs to be informed by both quantitative and qualitative information.

This multi-dimensional aspect of economic development led Stimson et al. (2006) to propose the following definition of regional economic development:

> .... Regional economic development is the application of economic processes and resources available to a region that result in the sustainable development of, and desired economic outcomes for a region and that meet the values and expectations of business, of residents and of visitors. (p. 6)

1.2 Changing Paradigms

Policy for economic development and regional planning strategy has undergone a series of evolutionary changes since World War II, driven by different paradigms of economic thought as shown in Fig. 1. Those paradigms have shaped the way regional and local communities and people think and plan for the future. But much thinking on regional economic development still remains embedded in the paradigms of the 1970s, because of an inherent reluctance of many regions and local communities to pro-actively embrace change.

Consequently, as suggested by Stimson et al. (2006):

> . . . many regions are not re-equipping themselves fast enough to compete effectively in the global age of business and technology of the post-industrial economy. To compete successfully in the global economy, regional organizations and businesses need to understand the implications of the paradigm shifts occurring in economic policy and strategy, and to build the flexible strategic infrastructure to do so. (p. 11)

A summary of the changing paradigms that have shaped regional economic development theory and planning strategy is presented in Fig. 1. It is, however, important to realize that time overlaps between these economic policy and the economic planning strategy paradigms are both deliberate and pragmatic, reflecting the reality of evolutionary changes in the paradigm approaches.

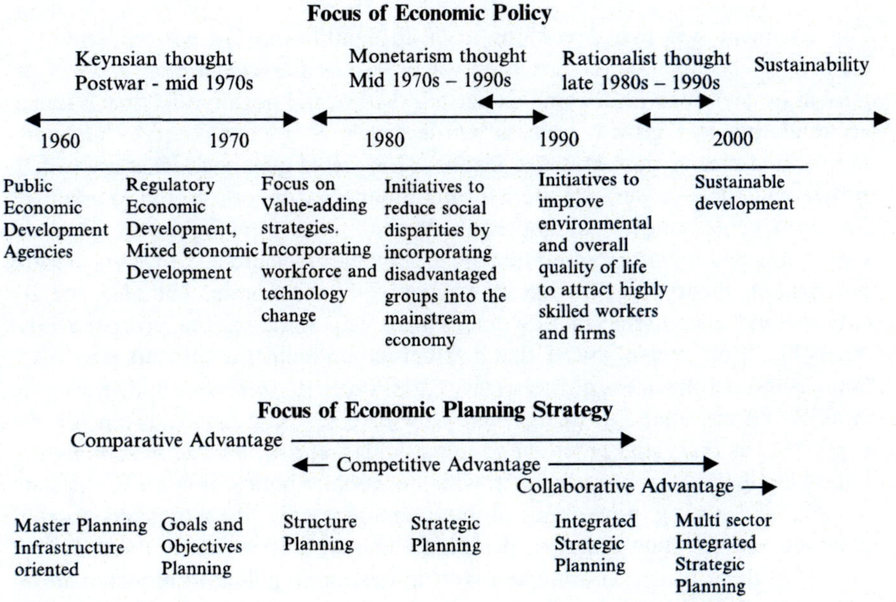

Fig. 1 Changing focus for economic development policy and planning strategy [Source: Stimson et al. (2006: p. 12)]

A detailed discussion of these changing paradigms is available in Stimson et al. (2006: pp. 11–17) and Stimson and Stough (2009). It focuses on a set of important issues all of which are evident in the context of a series of at least five paradigmatic evolutions:

1. *Neo-classical economic growth theory* had served as the basis of regional economic development theory predominantly through providing an understanding of the roles of labour and capital in the production function. Traditional neo-classical growth theory models assumed:

   - The homogeneity of production factors
   - Saw the price mechanism as the underlying adjustment mechanism of the model
   - Emphasized capital accumulation as the net product

   All of which lead to convergence thus eliminating inter-regional, inter-group and inter-sectoral differences over time in the long-run. But there was an inadequate explanation of the role of technology on productivity (Malecki 1991: p. 111).
2. Thus, counter-arguments began to arise, including *polarization theory* as represented early on by the work of Perroux (1950), Myrdal (1957) and Hirschman (1958), in which it was argued that:

   - Production factors are non-homogeneous
   - Markets are imperfect
   - The price mechanism is disturbed by *externalities* and *economies of scale*.

   The argument was that deviations from an equilibrium are *not* corrected by counter effects, but rather that they set off a circular *cumulative process* of growth or decline, with a complex set of positive and negative feedback loops accumulating to a growth process whose direction is fundamentally undetermined. In a *spatial* context, those feedback processes generated what are called *spread* and *backwash* effects, transferring impulses from one region to another. *Spatial structure* could be an important element in that growth process, generating *leading* and *lagging* regions that are highly interdependent. The advocates of polarization theory argued that it was not only economic, but also social, cultural, and institutional factors that explain why some regions prosper while others lag. More recently work that has focused on industrial districts (see, e.g., Scott 1988) and business clusters (see, e.g., Porter 1990; Feser 1998; Karlsson et al. 2005) has added to this evolution of theory which has been placing an explicit focus on spatial process and in particular on *agglomeration economies*.
3. Following the post-war era focus in which economic policy built on Keynesian thought and strong governmental interventions and the associated master planning and structure planning paradigms that were in vogue at the time, by the 1980s there had been a marked shift in economic policy to monetarism or economic rationalism and its focus on the dominance of markets, and associated with that was the rise in regional development policy to a focus on goals and objectives and strategic planning paradigms.

4. Paralleling these paradigm shifts was another shift in thinking on regional economic development from focus on *comparative advantage* associated with international trade theory in economics to a focus on the *competitive advantage* of regions as proposed by Porter (1985, 1986, 1990). And more recently the focus has been shifting to incorporate the notion of *collaborative advantage* (Huxham 1996).
5. Especially in the period since the mid-1970s the processes of *globalisation* have resulted in the emergence of an increasingly borderless economic world with increasingly unrestricted mobility of capital and labour and increasing freedom of trade in merchandise and services. Seemingly the influence of the nation state was reduced in a world where *cities* and particularly *mega-city regions* assumed increasing importance as *strategic hubs* and as the drivers of creativity, innovation and entrepreneurial activity and as they increasingly became the dominant engines of economic growth (Knight and Gappert 1989; Ohmae 1995; Prud-homme 1995; Florida 2002). That created new stresses for both nations and for regions and their governments in developing strategies to find a competitive edge in a globalized economy and a highly competitive and rapidly changing world. There was a considerable shift in regional development planning strategies towards the notions of enhancing regional self-help. More recently the emergence of concern for achieving sustainable development has diversified the goals for regional development and intensified competitive pressures. And it is presenting new challenges for institutional reform, leadership and governance.

1.3 The "New Growth Theory" Approaches

During the 1980s – by which time the focus in economic policy paradigms had shifted to monetarism and economic rationalism – there had been a shift from concerns about developing a regional *comparative advantage* to developing a regional *competitive advantage*, and there had been a shift in regional development planning strategy from *master planning* and *structural planning* to *strategic planning* paradigms and thus a new way of conceptualizing regional economic growth and development had begun to emerge which today is known as the "new growth theory". The evolution of the new growth theory and its focus on endogenous processes and factors as drivers of regional development and growth might be summarised as follows:

1. As early as the late 1970s, Rees (1979) had proposed that *technology* was a prime driver in regional economic development, and since then over the ensuing two to three decades the regional science literature has shown how technology is directly related to traditional concepts of *agglomeration economies* in regional economic development.
2. Economic theorists such as Romer (1986, 1990), Barro (1990), Rebelo (1991), Grossman and Helpman (1991), and Arthur (1994) sought to explain technical

progress in its role as a generator of economic development as an *endogenous* effect rather than accepting the neo-classical view of long term growth being due only to *exogenous* factors. In macroeconomic models of endogenous growth, technological progress was mainly seen as an endogenous process in an economic system, where knowledge is generally embedded in human capital that is enhanced through education, training, creativity, and R&D.

3. Thomas (1975) and later Erickson (1994), among others, showed how technological change was related to the competitiveness of regions. Norton and Rees (1979) and Erickson and Leinbach (1979) showed how the *product cycle*, when incorporated into a spatial setting, may impact differentially on regions through three stages, namely:

    - An innovation stage
    - A growth stage
    - A standardization stage

    Over the course of this transition, production shifts from the original high cost home region to a lower cost location – often one off-shore – which has been hastened through the evolution of the internationalization of the production process. Thus some regions were seen as the *innovators*, while others become the branch plants or recipients of the innovation, and those might even then become innovators via endogenous growth. Markusen (1985) extended the *product cycle theory* of regional development by articulating how profit cycles and oligopoly in various types of industrial organization and corporate development can magnify regional development differentials.

4. The concept of *innovative milieu* (Aydalot 1986; Camagni 1991; Maillat 1991) was formulated to explain the "how, when and why" of new technology generation. That notion linked back to the importance of agglomeration economies and localization economies that had been viewed as leading to the development of *new industrial spaces* (Scott 1988; Porter 1990). In particular Krugman's research (see Krugman 1991, 1995, 1996; Krugman and Venables 1996) led to a greater emphasis on *knowledge* as a tacit and primarily local good and the recognition of it as a driving endogenous self-reinforcing mechanism for regional development. But in discussing innovative industrial milieus, Castells and Hall (1994) had noted that:

    > ... despite all this activity ... most of the world's actual high-technology production and innovation still comes from areas that are not usually heralded as innovative milieus ... the great metropolitan areas of the industrial world. (p. 11)

    However there has been much emphasis on the importance of investment in human capital and its role in regional development (as emphasised by the OECD 2000; 2001).

5. Some theorists, have suggested that it is not just economic but also value and cultural factors – including *social capital* and *trust* – that are important in the rise of technology agglomerations as seen in the Silicon Valley phenomenon, where

collaboration among small and medium size enterprises through networks and alliances and links with universities forge a powerful R&D and entrepreneurial business climate.

6. There has also been a considerable emphasis on the role of *leadership* and *institutions* as factors that can enhance or even act as a catalytic effect in endogenous regional development as demonstrated by Stimson and Stough (2009). As Rees (2001) has pointed out, technology based theories of regional economic development need to incorporate the role of entrepreneurship and leadership, particularly as factors in the endogenous growth of regions, and it is the

> . . . link between the role of technology change and leadership that can lead to the growth of new industrial regions and to the regeneration of older ones. (p. 107)

Thus, the new growth theory models have allowed for and indeed have implied the importance of both agglomeration effects (economies of scale and externalities) and market imperfections, with the price mechanism not necessarily generating an optimal outcome through efficient allocation of resources. And there has been a considerable emphasis on intangible factors such as leadership, institutions, creativity, innovation and entrepreneurship, the endogenous "intangibles" that may enhance the performance of cities and regions.

The processes of capital accumulation and free trade have not necessarily led to convergence of wage and price levels between regions, with positive agglomeration effects tending to often concentrate activity in one or a few regions in many nations through the self-enforcing effects that attract new investment, and that process may be mediated positively by the endogenous "intangibles" we have referred to. The new growth theory actually has allowed for both *concentration* and *divergence* in regional development.

Most importantly, as the spatial distribution of knowledge and its spillovers are now considered to be important success factors in regional development, in framing and implementing regional development strategies it will be crucial for a city or region to fully understand the nature of the geographical patterns of knowledge diffusion and the barriers to access to knowledge as they relate to creativity, innovation and entrepreneurship as catalysts for employment and wealth generation (see, e.g., Keeble and Wilkinson 1999; Acs et al. 2002; Döring and Schnellenback 2006).

2 A Focus on Cities

Earlier in this chapter we made mention of the increasingly important role of cities – and in particular of mega-city regions – as economic hubs and as the magnets for innovation, creativity, entrepreneurship, and leadership for the generation of new business activity. That process is reinforcing the role of agglomeration forces. And it is also probably associated with the rise of an urban culture of increasing diversity and dynamism.

2.1 Analytic Frameworks

A number of analytic frameworks have been proposed and are discussed in the regional science and urban economics literature for understanding the urban dimensions of regional growth and development. They include the following seven (see Nijkamp 2008):

1. A *market-oriented view*, in which the urban rent gradient is the spatial-economic representation of the supply and demand for urban land by different categories of users, while taking into consideration density externalities, as advocated *inter alia* in the classical urban economics theory, in particular, land rent theory.
2. An *ecological socio-cultural view*, in which a blend of sociological and organistic urban viewpoints is offered to explain the structure of urban living and working patterns (advocated in particular by the so-called Chicago School).
3. A *clustering and industrial networks view*, in which urban dynamics is analysed from the perspective of a multiplicity of conflicting interests of urban stakeholders outlined by advocates of the so-called Los Angeles School (such as Scott 1988; Storper 1997).
4. A *politico-economic power view* of cities, in which in a globalizing world large cities act as global control and command centres with centripetal and centrifugal forces all over the world (advocated *inter alia* by Sassen 1994).
5. An *agglomeration advantage view*, in which urban agglomerations generate overwhelming advantages of scale and scope, so that cities become by necessity strong players in the space-economy (advocated *inter alia* by Glaeser 1994).
6. A *creativity view* on urban life, in which cities are the source of rejuvenation, innovation, radical breakthroughs and permanent change, as a result of the leading role of the creative class (see, e.g., Florida 2002).
7. A *virtual cities perspective*, in which in an emerging digital e-society cities act as key nodes in a virtual network and exploit all agglomeration benefits of their territory in a world-wide arena (advocated *inter alia* by Graham and Marvin 1996).

It would seem that there is a need for a more integrated theory of the role of cities and their regional economic development, and in this vein Nijkamp (2008) has proposed a "systems economics" approach which would:

- Offer a multi-disciplinary focus
- Be multi-actor oriented with an emphasis on interactions
- Cover economic systems from micro- to macro-analytical perspectives in a multi-layer way

- Be dynamic and based on evolutionary complexity
- Be analytical-quantitative in nature in order to map out key drivers and their impacts on complex systems

A "systems economics" approach (Antonelli 2011) to the role of cities would have merit because cities are characterized by three particular and distinct systems features, namely:

- Density and proximity externalities
- Dependence on their resource base (physical and cultural)
- Importance of interactive dynamics accruing from learning (including evolutionary and creativity) principles

Let us discuss those in turn:

1. *Density and proximity externalities* are particularly important because of the high degree of concentration of socio-economic and cultural advantages in large cities with their typical pools of skilled labour (particularly knowledge workers), high concentration of ICTs, and the role of knowledge transfers in creating an environment conducive to innovation and entrepreneurial activity. The associated agglomeration economies reduce transaction/interaction costs for firms, and in particular for start-ups.
2. The *resource base of cities*, nowadays is not just their traditional physical resources such as ports and airports that are important, but also their agglomerations of knowledge networks and cultural capital that are crucial. While firms may be increasingly footloose with respect to their city region, many are not so much so with respect to access to the concentrations of ICTs and logistics that city gateways proffer.
3. *Learning and creativity*, are increasingly the "intangibles" that cities possess that are the factors driving the economic growth of cities and mega-city regions. They are expressed in a city's:

   - Institutions
   - Culture
   - High degree of mobility of capital, of codified knowledge, and of (some) human capital

In such an environment learning means the capacity to adapt to rapidly changing competitive circumstances which requires institutional openness, dynamism and flexibility, especially through networks. This "learning regions" paradigm (which we briefly referred to earlier in this chapter) has been discussed extensively in the literature (see, e.g., Florida 1995; Simmie 1997; Camagni and Capello 2005; Camagni 2004; Crevoisier and Camagni 2000) which has emphasised the roles of innovation systems, technology complexes (including the knowledge spillover phenomena), post-Fordism new industry clusters, technology policy, local and regional institutions, and community action (see also Cooke 1998; Maskell and Malmberg 1999; Gertler and Wolfe 2002; Benner 2003). The OECD has been actively promoting this "learning concept" as a central plank in regional development strategy (OECD 2002), and it has in fact become more common for both national and city governments to embrace policies that seek to enhance high technology activity and investment in "smart infrastructure". Of course, this notion links to the importance of investment in education for human capital development and R&D enhancement as a path to national and regional economic growth and development in which the OECD (2001) had claimed that each extra year spent in full-time education with its corresponding approximately 10% rise in human capital would translate into about a 6% increase in per capita output.

2.2 *Creativity and Cities*

The importance of creativity as an economic driver in cities has been emphasised by Florida (2002) and Scott (2003), and there has been an avalanche of studies investigating the features and success conditions of creative urban environments (see, e.g., Heilbrun and Gray 1993; Pratt 1974; Vogel 2001; Hesmondhalgh 2002; Landry 2003; Power and Scott 2004; Markusen 2006). It is thus now common-place for urban development planning strategy to explicitly incorporate initiatives which focus on engendering "creative urban development", which might include a focus on design, culture and the arts as multi-faceted cornerstones for the innovative development of the city. As stated by Nijkamp (2009):

> ... it has become fashionable to regard cultural expressions like arts, festivals, exhibitions, media, communication and advertising, design, sports, digital expression and research as signposts for urban individuality and identity and departures for a new urban cultural industry. (p. 2)

Thus we see "old" cities like London, Liverpool, Amsterdam, Berlin, Barcelona, New York, San Francisco, Sydney or Hong Kong witnessing a profound transformation based on creative cultures. Nijkamp (2009) suggests that:

> .... This new orientation does not only provide a new dynamism for the city, it also has a symbolic value by showing the historical strength of these places as foundation stones for a new and open future. Clearly, blueprint planning of the city has become outdated. Hence, the creative sector has become an important signpost for modern urban planning and architecture, with major implications for both the micro structures of the city and its macro image towards the outer world. (p. 2)

Despite this increasing interest in the dynamics-enhancing impacts of creative activity, as yet an operational conceptualisation of creativity infrastructure and supra-structure has not been developed and that needs to be addressed in applied research. In doing so one is confronted with the considerable challenge of how to translate creative and cultural assets and expressions into commercial values and outcomes (such as value added, employment generation, visitor attraction, etc.). That means that private-sector initiatives are a *sine qua non* for effective and successful urban creativeness strategies. Consequently, critical success conditions for a flourishing urban creativeness strategy might be:

- An orientation towards local identity and local roots that is embedded in the notion of "a sense of place"
- A prominent commitment of economic stakeholders (particularly the private sector)
- The creation of a balanced and appealing portfolio of mutually complementary urban activities

Undoubtedly through their agglomeration advantages cities offer a broad array of business opportunities for creative cultures in which in particular self-employment opportunities and small- and medium-sized enterprises (SMEs) may play a central role in creating new urban vitality, including SMEs arising out of the entrepreneurial activity of a city's diverse ethnic groups and from new immigrants (see Dana 2008). Nijkamp (2009) has suggested that that may be helped through flanking and supporting urban conditions such as:

- Local identity
- An open and attractive urban "milieu" or atmosphere
- Using of tacit knowledge
- The presence of urban embeddedness of new business initiatives
- Access to social capital and networks

Those attributes may provide additional opportunities for a booming urban creativeness culture and an innovative, vital and open urban social ecology.

Certainly urban creativeness presupposes an open and multi-faceted culture and policy. However, the cultural and socio-ethnic pluriformity of modern cities might act to undermine the sense of a common identity with urban fragmentation possibly becoming a challenging new trend (e.g., in restaurants in Miami it is sometimes impossible to use English as a communication language). Nonetheless, cities have always been dynamic as the meeting places for people from diverse cultures and with varied ethnic origins, and with diverse educational backgrounds and talents. Nijkamp (2009) discusses how some cities more than others represent an open "agora" where ideas from diverse cultures and nations come together, and

> ... the challenge is how to turn possible tensions on such a multicultural "agora" into positive synergetic energy. (p. 4)

2.3 *The Critical Success Factors for Sustainable Innovative Development*

The urban fabric of modern cities thus forms a complex system that is influenced by many endogenous and endogenous forces. As Nijkamp (2009) has stated:

> .... In an open world dictated by global competitiveness, it is clear that cities are no longer islands of stable development, but are instead dynamic agglomerations operating in a force field where growth and decline are both possible. Cultural diversity may be a competitive

asset to improve the socio-economic performance of cities, but in case of ethnic-cultural tensions it may hamper a balanced development. (p. 4)

The issue is, then:

Which factors are decisive for a sustainable development of cities that is able to cope with both local and global forces?

It is possible to propose a production function for urban *sustainable innovative development* (SID) based on a *pentagon model* (see Fig. 2) – which has been used in other contexts by Nijkamp et al. (1994) and Capello et al. (1999) – with five critical success factors (CSFs):

1. The availability of *productive capital* (PC): This corresponds to neo-classical production theory where output is determined by the traditional production factors labour and capital.
2. The presence of *human capital* (HC): This refers to the quality of labour input obtained by means of education, training or new skills (e.g., in ICTs) and may be seen as a productivity-enhancing factor. Clearly a balanced distribution of human capital over people is of great importance.
3. The access to *social capital* (SC): This condition comprises interaction and communication between people, socio-economic bonds, social support systems, business networks (formal and informal), relations based on trust, and so on.
4. The usage of *creative capital* (CC): This may be seen as a great ability to cope with challenges and new opportunities, and is reflected in entrepreneurial spirit, new ways of thinking and acting, trend-setting artistic expressions, innovative foresights, and so on. Such a factor is often found in a multicultural urban melting pot.
5. The existence of *ecological capital* (EC): This condition takes for granted that a favourable quality of life, an ecologically benign condition in a city, presence of green space and water, or an attractive living climate (e.g., recreation and entertainment possibilities) contribute significantly to the innovative and sustainable potential of the city.

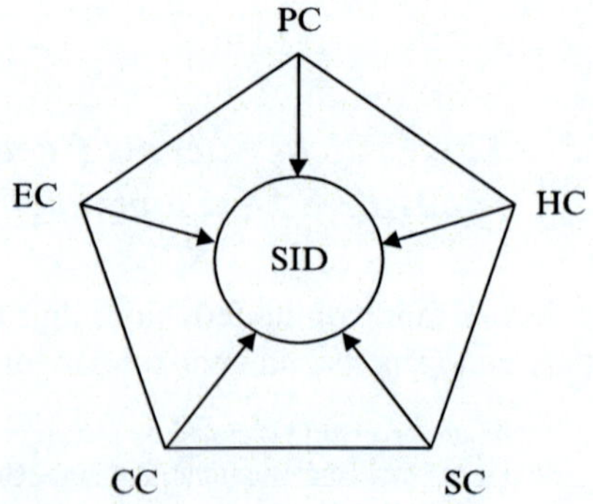

Fig. 2 A pentagon model presentation of urban sustainable creative forces [Source: Nijkamp (2009: p. 5)]

The various pentagon factors can, in principle, be measured and quantified, and next be put in an explanatory econometric model (for an empirical estimation see Capello et al. 1999).

3 Some Implications for Regional Development Planning

We certainly live in a rapidly changing and increasingly competitive world in which uncertainty and risk are considerable. As discussed by Stimson et al. (2006), the challenge facing economic development planners in contemporary times has been how to formulate economic policy that will respond to both:

- Global dynamics
- Sometimes (or often) a national vacuum in adoption of a regionally oriented macro policy in many countries

At one time regions were protected from outside competition, and to some extent their economies could be manipulated by national governments. But that ability has been overwhelmingly compromised as the economic rationalism pursued by many national governments left many cities and regions to fend for themselves. Many cities and regions continued to look to higher levels of government for support and resources to provide economic direction and investment to stimulate economic development. Unfortunately many cities and regions have failed to understand that globalization has left those higher levels of governments relatively weak when it comes to using their inherent power to apply economic and policy mechanisms to enhance the competitiveness of regional economies.

A number of key themes have emerged regarding what constitutes regional growth and development and what drives regional competitiveness. Not surprisingly there have been differences of views among regional economic development scholars, and some of those differences relate to the relative focus given to the roles of *exogenous* forces on the one hand and the roles of *endogenous* processes and factors on the other. But there does now seem to be an almost universal realisation of what Garlick et al. (2006) have referred to as the "institutional embeddedness" of *endogenous* processes and factors in regional development.

Of course *exogenous* factors are likely to remain important to a region's economic performance and how it develops over time; but increasing importance is being placed on *endogenous* forces as determinants of a region's competitiveness. However, regional economic development policy initiatives now tend to be more oriented – as they should be – towards measures that enhance local capacity and capability for a city or region to develop and cope with rapid change in an increasingly competitive global environment. While endogenous growth theory makes mention of leadership, entrepreneurship, and institutional factors, little systematic analysis has occurred to thoroughly conceptualize or, even more, measure their roles as endogenous factors in the development process.

But as discussed by Stimson et al. (2006), in the contemporary policy era of the last decade or two, it would seem that it has been more and more up to regions to develop and use their own devices to compete internationally in order to survive. Thus, it had become increasingly common in regional development planning strategy for there to be a reliance on endogenous processes, and typically that was espoused in regional economic development policy. To do that a region would need first have to have understood what the factors were that set the dynamics of the new economic age that had emerged the late twentieth century. In the wake of the current global financial crisis and recession conditions, it will be interesting to see whether these much changed macro circumstances will set the conditions for a rethinking of that regional self-reliance philosophy and usher in a new era of innovation in institutional arrangements which could incorporate more interventionalist policies in regional development strategy planning.

In the regional growth literature there is no doubt that the strategic importance of knowledge for innovation and entrepreneurship has been increasingly recognised. That has built on the notion of the "learning region" as proposed by Simmie (1997). As discussed by Capello and Nijkamp (2009), in a neoclassical framework of analysis long-range factors such as education, R&D, and technology, have played a critical structural role in the context of the spatial mobility of production factors, which could remove disparities (e.g., in terms of per capita income) in the long-run and, as a result, may equalise factor productivity across a nation's regions. And in the endogenous growth literature we have seen how *knowledge spillovers* and *institutional arrangements* in local regions are widely acknowledged as factors in explaining how knowledge spillovers are spread (as *growth spillovers*), with those knowledge spillovers representing pure externalities that produce non-compensating advantages for the receivers (Nijkamp and van Hemert 2009). But Capello (2009) has pointed to a discrepancy between the private and social optimum which creates the emergence for ad hoc policy interventions.

In the current economic climate of the global financial crisis and recession, Nijkamp and van Hemert (2009) have suggested that in trying to capture the catalytic effect of creativity, innovation and R&D in generating knowledge growth spillovers:

> . . . more than ever there is a role for government in focusing strong and directed efforts to boost the translation of scientific ideas into useful technologies, and to reinforce the base of science skills that drives this innovation. (p. 1)

They go on to say:

> . . .. Currently, there are different forces at play in the science domain that need attention and support from governments. Besides tensions between local and regional demands, the current crisis has highlighted the growing frictions between the individual and societal needs. (p. 1)

The challenges today include the need to revolutionise transport technologies, meet climate-change targets, and secure diversity of energy supply. On a national level, that will require more directed research, education and training innovation to develop the required skills to enact the new technologies, and the active participation of industry in government–science relations to help encourage innovation.

This changing socio-political environment, Hertz (2009) suggests, will require different research disciplines to work together more than ever.

In the context of regional development, Taylor (2009) has referred to the ability to capture ideas and discoveries that flow from research as the main test of whether the UK can recover growth and prosperity. He says that at present the UK does not have the workforce needed to enact new technologies to address the challenges just mentioned, and that is also the case across many if not all countries. While it is a major policy challenge, it does, nonetheless, represent an opportunity for local initiatives to be taken to boost investment in education and R&D, particularly in science and technology.

The notion is that, through what has been termed the "triple helix scenario" (Etzkowitz and Leydesdorff 1996, 1997, 2000), whereby investment in innovation and R&D inputs will lead to greater innovation outputs when they originate from local sources, cities and regions might be able to catalyse future economic growth. Such a notion affirms the existence of a spiral pattern of relations and links between, for example, three major institutional actors in a local environment – industry, university and research institutes, and government. In that relationship the education and research sector tends to have a critical part to play the context of economic growth and regional development in the contemporary knowledge-based economy and in helping societies to address the technological and policy challenges they face with respect to issues such as climate change and achieving more sustainable development. Thus, as Nijkamp and van Hemert (2009) say:

> .... Concentrations of outstanding scientific facilities and activities are very important to create challenging and attractive working conditions and opportunities for talented people. (p. 6)

That reinforces what Florida (2002) had suggested in his work on the "creative class" and the emergence of some cities as centres of creativity. Understanding the institutional barriers that mitigate against achieving this creativity and the associated economic dynamism of a city or region and how to unlock those barriers for the emergence of a "learning region" is an obvious priority in regional development strategy planning if the Pentagon model proposed by Nijkamp (2009) is to be pursued.

A significant issue will be the degree to which regional development and growth across regions will converge or diverge over time as a result of the "institutional embeddedness" of endogenous processes (Garlick et al. 2006). Another will be the nature of the "jumps and anomalies" Nijkamp (2008: p. 6) in urban and regional systems. Endogenous growth theory can help us to understand the complexities of a dynamic space-economy (including the shocks and bifurcations to which it is subject), but contextual drivers and government policies will continue to cause unexpected dynamics.

All these issues call for a solid analysis, based on a blend of theory, methodology, empirics, and policy analysis. The present volume offers a collection of refreshing contributions to modern regional economic growth theory against the background of innovation and entrepreneurship.

References

Acs ZJ, Anselin L, Varga A (2002) Patents and innovation counts as measures of regional production of new knowledge. Res Policy 31:1069–1085

Antonelli C (2011) The Dynamics of *Knowledge Externalities. Localized* Technological Change in Italy, Edward Elgar, Cheltenham

Arthur WB (1994) Increasing returns and path dependency in the economy. University of Michigan Press, Ann Arbor

Aydalot P (1986) Milieux innovateurs en Europe. GREMI, Paris

Barro RJ (1990) Government spending in a simple model of endogenous growth. J Polit Econ 98: S103–S125

Benner C (2003) Learning Communities in a learning region: the soft infrastructure of cross-firm learning networks in Silicon valley. Environ Plann A 35:1809–1830

Blakely EJ (1994) Planning local economic development: theory and practice, 2nd edn. Sage, Thousand Oaks, CA

Camagni R (ed) (1991) Innovation networks: spatial perspectives. Belhaven, New York

Camagni R (2004) Uncertainity, social capital and community governance: the city as a milieu. In: Capello R, Nijkamp P (eds) Urban dynamics and growth: advances in urban economics. Elsevier, Amsterdam, pp 121–152

Camagni R, Capello R (2005) Urban milieux: from theory to empirical findings. In: Boschma RA, Kloosterman RC (eds) Learning from clusters. Springer, Berlin, pp 249–274

Capello R (2009) Spatial spillovers and regional growth: a cognitive approach. Eur Plan Stud 17 (5):643–660

Capello R, Nijkamp P (eds) (2009) Handbook of regional growth and development theories. Edward Elgar, Cheltenham

Capello R, Nijkamp P, Pepping G (1999) Sustainable cities and energy policies. Springer, Berlin

Castells M, Hall P (1994) Technopoles of the world: the making of 21st century industrial complexes. Routledge, London

Cooke P (1998) Introduction. In: Braczyk HJ, Cooke P, Heidenreich M (eds) Regional innovation systems. UCL, London

Crevoisier O, Camagni R (2000) Les milieux urbains: innovation, systèmes de production et ancrage. EDES, Neuchatel

Dana LP (ed) (2008) Handbook of research on ethnic minority entrepreneurship. Edward Elgar, Cheltenham, UK

Döring Th, Schnellenback J (2006) What do we know about geographical knowledge spillover and economic growth? Reg Stud 40(3):375–395

Erickson RA (1994) Technology, industrial restructuring and regional development. Growth Change 25:353–379

Erickson RA, Leinbach T (1979) Characteristics of branch plants attracted to non metropolitan areas. In: Lonsdale R, Seyter HL (eds) Non metropolitan industrialisation. Winston, Washington, DC

Etzkowitz H, Leydesdorff L (1996) The future location of research: a triple helix of university-industry-government relations, II. EAAST Rev 15(4):20–25

Etzkowitz H, Leydesdorff L (1997) Universities and the global knowledge economy, a triple helix of university-industry-government. Pinter, London

Etzkowitz H, Leydesdorff L (2000) The dynamics of innovation: from National Systems and Mode 2 to a Triple Helix of university-industry-government relations. Res Policy 29:109–129

Feser EJ (1998) Enterprises, external economics and economic development. J Plann Lit 12:282–302

Florida R (1995) Toward the learning region. Futures 27(5):527–537

Florida R (2002) The rise of the creative class: and how it's transforming work, leisure, community and everyday life. Basic Books, New York

Garlick S, Taylor M, Plummer P (2006) An enterprising approach to regional growth: the role of VET in regional development, NCVER (National Council for Vocational Education Research). Australian National Training Authority, Adelaide
Gertler MS, Wolfe DA (eds) (2002) Innovation and social learning: institutional adaptation in an era of technological change. Macmillan, Basingstoke
Glaeser EL (1994) Cities, information and economic growth. Cityscape 1:9–47
Graham S, Marvin S (1996) Tele-communications and the city. Routledge, London
Grossman GM, Helpman E (1991) Innovation and growth in the global economy. MIT, Cambridge, MA
Heilbrun J, Gray CM (1993) The economics of art and culture. Cambridge University Press, New York
Hertz N (2009) Cooperation must rule. Nature 457:962–963
Hesmondhalgh D (2002) The cultural industries. Sage, London
Hirschman AO (1958) The strategy of economic development. Yale University Press, New Haven, CT
Huxham C (1996) Collaborative advantage. Sage, Thousand Oaks
Johansson B, Karlsson C, Stough RR (eds) (2001) Theories of endogenous regional growth. Springer, Berlin
Karlsson C, Johansson B, Stough RR (eds) (2005) Industrial clusters and inter-firm networks. Edward Elgar, Cheltenham, UK
Keeble D, Wilkinson F (1999) Collective learning and knowledge development in the evolution of regional clusters of high-technology SMEs in Europe. Reg Stud 33:295–303
Knight RV, Gappert G (eds) (1989) Cities in a global society. Sage, Newbury Park
Krugman P (1991) Geography and trade. MIT, Cambridge, MA
Krugman P (1995) Development, geography and economic theory. MIT, Cambridge, MA
Krugman P (1996) The self-organising economy. Blackwell, Oxford
Krugman P, Venables A (1996) Integration, specialization and adjustment. Eur Econ Rev 40:959–967
Landry C (2003) The creative city. Earthscan, London
Maillat D (1991) The innovation process and the role of the Milieu. In: Bergman EM, Maier G, Tödtling F (eds) Regions reconsidered: economic networks, innovation, and local development in industrialized countries. Mansell, London, pp 103–118
Malecki E (1991) Technology and economic development: the dynamics of local, regional and national competitiveness. Longman Scientific and Technical, Harlow
Markusen A (1985) Profit cycles, oligopoly and regional development. MIT, Cambridge, MA
Markusen A (2006) Urban development and the politics of a creative class. Environ Plann A 38 (10):1921–1940
Maskell P, Malmberg A (1999) Localised learning and industrial competitiveness. Camb J Econ 23(2):167–185
Myrdal G (1957) Economic theory and underdeveloped regions. Duckworth, London
Nijkamp P (2008) XXQ factors for sustainable urban development: a systems economic view. Rom J Reg Sci 2(1):1–34
Nijkamp P (2009) E Pluribus Unum, Region Direct 2(2):567–575
Nijkamp P, van Hemert P (2009) Knowledge Infrastructure and Regional Growth, *Regional Symbiosis* vol. 17, pp 1–10
Nijkamp P, Vleugel J, Maggi R, Masser I (1994) Missing transport networks in Europe. Ashgate, Aldershot
Norton RD, Rees J (1979) The product cycle and the decentralization of North American manufacturing. Reg Stud 13:141–151
OECD (2000) Links between policy and growth: cross-country evidence, draft paper for Working Party 1, Economics Department, OECD
OECD (2001) The well-being of nations: the role of human and social capital, centre for education research and innovation (CERI). OECD, Paris
OECD (2002) Innovative people. OECD, Paris

Ohmae K (1995) The end of the nation state: the rise of regional economics. Free Press, New York
Perroux F (1950) Economic space: theory and application. J Econ 4:90–97
Porter ME (1985) Competitive advantage: creating and sustaining superior performance. Free Press, New York
Porter ME (1986) Competition in global industries. Harvard Business School, Cambridge
Porter ME (1990) The competitive advantage of nations. MacMillan, New York
Power D, Scott A (2004) Cultural industries and the production of culture. Routledge, London
Pratt A (1974) The cultural industries production system. Environ Plann A 29:1953–1974
Prud-homme R (1995) On the economic role of cities, in cities and the new global economy. In: An international conference presented by the OECD and the Australian Government, Australian Government Publishing Service, Canberra, 3, pp 728–742
Rebelo S (1991) Long run policy analysis and long run growth. J Polit Econ 98:S71–S102
Rees J (1979) State technology programs and industry experience in the USA. Rev Urb Reg Dev Stud 3:39–59
Rees J (2001) Technology and regional development: theory revisited. In: Johansson B, Karlsson Ch, Stough RR (eds) Theories of endogenous regional growth. Springer, Heidelberg, pp 94–110
Romer PM (1986) Increasing returns and long-run growth. J Polit Econ 94(5):1002–1037
Romer PM (1990) Endogenous technological change. J Polit Econ 98:S71–S102
Sassen S (1994) Cities in a global economy. Pine Forge, Thousand Oaks
Scott AJ (1988) New industrial spaces: flexible production organization and regional development in North America and Western Europe. Pion, London
Scott AJ (2003) The cultural economy of cities. Sage, London
Simmie J (ed) (1997) Innovation, networks and learning regions? Jessica Kingsley, London
Solow RM (1956) A contribution to the theory of economic growth. Q J Econ 70:65–94
Solow RM (2000) Growth theory: an exposition. Oxford University Press, New York
Stimson RJ, Stough RR (2009) Regional economic development: methodology and analysis, ch. 9. In: Rowe JE (ed) Theories of local economic development: linking theory to practice. Ashgate, Aldershot, pp 169–192
Stimson RJ, Stough RR, Roberts BH (2006) Regional economic development: analysis and planning strategy, Revisedth edn. Springer, Berlin
Storper M (1997) The regional world: territorial development in a global economy. Guilford, New York
Taylor I (2009) Learn to convince politicians. Nature 457(19):958–959
Thomas MD (1975) Growth pole theory, technological change and regional economic growth. Pap Reg Sci Ass 34:3–25
Vogel H (2001) Entertainment industry economics. Cambridge University Press, New York

Interregional Knowledge Spillovers and Economic Growth: The Role of Relational Proximity

Roberto Basile, Roberta Capello, and Andrea Caragliu

1 Introduction

Standard neoclassical growth models (Solow 1956; Mankiw et al. 1992) implicitly assume that the technological progress is characterized by a worldwide global interdependence between economies without frictions. In contrast, recent mainstream contributions to the economic growth literature (Lòpez-Bazo et al. 2004; Ertur and Koch 2007) support the idea that technological interdependence is not homogenous across economies (countries or regions) and depends on their geographical connectivity scheme with other economies, which adds to reflections already envisaged in previous studies (Acs et al. 1994; Anselin et al. 2000). An important feature of technology is its aptitude to spread across borders (Coe and Helpman 1995, and Eaton and Kortum 1996, among others). However, the spatial diffusion of technological knowledge may be geographically bounded, so that the stock of knowledge in one region may spill over into other regions with an intensity which decreases with geographical distance (the so-called "spatial friction" hypothesis).

Based on these assumptions, spatial autoregressive reduced forms of the economic growth model have been derived, in which the growth rate of a region depends not only on its initial conditions and on its own structural characteristics (such as population growth rate and human and physical capital accumulation rates), but also on initial conditions, structural characteristics and growth rates of its neighbors. In particular, by assuming that technical progress depends on the stock of physical capital per worker and of human capital accumulated in other

R. Basile
ISAE (Institute for Studies and Economic Analyses), Piazza dell'Indipendenza, 4, 00185 Rome, Italy

R. Capello and A. Caragliu (✉)
Dipartimento BEST, Politecnico di Milano, Piazza Leonardo da Vinci 32, 20133 Milano, Italy
e-mail: andrea.caragliu@polimi.it

K. Kourtit et al. (eds.), *Drivers of Innovation, Entrepreneurship and Regional Dynamics*, Advances in Spatial Science, DOI 10.1007/978-3-642-17940-2_2,

countries and not merely in the home country, Ertur and Koch (2007) have obtained a growth equation characterized by parameter heterogeneity linked to the geographical location of the economies. In order to test these predictions, spatial econometric tools (such as spatial lag, spatial error and spatial Durbin models) have been largely used in the empirical literature (Lòpez-Bazo et al. 2004; Rey and Janikas 2005; Rey 2004; Rey and Montouri 2004). Some authors have also taken into account the possibility of parameter heterogeneity, using either spatial autoregressive local estimation methods (Ertur and Koch 2007) or spatial autoregressive semiparametric additive models (Basile 2008, 2009).

In this paper we test the hypothesis that geographical proximity is not the only dimension to be considered in order to capture the mechanisms governing knowledge spillovers. As already emphasized in the literature, other forms of proximity must be considered as complementary (or alternative) to physical distance: social proximity (Boschma 2005; Capello 2007, 2009a), organizational proximity (Bellet et al. 1993; Rallet and Torre 1995), institutional proximity (Lundvall and Johnson 1994), technological proximity (‘Cantner and Meder 2007) and specialization proximity (Ciccone 2002; Henderson 2003).

In this study we analyze the joint effect of relational and spatial proximity. The notion of relational proximity is based on a new concept of space, which accounts for the ways in which economic agents potentially interact and for the ways in which this interaction influences learning processes (Capello 2009a). Relational proximity is measured in terms of the difference between trust in two regions. Our assumptions are that knowledge spillovers depend on the presence of both geographical and relational proximity and that the simultaneous presence of geographical and relational proximity enhances the intensity of knowledge spillover.

We test these assumptions on a sample of 249 NUTS2 regions of the EU27 over the period 1990–2004. Along with a traditional spatial weights matrix, we introduce a matrix of inverse relational distance built on a measure of trust, defined as the capacity of economic agents in a regional context to act in cooperation with other actors, a capability which stems from a strong identity and sense of belonging, from shared trust and shared behavioral codes. Operationally, relational proximity is defined as the inverse distance of trust wealth among pairs of regions, normalized by the sum of the trust endowment in those areas. We find strong evidence of a positive role of relational proximity as a source of knowledge spillovers in the analyzed sample. We also produce evidence on the fact that geographical proximity enhances its positive external effects when regions are also close in terms of trust wealth.

The rest of the paper is organized as follows. In Sect. 2 we elaborate on the need to use the concept of relational distance to explain knowledge spillovers. Section 3 describes our dataset and the variables, while also providing an explanation of our measure of relational distance. Section 4 presents the results of an econometric analysis testing the assumptions of a positive role of relational proximity as a source of knowledge spillovers and of the super-additive effect of relational and spatial proximity. Section 5 concludes.

2 Knowledge Spillovers: The Role of Physical and Relational Distance

Economic theory is increasingly aware of the strategic role played by – voluntary or unintended – technological interdependence among economic actors. In particular, the interest in knowledge spillovers lies in the fact that they represent pure externalities, producing non-compensated advantages for receivers; a discrepancy between private and social optimum generates the need for specific policy interventions.

The concept of knowledge spillovers has stimulated interest in economic theories with rather different approaches, from mainstream to heterodox views. In neoclassical growth models, the dominant paradigm is that national growth rates depend on the growth rates and income levels of other countries. Stylized facts demonstrate that economic activity is concentrated at different spatial levels – countries, regions, cities (Easterly and Levine 2001), the reason lying in the strong global interdependence of technological progress. Knowledge accumulation affects the technological development (à la Solow 1956), the physical capital accumulation (à la Romer 1986) and the human capital accumulation (à la Lucas 1988) in the home country; what is new is the idea that knowledge accumulated in one country affects technological development and growth *of other countries*. In these models the intensity of the knowledge spillover effect depends on socio-economic or institutional proximity, measured by an exogenous variable, namely the geographical proximity of countries (Ertur and Koch 2007).

Regional economists and economic geographers achieve the same result, developing the concept at a more spatially disaggregated level of analysis. Knowledge spillovers imply that knowledge created by an organisation generates positive effects not only within it, but also for other organisations located in neighbouring regions (Fischer et al. 2006). This literature differentiates with mainstream economics as knowledge spillovers are interpreted as a spatially-bounded phenomenon: they take place mainly among regions or cities, rather than countries. This in turn would facilitate the exchange of information, face-to-face contacts, trade and market relationships, all within a pure gravity type logic. Such explanations date to Marshall's identification of high flows of information and ideas between firms of a region – what is "in the air" – as one of the main reasons for concentration of activities in space (Marshall 1920). In a pure spatial/geographical approach, the knowledge transmission channels are epidemiological contacts among local agents (Capello 2007).

More recently, doubts have been expressed on the idea that the mere geographical proximity is able to interpret all mechanisms behind knowledge spillovers (Boschma 2005; Capello 2007, 2009a, b). Geographical proximity justifies knowledge spillover effects through simple gravity-type processes, that hold at country, region or city levels, which limit the interpretation of the spillover effects under two perspectives (1) on the one hand, its validity at different geographical levels makes the spillover an a-spatial concept; (2) on the other, a pure geographical, gravity-type approach does not explain the learning processes of agents and contexts: learning on how to translate knowledge into innovation, learning on how to get the highest benefits from the presence of a multinational enterprise, learning to attract resources at the local level and to apply them in a creative and innovative way. Different learning processes explain why two regions at the same distance from a third highly innovative region may have a completely different absorption capacity of knowledge spillovers.

This implies that as other regions face endogenous growth processes, the extent to which a region can benefit from the external stimulus depends also, in our conceptual framework, on the relative differences in trust between the regions. As differences in trust decline, the ease with which knowledge travels and can be understood, decoded and efficiently exploited increases (Capello et al. 2010). Lower differences in trust between regions implies therefore higher absorptive capacity of firms, individuals and institutions, as well as lower transaction costs in the process of knowledge decoding and transfer.

Recent work has taken the need for non-geographical notions of distance seriously. In Maggioni et al. (2007) the effect of relational proximity, along with more traditional geographical proximity on growth spillovers is explored, with the use of data on research networks built up with EU Fifth Framework Programme and EPO co-patent applications. Ponds et al. (2010) proceed a step beyond and use the geographical and relational spatial lag of the performance measure as an independent variable simultaneously.

In order to introduce learning mechanisms in the explanation of knowledge spillovers, a relational approach is required, that explains the ways in which agents and contexts learn: this approach mainly interprets knowledge accumulation as the accumulation of knowledge through cooperative learning processes (Camagni 1991; Keeble and Wilkinson 1999, 2000), nourished by spatial proximity ("atmosphere" effects), network relations (long-distance, selective relationships), interaction, creativity and recombination capability.

This approach entails a relational definition of space. Functional/hierarchical, economic and social interactions take place in this space and are in turn embedded into geographical space (Camagni 1991; Camagni and Capello 2009). Relational space plays a role in learning processes. It develops and reinforces interactive processes between actors at the local level. It forms the set of shared behavioural codes, common culture, the capital of trust among agents and the sense of belonging. In turn, it depends on the social glue that is present in the region, which represents a pre-requisite for a creative interaction. These characteristics act on the capacity of firms to engage in market interactions. They develop and enhance collective learning processes by means of specific territorial channels through which knowledge flows by virtue of (a) the huge mobility of professionals and skilled labour – among firms but internally to the local labour market defined by the district or the city, where mobility of this kind is highest, and (b) intense co-operative relations among local actors and, in particular, customer-supplier relationships in production, design, research and, finally, knowledge creation (Camagni and Capello 2002).[1]

Territorial channels of knowledge flows are typical of production contexts characterised by the presence of small and medium sized firms (SMEs). The average dimensions of firms fostering the exchange of knowledge and the mutual transcoding of tacit information is not sterile: in fact, previous streams of literature, such as the milieu innovateur and the industrial district theory, have suggested that tacit knowledge exchange is maximized in SMEs. However, they are also relevant in contexts where large firms develop their own internal knowledge, culture and know-how by enhancing internal interactions and boosting selective external interaction with industrial partners, universities, professionals and research centres. In this view, the channels through which knowledge spreads are *territorialized*, embedded into the socio-cultural structure of a local system and, therefore, anchored by definition to the local area. Thus, the territorial reasons for a spatially-bounded effect of knowledge spillovers are identified and the limits of a-spatial theories in which knowledge spillovers concept is applied indifferently to countries, regions or cities are overcome.

Relational proximity – defined as the similarities of two areas in terms of shared behavioural codes, common culture, mutual trust, sense of belonging and cooperation capabilities – plays an important role in the capacity of a region to absorb knowledge spillovers. Cooperative learning processes are nourished by spatial proximity ("atmosphere" effects), network relations (long-distance, selective relationships), interaction and cooperation. Therefore, while geographical proximity is a good proxy for the "atmosphere effect", relational proximity measures the potential interaction and cooperation capabilities in knowledge accumulation. Relational proximity is therefore at least as important as geographical proximity in order to understand the micro-foundations of knowledge spillovers and the channels through which knowledge diffuses. Being geographically close to a region with similar relational capacity reinforces knowledge diffusion between the two areas. By the same token, relational proximity reinforces the effects generated by geographical closeness thanks to synergies and increasing returns.

From this conceptual framework we obtain two testable assumptions: *H1*. "knowledge spillovers depend on the presence of both geographical and relational proximity" and *H2*. "the simultaneous presence of geographical and relational proximity enhances the intensity of knowledge spillovers". Hypotheses *H1* and *H2* will then be empirically tested in Sect. 4 by introducing a relational distance effect within the Ertur and Koch's (2007) approach.

[1] A collective learning process of this kind was first hypothesized by the GREMI group (Camagni 1991; Perrin 1995) and subsequently widely adopted as a sound theoretical concept for the interpretation of knowledge-based development and innovation (Keeble and Wilkinson 1999, 2000; Capello 1999; Cappellin 2003).

3 Data and Variables

3.1 Basic Variables

We test our two hypotheses by estimating growth regression models on a sample of 249 NUTS2 regions belonging to the enlarged Europe (EU27). The dependent variable is the labour productivity growth rate computed for the period 1990–2004, $\gamma_y = T^{-1}(\ln y_T - \ln y_0)$. Basic data come from EUROSTAT Regio and Cambridge Econometrics databases, which include information on real gross value added, employment, investment, secondary education attainment and R&D investments. We measure labour productivity, y, as the ratio between total real value added and total employment; the saving rate, s_k, as the average share of gross investments on real gross value added; the human capital accumulation rate, s_h, as the average percentage of a region's working population in secondary school. Finally, n is the average growth rate of total employment.

In the last set of estimates we carry out a robustness check, controlling for other variables that modern regional growth theory considers as potentially relevant in explaining regional performance: sectoral composition (Perloff et al. 1960), agglomeration externalities (Ciccone and Hall 1996; Ciccone 2002), externalities associated to sectoral diversity (Jacobs 1969; Glaeser et al. 1992; Beaudry and Schiffauerova 2009) and R&D intensity (Sterlacchini 2008).

Sectoral composition is measured by the share of agricultural employment on total regional employment, $Share(agr)$, assuming that a higher share of agriculture may subsequently reduce economic performance. Agglomeration externalities, *dens*, are measured by the density of employment (ratio between total employment and regional surface in km^2) Jacobs externalities are measured as the median of Balassa indices, $Jacobs = median\left(\frac{E_{is}/E_i}{E_s/E}\right)$, where i denotes the region and s indexes the sector, E_{is} stands for average employment in the s-th sector (at two-digit level of the classification of economic activity)[2] for the i-th region, E_i is the average overall employment in the i-th region, E_s indicates the employment in the s-th sector in Europe, while E is the overall European employment.[3] Finally, R&D intensity, *r&d*, is measured by the percentage of total intramural R&D expenditure on gross value added.

---

[2]Namely, sector DA (food products, beverages and tobacco), DB (textile and textile products), DC (leather and leather products), DD (wood and wood products), DE (pulp, paper and paper products; publishing and printing), DF (coke, refined petroleum products and nuclear fuel), DG (chemicals, chemicals products, and man-made fibres), DH (rubber and plastic products), DI (other non-metallic mineral products), DJ (basic metals and fabricated metal products), DK (machinery and equipment n.e.c.), DL (electrical and optical equipment), DM (transport equipment) and DN (manufacturing n.e.c.).

[3]Since the Balassa index follows an asymmetric distribution (with a fixed lower bound, 0, and a variable upper bound, E/E_i), its median turns out to be the most appropriate indicator of the distribution position. When the median is low, an economy shows a comparative advantage in a large share of sectors and its productive structure is therefore diversified, and vice versa. So, we use the median as a direct measure of diversification.

Spatial lags of residuals and variables are computed using different distance-based spatial weights matrices. More precisely, for the diagnostics of residuals from the estimates, we use a binary spatial weights matrix with a distance based cut-off, whose elements w_{ij} assume value of 1 if the distance between the centroids is lower than 424 km (the minimum distance which allows all regions to have at least one neighbour) and zero otherwise:

$$w_{ij} = \begin{cases} 1 & \textit{if } d_{ij} \leq 424\,km \\ o & \textit{otherwise} \end{cases}$$

Elements w_{ii} on the main diagonal are set to zero by convention, whereas elements w_{ij} indicate whether region i is spatially connected to region j. For the computation of spatial lag variables, we compute a more general inverse-distance spatial weights matrix

$$w_{ij} = \begin{cases} d_{ij}^{-1} & \textit{if } d_{ij} \leq 424\,km \\ o & \textit{otherwise} \end{cases}$$

where, again, $w_{ii} = 0$. In order to normalize the outside influence upon each region, the weights matrix is row-standardized, so that the spatial lag of a variable is simply the weighted average of the neighbors' observations.

The chosen time span (1990–2004) encompasses, among others, two major breaks in the European history. First, it starts with the fall of the Communist regimes in Eastern countries. Then, it ends with the years of the biggest wave of enlargements (2004 and 2007, respectively) of the European Union, which coincided with the inclusion of 12 more countries in the EU. Although the enlargements themselves have physically taken place after the period surveyed in the present paper, the theory of rational expectations offers support to the idea that most of the effects of the 2004 enlargement may be already captured in the final years of the sampled period.

While these two major events are likely to have influenced our results, in particular in terms of growth rates, we believe that this study may shed further light precisely on the reasons of different economic performance of European regions. The switch to a competitive, market-based economic regime, and the announcement effect of the 2004 enlargement may in fact have boosted New Member States (NMS) economies much more than what actually happened. In fact, initial trust differences among regions may contribute to the explanation of growth differentials in EU regions beyond the cyclic effects present in any sample.

3.2 A Measure of Relational Distance

The core of our tests entails the definition and computation of relational distance. As stated in Sect. 2, we believe collective learning to be enhanced not only by the physical proximity of relevant actors (individuals, firms and institutions), but also by relational proximity. Holding physical distance constant, knowledge flows more easily when people face low transaction costs in the process of exchanging information: this requires a high level of trust within organizations (La Porta et al. 1997), which in turn facilitates the effectiveness of weak ties (Granovetter 1973).

Among the several dimensions along which we may measure relational proximity, we select the one we believe to have the highest impact on knowledge flows: trust. As levels of trust rise, individuals and firms are more prone to exchange knowledge. Trust enhances local channels of knowledge transmission, especially cooperation among local actors, local firms, clients and suppliers.

We measure trust in the most direct way, exploiting information collected by the EVS.[4] In particular, citizens have been asked "How much do you trust people?" The scale of possible answers ranges from 1 ("I trust them completely") to 5 ("I don't trust them at all"). For each region we calculate the percentage of answers 1 and 2 over the subsample of EVS individuals that answered this question.[5] Thus, for each region we have a measure of the average percentage of people who trust others "completely" or "enough": the vector's domain is $trust \in [0, 1]$.[6] Trust distance between regions i and j is therefore calculated as $w_{rel}^{ij} = d_{ij}^{-1}$, with $d_{ij} = |trust_i - trust_j| / (trust_i + trust_j)$. The inverse-distance trust weights matrix is finally created as the inverse of the absolute distance between trust levels in each region.[7]

Figure 1 plots the values of our relational distance indicator as a function of its numerator and denominator. The indicator increases monotonically with the numerator and decreases monotonically with the denominator. This allows us to conclude that, if two regions display a minor difference between their level of trust, while both having high values of trust, our indicator signals a lower distance than in the case of two regions with low difference between their level of trust and low values of trust. By the same token, our indicator assigns a higher distance between two regions with a high difference in their trust levels and a low amount of total trust

---

[4]EVS is among the widest surveys comprising statistical information from individual questionnaires on the values of European citizens. This paper uses its 1990 wave, which perfectly matches the initial year of our estimations. More information can be found on www.europeanvalues.nl

[5]This amounts to 37,107 cases, with just 1,106 individuals missing; hence, the question had a reply rate of about 97% of the individuals interviewed. The EVS sample was drawn from the population of adult citizens over 18 years of age. In some countries, random sampling was applied, in others quota sampling. The samples were weighted to correct for gender and age: the survey, therefore, correctly represents the population of each region.

[6]The actual range of the variable goes from 0.03 (recorded for Sardinia) to 0.64 (corresponding to Sydsverige).

[7]Glaeser et al. (2000) reviews the use of the EVS trust question. They find the question may also capture the level of trustworthiness of individuals, while also detecting high correlations among the EVS trust level measured *within the survey* and the outcome of two experiments aiming at identifying trust *behaviors*.

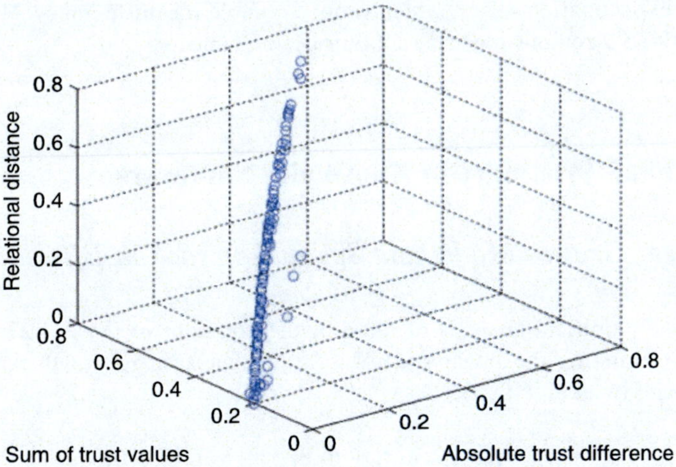

Fig. 1 Relational distance indicator as a function of its components

than between two regions with a high difference in their trust levels and a high amount of total trust (and vice versa).[8,9]

As for the correlation between physical and relational distance, Pearson's correlation index is equal to −0.18, significant at the 99% confidence level. Although in absolute terms not particularly high, the correlation coefficient implies that pure geographical proximity does not necessarily imply relational proximity. Spatially neighbouring regions may actually enjoy significantly different levels of trust, which therefore contributes to our understanding of why, ceteris paribus, regions with similar endowments of physical factors and with analogous locations display significantly different growth rates.

The issue of missing values is quite relevant for two reasons. First, as the data coverage starts from 1990 and includes former-communist regions, several statistics are missing for the first years after the fall of communist regimes. Second, we exclude Slovenija, Cyprus, Latvija and Malta in order to avoid problems with weights matrices (either spatial proximity is difficult to define or else EVS data are missing, as is the case for Cyprus). We also exclude Bulgarian regions in order to fully exploit the EUROSTAT Regio database and, in particular, its ATECO

[8]Notice that, as easily detectable from the Figure, and clear from the social distance formula, the indicator takes on value zero (whatever the sum of trust levels in the regions) when the numerator is zero. This may, however, happen for all the regions sharing the same level of trust. For such regions, it does not matter whether they share a high, medium or low level of trust: our indicator scores zero anyway. This is shown in Figure 1 with the dots on the *xy* plane.

[9]The same conclusions could be obtained mathematically. In fact, both the numerator as well as the denominator of the social distance measure are of first order, thus converge asymptotically with the same speed; besides, they both map on the positive half of the real numbers. This line of reasoning is behind the shape of the plot depicted in Fig. 1.

2-digits level occupation series, which is not available for this country. We end up with 249 NUTS 2 regions covering 22 European Countries.

4 Relational Proximity in Knowledge Spillovers

4.1 Linear Growth Model and Spatial Interaction Effects

We begin the empirical analysis of the growth behaviour of the EU-27 NUTS-2 regions by estimating the linear specification of neoclassical growth model proposed by Mankiw et al. (1992).

$$\gamma_y = \beta_0 + \beta_1 \ln y_0 + \beta_2 \ln s_k + \beta_3 \ln s_h + \beta_4 \ln(n + 0.05) + \varepsilon \tag{1}$$

The estimation results (not reported, but available upon request), obtained using heteroskedasticity-corrected variance-covariance matrices as suggested by Cribari-Neto (2004), confirm the theoretical predictions: regional productivity growth rates are positively affected by physical and human capital accumulation rates and negatively influenced by employment growth rates and initial productivity levels. Thus, the conditional convergence hypothesis cannot be rejected, but the speed of convergence (equal to 0.427%) is rather slow and the corresponding half-life is 162 years (almost three times the one estimated for Western Europe regional samples by Le Gallo et al. 2003, among others).

Even though OLS estimates tend to corroborate the hypotheses suggested by Mankiw et al. (1992), the diagnostics of the residuals reveal that the linear augmented Solow model is mis-specified due to (a) the assumption of homogenous behaviour (the RESET test raises doubts on the capacity of the linear functional form to properly capture the data generating process) and (b) the omission of variables that capture technological interdependence (Moran's I tests yields to reject the assumption of spatial independence of the residuals).[10] Given these results, we relax the hypothesis of linearity and spatial independence and estimate a semiparametric spatial Durbin model (SDM) as well as a semiparametric spatial lag model (SAR).

The semiparametric spatial Durbin growth regression model can be specified as (Basile 2008, 2009).

[10]Moran's I tests have been performed using distance-based binary spatial weights matrices. Many distance cut-offs, ranging from 420 km (the minimum distance which allows all regions to have at least one neighbour) to 1,020 km with a step of 50 km, have been adopted. All corresponding spatial weights matrices yield significant values of Moran's *I*. The highest standardized Moran's *I* value occurred in correspondence to the minimum distance.

$$\begin{aligned}\gamma_y = \beta_0 + f_1(\ln y_0) + f_2(W_{dist} \ln y_0) + f_3(\ln s_k) + f_4(W_{dist} \ln s_k)\\ + f_5(\ln s_h) + f_6(W_{dist} \ln s_h) + f_7(\ln(n + 0.05)) + f_8(W_{dist} \ln(n + 0.05))\\ + \rho_{dist} W_{dist} \gamma_y + \varepsilon\end{aligned} \tag{2}$$

where $f_j(.)$ are unknown smooth functions of the covariates, W_{dist} is a spatial weights matrix, the smooth terms $f_2(\cdot), f_4(\cdot), f_6(\cdot)$ and $f_8(\cdot)$ capture the effect of the spatial lags of the exogenous variables, ρ_{dist} is a parameter measuring the amount of global spatial externalities (or spatial technological interdependence) and ε is a vector of independently distributed errors. This specification is consistent with the economic growth model developed by Ertur and Koch (2007) based on the assumption that technological knowledge spread across regions/countries with an intensity which decreases with geographical distance. The matrix W_{dist} used to estimate this model has been described in the previous section.

LeSage and Pace (2009) suggest that the SDM specification can also be derived from a data generating process characterized by unobserved heterogeneity and that the SDM nests both SAR and SEM (spatial error model). However, the SDM specification implies an inflation of smooth terms (especially when two different weight matrices are used in the same model and the number of exogenous variables is not negligible). An alternative method to control for unobserved spatial heterogeneity, rather diffused in spatial statistics (Venables and Ripley 2002), consists of including in the model a spatial trend surface, that is a bi-dimensional smooth function of northing (no) and easting (e), $f(no, e)$, instead of spatially lagged exogenous variables:

$$\begin{aligned}\gamma_y = \beta_0 + f_1(\ln y_0) + f_2(\ln s_k) + f_3(\ln s_h)\\ + f_4(\ln(n + g + \delta)) + f_5(no, e) + \rho_{dist} W_{dist} \gamma_y + \varepsilon\end{aligned} \tag{3}$$

Both models (2) and (3) include the endogenous term $W_{dist}\gamma_y$.[11] In order to deal with endogeneity problems in a nonparametric framework, Blundell and Powell (2003) have proposed to use the "control function" approach which consists of two steps. In the first one, an auxiliary nonparametric regression $W_{dist}\gamma_y = \beta_0 + f_1(.) + ... + h(Z) + v$ is considered, with Z a set of conformable instruments and v a sequence of random variables satisfying $E(v|Z) = 0$. The second step consists of estimating an additive model of the form $\gamma_y = \beta_0 + f_1(.) + ... + \rho W_{dist}\gamma_y + f_J(\hat{v}) + \varepsilon$.

We employ the methodology proposed by Wood (2006) to estimate models (2) and (3) with spline-based penalized regression smoothers which allows for automatic and integrated smoothing parameters selection via GCV. For the two spatial dimensions no and e, an isotropic thin plate regression spline basis function is used, as suggested by Augustin et al. (2009). The econometric results reveal that the two

[11] In linear spatial regression analysis, Kelejian and Prucha (1998) have proposed a 2SLS procedure to estimate the spatial autocorrelation regression model and have suggested using spatial lags of the strictly exogenous variables as instruments.

models perform quite similarly in terms of adjusted R^2, percentage of explained deviance, GCV score and AIC. Both specifications allow to predict the spatial variability in growth behaviour better than non-spatial linear and nonlinear models and to solve the issue of spatial autocorrelation in the residuals (Moran's I statistics are no more significant). Given the similarity between the two models, and due to the lower number of smooth terms included in the augmented spatial lag model (3), we will keep this specification as the preferred one to continue our from-particular-to-general estimation strategy.

Two-step estimation results of model (3) are reported in Table 1. The F-tests for the overall significance of the smoothed terms have p-values lower than 0.05 in all cases, while the number of effective degrees of freedom (*edf*) suggests that the relationship between regional growth and its determinants is far from being linear. Figure 2a–d show the fitted univariate smooth functions (solid lines), alongside Bayesian confidence intervals (shaded grey areas) at the 95% level of significance, computed as suggested by Wood (2006). In each plot, the vertical axis displays the scale of the expected values of regional growth, while the horizontal ones report the scale of each determinant.

An inverted U-shaped relationship between growth and initial conditions emerges (Fig. 2a), with a clear downward pattern in $\hat{f}_1(\ln y_0)$ only after a certain threshold of the relative level of GDP per worker in 1990 ($\ln y_0$): specifically, a diverging behaviour characterizes the group of Eastern regions (45 regions), while Western regions maintain a conditional predicted convergence path. The assumption of identical speed of convergence is consequently rejected.

Nonlinearities in the effects of gross physical investment, $\hat{f}_2(\ln s_k)$, and of secondary school enrolment ratio, $\hat{f}_3(\ln s_h)$, are clearly detected. Specifically, an increase in the saving rate is associated with an increase in growth rate only when $\ln s_k$ is above the EU average. The existence of a threshold in the effect of $\ln s_k$

Table 1 Nonparametric estimation results of the additive nonlinear model

Variables	Unconstrained nonlinear growth model	
	F-tests and p-values	Edf
$f_1(\ln y_0)$	14.418 [0.000]	2.543
$f_2(\ln s_k)$	16.911 [0.000]	2.424
$f_3(\ln s_h)$	3.165 [0.013]	3.649
$f_4(\ln(n+0.05))$	5.721 [0.000]	2.607
$R^2_{adj.}$	0.533	
Deviance	55.4	
AIC	740	
GCV	1.137	
Moran's I	3.848 [0.000]	

Notes: Dependent variable: productivity growth rate. F test and p-value (in squared brackets) for the overall significance of smooth terms are reported in Column 2. Edf are the effective degrees of freedom. Deviance is the percentage of explained deviance. AIC is the Akaike Information Criterion. GCV is the generalized Cross Validation. Moran's I standard deviates and p-values are computed using a great-circle distance-based binary weights matrix with a threshold distance of 424 km. The number of observations in the sample is 249

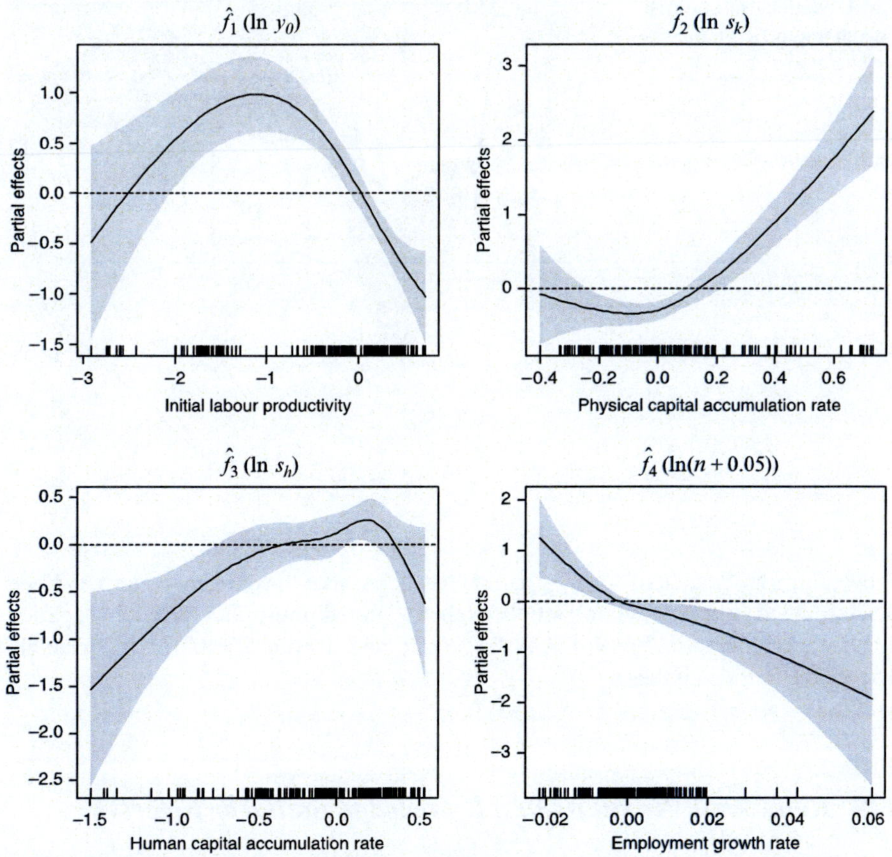

Fig. 2 Model 2: partial effects of univariate smooth terms

(Fig. 2b) is in line with Azariadis and Drazen (1990) theoretical prediction. Quite differently, the rate of schooling has a positive effect on regional growth only up to a threshold, then a downward pattern comes out (Fig. 2c). To explain this odd result, it may be useful to observe that most of the regions with levels of $\ln s_h$ higher than the threshold belong to Eastern countries. It is recognised that, despite the high enrolment rates in primary and secondary schooling, a decline in the quality of education is of particular concern in these countries. Finally, the influence of the employment growth rate on regional growth is monotonically negative, albeit the marginal effect is not homogenous across the sample (Fig. 2d). Finally, Fig. 3 displays the partial effect of the smooth interaction between latitude and longitude, $f(no, e)$.

The value of the spatial autocorrelation parameter $\hat{\rho}_{dist}$ is equal to 0.58 and statistically significant at the 5%, confirming the role of spatial frictions in the interregional diffusion of technological spillovers. The endogeneity of the lag term $W_{dist}\gamma_y$ is confirmed by significance of the control function $f_J(\hat{v})$. As pointed out

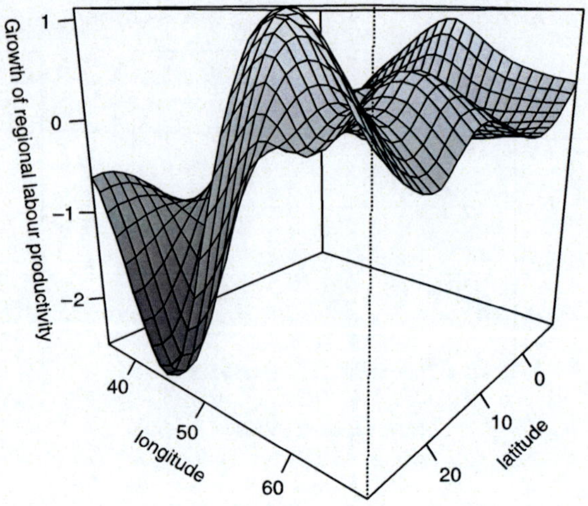

Fig. 3 Partial effects of the smooth interaction term $f_{10}(lat, long)$

above, the significance of the $\hat{\rho}_{dist}$ parameter means that the exogenous terms affect the left hand side of the model through a "global multiplier effect" ("spatial diffusion with friction") (see Anselin 2004; and Basile 2008, for a thorough discussion of these issues).

4.2 Relational Proximity and Knowledge Spillover Effects

In this section we present the results of an econometric analysis aimed at testing the two hypotheses illustrated in Sect. 2 (statements *H1* and *H2*). We first analyze proposition *H1*, according to which knowledge spillovers depend on the presence of both geographical and relational proximity. This assumption is tested by including on the right-hand side of the growth regression model a linear term measuring the relational spillover variable, $W_{rel}\gamma_y$, along with the linear term measuring the spatial spillover effect, $W_{dist}\gamma_y$, while all other variables are treated as nonlinear smooth terms:

$$\begin{aligned} \gamma_y = \beta_0 + f_1(\ln y_0) + f_2(\ln s_k) + f_3(\ln s_h) + f_4(\ln(n+g+\delta)) \\ + f_5(lat, long) + \rho_{dist} W_{dist}\gamma_y + \rho_{rel} W_{rel}\gamma_y + f_6(\hat{v}_{dist}) + f_7(\hat{v}_{rel}) + \varepsilon \end{aligned} \tag{4}$$

Spatial interactions are based on the inverse distance weights matrix, while relational interactions are modelled with the relational inverse distance matrix W_{rel} illustrated in Sect. 3. A control function approach is adopted in order to control for the endogeneity of both variables. "Relational lags" of the exogenous explanatory variables are therefore considered as further instruments. The smooth interaction between latitude and longitude, $f(lat, long)$, accounts for spatial trends in the DGP.[12]

Estimation results of model (4) are shown in Table 2. The ρ_{dist} and the ρ_{rel} parameters measure the degree of spatial spillovers and of relational spillovers, respectively. The magnitude of $\hat{\rho}_{dist}$ is rather high (0.788) but in line with that reported in previous analyses (Basile 2009), while the coefficient $\hat{\rho}_{rel}$ is equal to 0.219 and significant at the 1% level. Apparently, a relevant role for geographical distance is maintained when the effect of relational distance is accounted for. The economic interpretation of a significant $\hat{\rho}_{rel}$ parameter is similar to the one usually adopted with respect to the $\hat{\rho}_{dist}$ parameter. Thus, we can say that a random shock in regions i as well as a change in the level of an exogenous variable (such as human or physical capital investments) in regions i influence not only the growth outcome of that region, but also the growth outcome of all other regions with a strength decreasing with the relational distance between the regions (along with the geographical distance between regions).

While the control function associated to the geographical lag – $f_6(v_{dist})$ – is highly significant, confirming the endogeneity of this process, trust lag does not turn out to be endogenous. We interpret this result as a further demonstration of the slow pace at which soft forms of capital accumulate over time (Putnam 2000). Trust capital is as easy to spoil as difficult to accumulate. Synergies among local actors crucially depend on mutual understanding, which in turn thrives on high education levels, cultural homogeneity and sharing similar values.

As mentioned above, all other terms enter the model nonlinearly. Nonlinearities in economic growth regression usually come out from three possible reasons (1) the existence of multiple steady-states in the DGP, (2) the omission of relevant growth determinants, (3) nonlinearity in the production function. The number of effective degrees of freedom (*edf*) associated to each smooth term is always higher than one suggesting that in fact the relationship between regional growth and its determinants is far from being linear. Specifically, the estimation results suggest that an increase in the saving rate is associated with an increase in growth rate only when $\ln s_k$ is above the EU average. Quite differently, the rate of schooling has a positive effect on regional growth only up to a threshold, then a downward pattern comes out. Finally, the influence of the employment growth rate on regional growth is monotonically negative, albeit the marginal effect is not homogenous across the sample.[13]

---

[12]Prior to the main tested hypotheses, we adopted a from-particular-to-general specification strategy to choose the most suitable specification. The first step entails estimating a basic human-capital augmented neoclassical model *à la* Mankiw et al. (1992). Next, the hypothesis of linearity and of spatial independence is relaxed, as residuals of the first OLS estimates display spatial autocorrelation. We therefore estimated a spatial Durbin model (and a spatial lag) model, *à la* Ertur and Koch (2005, 2007) and Basile (2008, 2009). Finally, we augment the spatial lag specification by incorporating social proximity effects. This section presents the econometric results of the preliminary two steps, while the effect of social proximity is analyzed in Sect. 4.2.

[13]The plots of these smooth terms are not reported in the paper, but they are available upon request.

Finally, although the effect of the other variables remains qualitatively similar to those obtained in a linear setting, the inclusion of relational knowledge spillovers consistently improves all relevant fit statistics and choice criteria: the adjusted R^2 is equal to 0.733 (it is equal to 0.37 in a linear setting), the percentage of explained deviance and the Generalized Cross Validation (GCV) score are equal respectively to 76.9 and to 0.719, while the Akaike criterion decreases with respect to linear estimates to 621. The Moran's I statistics are not statistically significant.

However, one point in this relationship is not yet fully clear. How do geographical and relational distance interact with each other? The answer to this question is the object of the next section.

4.3 Geographical and Relational Proximity: Synergies in Knowledge Spillover Effects

The growth model presented in (4) can be further adapted in order to take possible (nonlinear) interactions between the two global spillover effects into account. Column 2 in Table 1 reports the estimates of a fully nonlinear econometric model, where the extent to which knowledge spills over surrounding regions is accounted for by a smooth term. In this case non linear structure is ex ante imposed for the distribution of geographical and relational frictions among European regions in the emergence of knowledge spillovers.

This statement corresponds to testing the second research question (*H2*) presented in Sect. 2. This translates in the following testable nonlinear equation:

$$\begin{aligned}\gamma_y = \beta_0 + f_1(\ln y_0) + f_2(\ln s_k) + f_3(\ln s_h) + f_4(\ln(n + g + \delta)) \\ + f_5(lat, long) + f_6\left(W_{dist}\gamma_y, W_{rel}\gamma_y\right) + f_7(\hat{v}_{dist}) + f_8(\hat{v}_{rel}) + \varepsilon\end{aligned} \quad (5)$$

Results of estimating (5) are shown in Table 2. The smooth term is highly significant at all conventional levels. From this analysis and provided estimates for relevant controls that do not differ consistently from similar linear models, we can infer that indeed not only geographical distance plays a role in mitigating the extent to which knowledge spillovers travel. Also, relational distance co-determines the geography of knowledge spillovers.

This statement can also be seen graphically. Figure 4 plots the joined effect of $W_{dist}\gamma_y$ and $W_{rel}\gamma_y$ from two different perspectives. The vertical axis displays the scale of the expected values of regional growth, while the two axes of the horizontal plane report the scale of $W_{dist}\gamma_y$ and $W_{rel}\gamma_y$.

Not only does a smaller distance – both in terms of geographical and relational space – increases the magnitude of estimated knowledge spillovers; but also the effects of the two measures of proximity mutually reinforce. We can therefore infer that knowledge flows more easily between regions that are not too distant either from a geographical or a relational point of view. We can conclude that, *ceteris*

Table 2 Spatial and relational spillovers

Variables	Linear externalities		Nonlinear interaction between spatial and relational externalities	
	F-tests and p-values	Edf	F-tests and p-values	Edf
$f_1(\ln y_0)$	6.002 [0.000]	3.028	10.718 [0.000]	3.259
$f_3(\ln s_k)$	18.912 [0.000]	2.483	19.239 [0.000]	2.079
$f_5(\ln s_h)$	7.505 [0.000]	3.942	9.528 [0.000]	3.938
$f_4(\ln(n+0.05))$	3.022 [0.038]	2.555	5.572 [0.019]	1.000
$f_5(lat, long)$	2.387 [0.002]	15.649	3.068 [0.010]	4.988
$f_6(v_{dist})$	23.797 [0.000]	2.894	37.230 [0.000]	2.847
$f_7(v_{soc})$	1.175 [0.279]	1.000	0.187 [0.665]	1.000
ρ_{dist}	0.788 (0.245) [0.001]			
ρ_{soc}	0.219 (0.067) [0.001]			
$f_8(W_{dist}\gamma_y, W_{soc}\gamma_y)$			17.823 [0.000]	4.567
$R^2_{adj.}$	0.733		0.747	
Deviance	76.9		77.1	
GCV	0.719		0.650	
AIC	621		599	
Moran's I	−0.229 [0.590]		0.199 [0.421]	

Notes: Dependent variable: productivity growth rate. F test and p-value (in squared brackets) for the overall significance of smooth terms are reported in Column 2. Edf are the effective degrees of freedom. Deviance is the percentage of explained deviance. AIC is the Akaike Information Criterion. GCV is the generalized Cross Validation. Moran's I standard deviates and p-values are computed using a great-circle distance-based binary weights matrix with a threshold distance of 424 km. The number of observations in the sample is 249

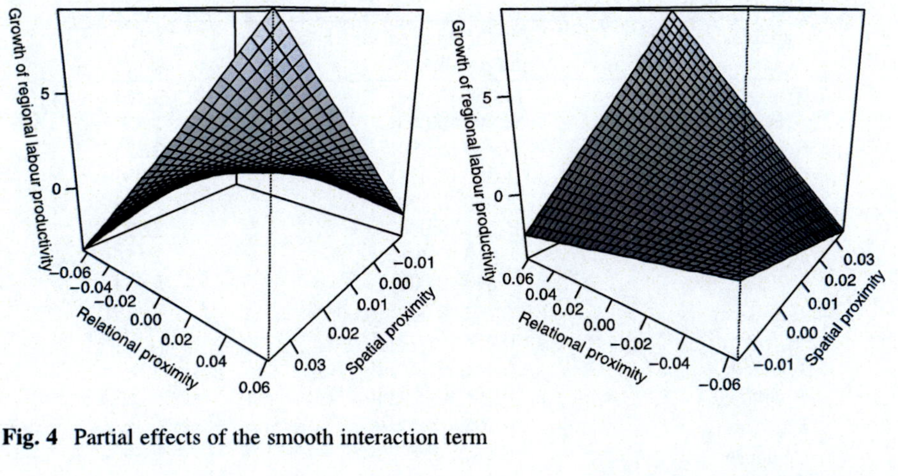

Fig. 4 Partial effects of the smooth interaction term

paribus, both spatial and relational proximity co-determine knowledge spillovers and their impact is maximized when regions are both physically as well as relationally proximate.

A final improvement in our estimates entails a richer semi-nonlinear econometric model, where we also control for selected growth-enhancing factors that have been previously found to be relevant in the regional growth literature. This is done in the next sub-section.

4.4 Robustness Checks

To test the robustness of our results, we control for the omission of possibly relevant growth determinants. Specifically, we take agglomeration and Jacobs externalities as well as sectoral composition and R&D intensity as controls entering the model linearly whereas we allow the other variables to make up the nonlinear component of the semiparametric model:

$$\begin{aligned}\gamma_y = \beta_0 &+ f_1(\ln y_0) + f_2(\ln s_k) + f_3(\ln s_h) + f_4(\ln(n+g+\delta)) \\ &+ f_5(lat, long) + f_6(W_{dist}\gamma_y, W_{rel}\gamma_y) + f_7(\hat{v}_{dist}) + f_8(\hat{v}_{rel}) \\ &+ \beta_1 \ln sh(agr) + \beta_2 \ln(dens) + \beta_3 \ln(Jacobs) + \beta_4 \ln(R\&D) + \varepsilon\end{aligned} \quad (6)$$

Results of estimating (6) are shown in Table 3.

The addition of the above mentioned control variables does not substantially change the main conclusions of this paper on the role of relational spillovers. Spatial and relational distance both mediate in the ease with which knowledge

Table 3 Robustness checks

Variables	Coefficients, std.err., F-tests and p-values	Edf
$f_1(\ln y_0)$	16.456 [0.000]	3.414
$f_3(\ln s_k)$	30.990 [0.000]	1.850
$f_5(\ln s_h)$	7.879 [0.000]	3.851
$f_4(\ln(n+0.05))$	5.389 [0.021]	1.000
$f_5(lat, long)$	1.050 [0.379]	3.699
$f_6(v_{dist})$	35.523 [0.000]	2.727
$f_7(v_{soc})$	0.913 [0.423]	2.556
$f_8(W_{dist}\gamma_y, W_{soc}\gamma_y)$	14.613 [0.000]	5.484
$\ln Sh(agr)$	−0.263 (0.093) [0.005]	
$\ln dens$	0.056 (0.074) [0.447]	
$\ln Jacobs$	−0.461 (0.259) [0.076]	
$\ln r\&d$	0.208 (0.083) [0.013]	
$R^2_{adj.}$	0.767	
Deviance	79.3	
GCV	0.610	
AIC	582	
Moran's I	−0.800 [0.782]	

Notes: see Table 1

spreads. Also, their effect is higher when regions are proximate both spatially as well as relationally.

Nevertheless, with the inclusion of these important control variables, the nonlinear spatial trend – $f_5(lat, long)$ – loses its significance. We can therefore conclude that unobserved spatial heterogeneity in growth behaviour displays no more relevant patterns. Moreover, with the inclusion of additional control variables, the precision of our estimates increases further: the R^2 and the percentage of explained deviance increase respectively to 0.767 and 79.3, the GCV score abates to 0.610 and the Akaike criterion decreases to 582. The significance level associated to Moran's I statistics also further reduces.

Additional controls are significant in three out of four cases. Signs associated to the additional controls are all in line with the literature. In particular, there is evidence of *negative* Jacobs externalities, which may be linked to the level of spatial aggregation of our data. Jacobs externalities were deemed to play a major role in large, diversified and creative cities (in regions at the EU NUTS3 level definition) (Jacobs 1969; Beaudry and Schiffauerova 2009), while sectoral specialization may actually foster productivity growth at the NUTS2 level. However, more research may shed light on this highly debated issue.

5 Conclusions

Regional spillovers are growth enhancing elements of a region which, as pure public goods, exert positive (negative) effects on other regions, with remarkable distance-decay effects. The reasons behind the spatially-bounded nature of spillovers may be found in spatial proximity (following a pure spatial-geographical approach) as well as in other notions of proximity. In this paper, we test the hypothesis that relational proximity, intended as the proximity between pairs of regions in developing collective learning processes, co-determines knowledge spillovers. Specifically, we test the hypothesis that both geographical and relational proximity explain the mechanisms behind knowledge spillovers. Space here is therefore defined along two axes: a relational space, where functional, hierarchical, economic and relational interactions take place, and the geographical space in which these relations are embedded.

We test the role of relational proximity as determinant of knowledge spillovers using a sample of 249 EU27 NUTS2 regions over the period 1990–2004. The evidence strongly supports the idea that relational space adds information on the way agents interact and on how knowledge spillovers are generated. Thus, relational as well as physical proximity are found to be key determinants for knowledge spillover exploitation. Also, we find that the effects of geographical and relational proximity on knowledge spillovers reinforce each other; data clearly show that, *ceteris paribus*, regions closer in spatial terms exchange knowledge more easily when their levels of trust is similar. These two main results are robust both to different choices of models, allowing for spatial heterogeneity of the estimated parameters and controlling for endogeneity of the processes explained by the model, as well as to the inclusion of other relevant growth determinants.

Our results call for further research on the topic. Only recently attention has been paid to the different definitions of space that might determine the extent to which knowledge travels. Empirical assessments of these theories are quite rare and more empirical research, supported by strong theories, might help in accounting for more complex and satisfactory definitions of space.

This paper has also relevant policy implications. EUs DG Regio, i.e. the European Regional Authority, is in charge of regional policy for the EU27 Member States, and explicitly focuses mainly on territorial cohesion,[14] with a clear commitment to reducing spatial disparities between European regions, in terms of economic wealth, and, consequentially, of future opportunities (European Commission 1996, 1999; European Council 1999a, b). Soft policies are part of the policy bundle for this Authority; however, seldom have more comprehensive context policies been attempted.

In fact, there is evidence that social capital can be accumulated, thus enhancing relational proximity. Books like Putnam (2000) are replete with examples of local US communities feeding their wealth of social values, trust and norms, laying the basis for future socio-economic improvements. Investing in social and relational capital is costly and expensive. Rules and norms, trust and values have typically long accumulation time, while also presenting very short spoiling periods. Social capital, therefore, seems to accumulate at a slow pace and risks to dissipate at a fast rate. However, regions may significantly benefit from such investment.

The propensity to cooperate is for instance the object of some cooperation-enhancing research policies. A few recent examples include the voucher issued by the province of Limburg (Netherlands) and that released by the region of Lombardy. In the first case, Limburg started a pilot project in 1998, randomly assigining vouchers to 20 SMEs in order to foster cooperative behaviour aiming at R&D activities. Similarly, in 2005 Lombardy released R&D cooperation vouchers to firms and Technology Transfer Centres for improving technical contents of an innovation or for patenting. Target firms included SMEs, on the premise that this is the segment of the market that faces the biggest constraints to cooperative behaviour in research and patenting. In both cases, evidence suggests that cooperative behaviour indeed increased among SMEs after the introduction of the vouchers.

These examples present the case for similar policies, for instance in the form of tax reductions or rebates, for firms and institutions lacking, fully or partially, the capability to cooperate. In presence of a typical market failure, these measures may actually build up the stock of trust needed to foster cooperation between distant areas, thus causing faster and more efficient growth spillovers between regions, and therefore an increase in the long run equilibrium growth rate for EU areas.

[14]Keywords on DG Regios' website as of May 5, 2010 include the following terms: "Beneficiaries", "Future of Cohesion Policy", "Territorial Cohesion", "Territorial Co-operation", "Closure 2006", "RegioStars", "Economic crisis", "Cohesion reports", "Danube strategy", and "Ex Post Evaluation 2000–2006".

References

Acs Z, Audretsch D, Feldman M (1994) R&D spillovers and recipient firm size. Rev Econ Stat 76:336–340

Anselin L (2004) Spatial externalities, spatial multipliers and spatial econometrics. International Regional Science Review 26:153–166

Anselin L, Varga A, Acs Z (2000) Geographic and sectoral characteristics of academic knowledge externalities. Pap Reg Sci 79(4):435–443

Augustin, N., Musio, M., von Wilpert, K., Kublin, E., Wood, S. and Schumacher, M. (2009). "Modelling spatio-temporal forest health monitoring data", Journal of the American Statistical Association, 104(487):899–911

Azariadis C and Drazen A (1990) Threshold externalities in economic development. Quarterly Journal of Economics 105:501–526

Basile R (2008) Regional economic growth in Europe: a semiparametric spatial dependence approach. Pap Reg Sci 87:527–544

Basile R (2009) Productivity polarization across regions in Europe. The role of nonlinearities and spatial dependence. Int Reg Sci Rev 32(1):92–115

Beaudry C, Schiffauerova A (2009) Who's right, Marshall or Jacobs? The localization versus urbanization debate. Res Policy 38:318–337

Bellet M, Colletis G, Lung Y (1993) Introduction au numéro special sur l'économie de proximité. Révue d'économie régionale et urbaine 3:357–364

Blundell R and Powell J (2003) Endogeneity in nonparametric and semiparametric regression models. In Dewatripont M, Hansen L and Turnsovsky SJ (eds.) Advances in Economics and Econometrics, Cambridge: Cambridge University Press

Boschma RA (2005) Proximity and innovation: a critical assessment. Reg Stud 39(1):61–74

Camagni R (1991) Local milieu, uncertainty and innovation networks: towards a new dynamic theory of economic space. In: Camagni R (ed) Innovation networks: spatial perspectives. Belhaven-Pinter, London, pp 121–144

Camagni R, Capello R (2002) Milieux innovateurs and collective learning: from concepts to measurement. In: Acs Z, de Groot H, Nijkamp P (eds) The emergence of the knowledge economy: a regional perspective. Springer, Berlin, pp 15–45

Camagni R, Capello R (2009) Knowledge-based economy and knowledge creation: the role of space. In: Fratesi U, Senn L (eds) Growth and innovation of competitive regions: the role of internal and external connections. Springer, Berlin, pp 145–166

Cantner U, Meder A (2007) Technological proximity and the choice of a cooperation partner. J Econ Interact Coord 2:45–65

Capello R (1999) Spatial transfer of knowledge in high-technology milieux: learning vs. collective learning processes. Reg Stud 33(4):353–365

Capello R (2007) Regional economics. Routledge, London

Capello R (2009a) Spatial spillovers and regional growth. Eur Plan Stud 17(5):639–658

Capello R (2009b) Indivisibilities, synergy and proximity: the need for an integrated approach to agglomeration economies. Tijdschrift voor Economische en Sociale Geographie 100(2):145–159

Capello R, Caragliu A, Nijkamp P (2011) Territorial captial and regional growth: increasing returns in knowledge use Tijdschrift voor economische en sociale geografie 101(2):1–17

Cappellin R (2003) Territorial knowledge management: towards a metrics of the social dimension of agglomeration economies. Int J Technol Manage 26(2–4):303–325

Ciccone A (2002) Agglomeration effects in Europe. Eur Econ Rev 46:213–227

Ciccone A, Hall RE (1996) Productivity and the density of economic activity. Am Econ Rev 86 (1):54–70

Coe DT, Helpman E (1995) International R&D spillovers. Eur Econ Rev 39(5):859–887

Cribari-Neto F. (2004), Asymptotic inference under heteroskedasticity of unknown form. Computational Statistics & Data Analysis 45, 215–233

Easterly W, Levine R (2001) It's not factor accumulation: stylized facts and growth models. World Bank Econ Rev 15:177–219
Eaton J, Kortum S (1996) Trade in ideas. Patenting and productivity in the OECD. J Int Econ 40(3–4):251–278
Ertur C, Koch W (2007) Growth, technological interdependence and spatial externalities: theory and evidence. J Appl Econometrics 22:1033–1062
European Council (1999a) Presidency conclusions, European Council, 24 and 25 March
European Council (1999b) Council regulation (EC) No. 1260/1999 laying down general provisions on the Structural Funds, Official Journal of the European Communities, 26 June 1999
European Commission (1996) First cohesion report, Commission of the European Community, Brussels, November
European Commission (1999) Sixth periodic report on the social and economic situation and development of the regions of the community, Commission of the European Communities, Brussels, February
Fischer M, Scherngell T, Jansenber E (2006) The geography of knowledge spillovers between high-technology firms in Europe: evidence from a spatial interaction modeling perspective. Geogr Anal 38:288–309
Glaeser EL, Kallal HD, Scheinkman JA, Shleifer A (1992) Growth in cities. J Polit Econ 100(6): 1126–1152
Glaeser EL, Liaison DI, Scheinkman JA, Soutter CL (2000) Measuring trust. Q J Econ 115(3): 811–846
Granovetter M (1973) The strength of weak ties. Am J Sociol 78(6):1360–1380
Henderson JV (2003) Marshall's scale economies. J Urban Econ 43:1–28
Holod D, Reed R (2004) Regional spillovers, economic growth, and the effects of economic integration. Econ Lett 85(1):35–42
Jacobs J (1969) The economy of cities. Vintage, New York
Keeble D, Wilkinson F (1999) Collective learning and knowledge development in the evolution of regional clusters of high-technology SMS in Europe. Reg Stud 33:295–303
Keeble D, Wilkinson F (2000) High technology clusters, networking and collective learning in Europe. Ashgate, Aldershot
Kelejian HH, Prucha IR (1998) A generalized spatial two-stage least squares procedure for estimating a spatial autoregressive model with autoregressive disturbances. Journal of Real Estate Finance and Economics 17:99–121
LeSage J and Pace RK (2009) Introduction to Spatial Econometrics, Taylor & Francis Group, LLC
La Porta R, Lopez-De-Silanes F, Shleifer A, Vishny R (1997) Trust in large organizations. Am Econ Rev Pap Proc 87:333–338
Le Gallo J, Ertur C and Baumont C (2003) A spatial econometric analysis of convergence across European regions, 1980–1995. In Fingleton B (ed.) European regional growth. Berlin: Springer-Verlag
Lòpez-Bazo E, Vayà E, Artìs M (2004) Regional externalities and growth: evidence from European regions. J Reg Sci 44:43–73
Lucas R (1988) On the mechanics of economic development. J Monetary Econ 22:3–42
Lundvall B-A, Johnson B (1994) The learning economy. J Ind Stud 1(2):23–42
Maggioni MA, Nosvelli M, Uberti TE (2007) Space versus networks in the geography of innovation. A European analysis. Pap Reg Sci 86(3):471–494
Mankiw NG, Romer D, Weil DN (1992) A contribution to the empirics of economic growth. Q J Econ 107:407–437
Marshall A (1920) Principles of economics. Macmillan, London
Perloff H, Dunn E, Lampard E, Muth R (1960) Region, resources and economic growth. John Hopkins, Baltimora, MD
Perrin JC (1995) Apprentissage collectif, territoire et milieu innovateur: un nouveau paradigme pour le développement. In: Ferrão J (ed) Políticas de Inovação e Desenvolvimento Regional

et Local, Edição do Instituto de Ciencias Sociais de Universidade de Lisboa, republished in Camagni R, Maillat D (2006) Milieux innovateurs, Economica-Anthropos, Paris, pp 99–128
Ponds R, van Oort F, Frenken K (2010) Innovation, spillovers and university–industry collaboration: an extended knowledge production function approach. J Econ Geogr 10(2):231–255
Putnam RD (2000) Bowling alone. Simon and Schuster, New York, NY
Rallet A, Torre A (eds) (1995) Économie industrielle et économie spatiale. Economica, Paris
Rey SJ (2004) Spatial dependence in the evolution of regional income distributions. In: Getis A, Mur J, Zoeller H (eds) Spatial econometrics and spatial statistics. Palgrave, Hampshire, pp 194–213
Rey SJ, Janikas MV (2005) Regional convergence, inequality and space. J Econ Geogr 5:155–176
Rey SJ, Montouri BD (2004) U.S. regional income convergence: a spatial econometric perspective. In: Cheshire P, Duranton G (eds) Recent developments in urban and regional economics. Edward Elgar, Cheltenham, pp 389–404
Romer P (1986) Increasing returns and long-run growth. J Polit Econ 94(5):1002–1037
Solow R (1956) A contribution to the theory of economic growth. Q J Econ 70:65–94
Sterlacchini A (2008) R&D higher education and regional growth: uneven linkages among European regions. Res Policy 37:1096–1107
Venables W and Ripley B (1994) Modern Applied Statistics with S-Plus, New York (NY): Springer-Verlag
Wood SN (2006) Generalized additive models. an introduction with R, Boca Raton (FL): Chapman & Hall/CRC

Agglomeration and New Establishment Survival: A Mixed Hierarchical and Cross-Classified Model

Martijn J. Burger, Frank G. van Oort, and Otto Raspe

1 The Firm in the "New Economics of Urban and Regional Growth"

Innovative economic developments take place in urban areas and industrial clusters. Urban and regional planners, geographers and economists are interested in the forces that create, shape and maintain these concentrations of economic activities (Van Oort 2004). Since the early 1990s, a growing empirical literature has emerged in the field of regional science and urban economics. It examines whether spatial circumstances give rise to agglomeration economies – external economies from which firms can benefit through co-location – that endogenously induce localized economic growth (Glaeser et al. 1992; Henderson et al. 1995; Combes 2000; Rosenthal and Strange 2003; Brülhart and Mathys 2008). As this literature tends to combine the traditional urban economics and regional science literature with new growth theory (Romer 1986; Lucas 1988), Glaeser (2000) has dubbed this line of research the "New Economics of Urban and Regional Growth." Many of the empirical studies under this heading show that agglomeration economies may be one source of the uneven distribution of economic activities and growth across cities and regions. In their survey of empirical literature on the benefits of agglomeration, Rosenthal and Strange (2004) point out that the elasticity of productivity to city and industry size typically ranges between 3 and 8%. However, the effects of agglomeration economies on localized economic growth generally differ across

M.J. Burger (✉)
Department of Applied Economics, Erasmus University Rotterdam, Room H13-27, P.O. Box 1738, 3000 DR Rotterdam, The Netherlands
e-mail: mburger@ese.eur.nl

F.G. van Oort
Department of Economic Geography, Utrecht University, P.O. Box 80115, 3508 TC Utrecht, The Netherlands

O. Raspe
Urban and Regional research centre Utrecht (URU), Utrecht University, P.O. Box 80115, 3508 TC Utrecht, The Netherlands

K. Kourtit et al. (eds.), *Drivers of Innovation, Entrepreneurship and Regional Dynamics*, Advances in Spatial Science, DOI 10.1007/978-3-642-17940-2_3,

sectors, space, and time (Rosenthal and Strange 2004; Van Oort 2007; De Groot et al. 2009; Melo et al. 2009; Neffke 2009; Burger et al. 2010).

At the same time, relatively little is known about the importance of agglomeration economies for the performance of firms (Acs and Armington 2004; Martin et al. 2011). Many empirical studies on agglomeration use aggregated data with cities or city-industries as the basic reference unit. Hence, these studies provide only limited insights and weak support for the effects of agglomeration economies on firm performance. Regional-level relationships are not necessarily reproduced at the firm level because information on the variance between firms is lost when using aggregated data. Hence, even if regions endowed with a greater number of agglomeration economies grow faster, this conclusion cannot be generalized to firms. In the social sciences, this problem is referred to as the "ecological fallacy" (Robinson 1950) or the "cross-level fallacy" (Alker 1969).

In addition, agglomeration effects found in area-based studies can be compositional (Macintyre et al. 1993). For example, articles in the economic and industrial organization literature often argue that large firms are more likely to grow compared to small firms due to internal economies of scale. Hence, a location may be fast-growing due to the concentration of large firms rather than the localization of externalities or the external economies of scale present. This has been shown in the work of Combes et al. (2008) and Mion and Naticchioni (2009) on spatial sorting and spatial wage disparities. In addition, Baldwin and Okubu (2006) show that the agglomeration of productive firms may simply be a result of a spatial selection process in which more productive firms are drawn to dense economic areas. For this reason, it remains unclear whether geographical differences are an artifact of location characteristics (e.g., agglomeration economies) or simply caused by differences in business and economic composition. This endogeneity problem makes it even more difficult to draw inferences about firms when using cities or regions as the lowest unit of analysis.

Thus far, only a few studies have used firm-level data to assess the effect of agglomeration economies on firm performance. Audretsch and Dohse (2007) find that German firms located in a knowledge-based cluster grow faster than firms located in a region less endowed with knowledge resources. Henderson (2003) considers the productivity effect of employment density in a plant's own county versus neighboring counties. Using industry and time dummies, he finds that a 10% increase in employment in a plant's own county increases the productivity of a plant by 0.8% in the high-tech industry. Using French firm-level data (both manufacturing and services), Martin et al. (2011) find that doubling the size of the firm's sector increases firm productivity by 4–5%. Baldwin et al. (2010) find similar results for the effect of own industry size (in terms of buyer and supplier networks, labor market pooling and knowledge spillovers) on firm productivity in five broad manufacturing sectors in Canada. These studies find no effect of city size on firm productivity.

Although the relative shortage of firm-level evidence in the agglomeration economics literature can mainly be ascribed to data limitations and confidentiality restrictions, this is remarkable nevertheless because the theories that underlie agglomeration economies are microeconomic in nature (Martin et al. 2011). In other words, agglomeration economies do not directly foster regional economic growth, but do so only indirectly, through their effect on firm performance. In this chapter, we focus on the determinants of the survival of new establishments in the advanced producer services sector in the Netherlands. Employing a mixed hierarchical and cross-classified probit regression, we introduce a model of firm survival that is specific to characteristics of the internal and external environment of the firm. This external environment may consist of several components, such as its location, sector or club (location-by-sector). We add to previous studies in three ways. First, we explicitly disentangle the location effect from the firm and sector effect. Second, we analyze whether firms benefit from agglomeration economies asymmetrically in relation to their size. Third, we focus on new establishment survival in advanced producer services; an industry not researched much yet. Controlling for firm and sector characteristics, we find that location accounts for about 4% of the variance in the probability of survival of new establishments. We find that localization and urbanization economies have a positive effect on the survival of new establishments. However, new establishments with large start-up sizes tend to profit more from agglomeration economies than new establishments with small start-up sizes.

2 The Macro to Micro Link in Agglomeration Economics

2.1 Agglomeration Economies

The origin of the agglomeration economies concept can be traced back to the end of the nineteenth century. At the fin de siècle, the neoclassical economist Alfred Marshall aimed to overturn Malthus' and Ricardo's pessimistic (but influential) predictions on the co-evolution of economic and population development. He introduced a form of localized aggregate increasing returns to scale for firms. In his seminal work, *Principles of Economics* (Book IV, Chapter X), Marshall (1890) mentioned a number of cost-saving benefits or productivity gains external to a firm. He argued that a firm could benefit from co-location with other firms engaged in the same sort of business. Marshall considered these *agglomeration economies* to be uncontrollable and difficult to regulate, as well as immobile or spatially constrained.

Marshall (1890) focused on a local specialist labor pool, the role of local knowledge spillover, and the existence of non-traded local inputs. In contrast, Hoover (1948), Ohlin (1933) and Isard (1956) allocated the sources of agglomeration advantages into internal economies of scale and external economies of scale in the form of localization and urbanization economies. Production cost efficiencies realized by serving large markets may lead to increasing returns to scale in a single firm. There is nothing inherently spatial in this concept, other than that the existence of a single large firm in space implies a large local concentration of employment (Van Oort 2004). However, external economies are qualitatively very different.

Whether due to firm size or a great number of local firms, a high level of local employment may allow for the development of external economies within a group of local firms in a sector. These are known as *localization economies*. The strength of these local externalities is assumed to vary, implying that they are stronger in some sectors and weaker in others (Duranton and Puga 2000). The associated economies of scale comprise factors that reduce the average cost of producing outputs in that locality. Following Marshall (1890), a spatially concentrated sector can exert a pull on (and uphold) a large labor pool that includes workers with specialized training in the given industry. Obviously, this reduces search costs and increases flexibility in appointing and firing employees. Moreover, a concentration of economic activity in a given sector attracts specialized suppliers to that area, which in turn reduces transaction costs. Finally, agglomerated firms engaged in the same sector can profit from knowledge spillover, as geographic proximity to other actors facilitates the diffusion of new ideas or improvements related to products, technology and organization.

On the other hand, *urbanization economies* reflect external economies passed to enterprises as a result of savings from the large-scale operation of the agglomeration or city as a whole. Thus, they are independent of industry structure. Relatively more populous localities, or places more easily accessible to metropolitan areas, are also more likely to house universities, industry research laboratories, trade associations and other knowledge-generating institutions. The dense presence of these institutions, which are not solely economic in character but also social, political and cultural, supports the production and absorption of knowledge, stimulating innovative behavior and differential rates of interregional growth (Harrison et al. 1997). However, areas that are too densely populated may also result in a dispersion of economic activities due to pollution, crime or high land prices. In this respect, one can speak of urbanization diseconomies.

Agglomeration economies are more complex than Marshall originally presented. Quigley (1998), for instance, describes additional features embedded in the categorization but not recognized for their individual value. These include scale economies or indivisibilities within a firm, the historical rationale for the existence of productivity growth in agglomerated industries in the first place (Isard 1956). In consumption terms, the existence of public goods leads to urban amenities. Cities function as ideal institutions for the development of social contacts, which correspond to various kinds of social and cultural externalities (Florida 2002). Moreover, agglomeration economies may provide greater economic efficiency growth due to potential reductions in transaction costs (Martin and Ottaviano 2001). The growing importance of transactions-based explanations of local economic productivity growth is a logical outcome of the interaction between urban economies and knowledge-based service industries (Castells 1989). They have also become more important recently. Lastly, Quigley (1998) points to the law of large numbers regarding the possibility of fluctuations in the economy. Fluctuations in purchases of inputs are usually imperfectly correlated across firms, as are the sales of outputs across buyers. As such, opportunities to pool supplies imply that firms are required to hold fewer inventories.

2.2 *From Macro to Micro*

The features of agglomeration economies described above may explain why regions characterized by an agglomeration of economic activities tend to exhibit higher economic growth (McCann and Van Oort 2009). Despite the focus in the empirical literature on the relationship between agglomeration economies and regional growth as a macro-level phenomenon, the underlying theory of agglomeration contains both macro- and micro-level propositions (see Rosenthal and Strange 2004). Although these propositions begin and end at the urban or regional level, they recede at the level of the individual firm. Coleman (1990) explored this fact in his bathtub model (also known as the "Coleman boat"), concluding that system-level phenomena (e.g., agglomeration) influence system outcomes (e.g., regional economic performance) through their effect on firms' orientations and performance. In this respect, performance differences between regions cannot be perceived as a direct result of macro-economic differences between regions. Instead, they are by-products of firms' individual behaviors.

Firms are interested in seeking agents whose production function is partly determined by the region or city in which they are embedded. This is influenced by the opportunities (agglomeration economies) and constraints (agglomeration diseconomies) present in this external environment (Granovetter 1985; Grabher 1993). In turn, differences in opportunities and constraints across regions generate differences in firm performance and, hence, in regional performance. Firms optimize their own performance but do not strive for regional growth. This phenomenon is more explicitly described as follows (see Fig. 1):

1. The region in which a firm is embedded generates opportunities and economic constraints for firms located in that region through agglomeration economies and agglomeration diseconomies (macro-to-micro transition).

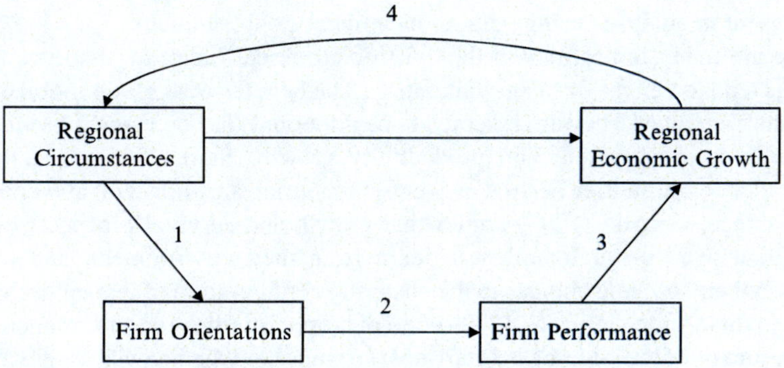

Fig. 1 Macro- and micro-level propositions: effects of regional circumstances on regional economic growth

2. Firms with more economic opportunities and less economic constraints (Proposition 1) tend to perform better in terms of survival chances, employment growth or productivity growth (purposive action).
3. Regions containing successful firms (Propositions 1 and 2) exhibit higher economic growth. Regional performance is here conceptualized as the weighted sum of the firms' performances (micro-to-macro transition).
4. Regional performance affects regional circumstances, resulting in a feedback loop. In this fashion, the model can be linked to the evolutionary development of regions.

Two features of this theoretical model call for clarification. First, the firm's external environment consists not only of the location (physical environment), but of other components, such as the sector in which the firm is embedded (functional environment, Lambooy 1993). For example, firms nested within the same sector share the same technologies and are affected by the same labor market policies and product life cycle. Second, not all opportunities and constraints facing a firm are related to macro-level properties, such as initial firm size, age or entrepreneurship (Santarelli and Vivarelli 2007). However, even when constraints and resources are firm-based, to what extent their effect is independent of the external environment often remains debatable. In this chapter, we focus on the first two propositions and examine to what extent the macro–micro link exists in agglomeration economics.

3 New Establishment Survival in Advanced Producer Services

In order to examine the relationship between agglomeration economies and firm performance, we concentrate on the survival of new establishments in the advanced producer services sector in the Netherlands. An obvious advantage of focusing on new establishments is that these are less constrained by previous decisions, such as past capital installments, which influence how they value the marginal worker and whether new employment is created (Rosenthal and Strange 2003). In the absence of many establishment-level variables, we avoid the endogeneity problems that are often present in analyses using data on incumbent establishments.

There are many indications in the existing empirical literature that new establishments tend to benefit from agglomeration. The benefits of agglomeration extend beyond start-up rates. They are assumed to be important for processes subsequent to entry, such as employment and productivity growth (Stam 2005). Questioning whether these externalities bestow new entrepreneurial start-ups with any competitive advantage, Geroski (1995) argues that growth and survival prospects of new firms will depend on their ability to learn from their environment, and to link changes in their strategic choices to the changing configuration of that environment. Related to this, Audretsch et al. (2006) find that opportunities for entrepreneurship, and therefore of knowledge-based start-ups, are superior when new firms are able to access externalities through geographic proximity to knowledge sources. The underlying argument is that a new firm that must generate its own knowledge capital will be limited by scale and time. It has neither the resources nor the experience to generate ideas. However, a new firm that uses external knowledge and ideas can leverage its own knowledge capital by standing on the shoulders of giants. In addition, the processes subsequent to entry are important, and survival is one of the main goals of a new firm. In line with Audretsch and Mata (1995), we argue that survival (and, later on, growth processes) subsequent to the entry is at least as important as the entry process itself. The post-entry performance of establishments reveals the selection process of markets. The section process enables some of the new entrants to survive and prosper, while others stagnate and ultimately exit.

Our selection of economic activities focuses on new establishments in 19 advanced producer services sectors (see Table 1). Although we realize that agglomeration theory is originally based on the concentration of manufacturing, so most empirical research has focused on this sector (Melo et al. 2009), we argue that advanced business services can profit extensively from agglomeration externalities, as advanced business services are among the most concentrated economic sectors in Europe (Brülhart and Traeger 2005) and these kinds of activities involve the creation, accumulation and dissemination of knowledge (Miles et al. 1995). Advanced producer services are characterized by their heavy reliance on professional knowledge, both codified (explicit) and tacit (implicit). These can be considered a primary source of information and external knowledge; they can use their knowledge to produce intermediary services for their clients' production processes and they are typically supplied to business through strong supplier user interactions (Illeris 1996; Muller and Zenker 2001). Bennett and Smith (2002) find that customers of advanced producer services search for a supplier within a radius of 25 km.

Table 1 Sectors in advanced producer services used in the analysis of new establishment

#	Sector
1	Warehousing and support activities for transportation
2	Publishing
3	Banks and insurance
4	Financial services
5	Real estate activities
6	Rental and leasing activities
7	Computer services activities
8	Information services activities
9	Legal services
10	Accounting
11	Market research
12	Advertising
13	Management consultancy activities
14	Architectural and engineering activities
15	Scientific and research activities
16	Employment activities
17	Office administrative, office and business support activities
18	Services to buildings
19	Telecommunication

4 Putting Theory into Practice: A Multi-level Framework

4.1 Exploring the Macro to Micro Link

Hierarchical or multi-level modeling, which allows the micro- and macro-levels to be modeled simultaneously, is becoming an increasingly common practice in the social sciences. Goldstein (2003) and Moon et al. (2005) summarize recent overviews of area-based studies in relation to multi-level modeling.

Following Jones (2004), there are two distinct advantages to multi-level models. First, multi-level models offer a natural way to assess *contextuality,* or to what extent a link between the macro-level and micro-level exists. Applying multi-level analysis to empirical work on agglomeration starts from the simple observation that firms sharing the same external environment are more similar in their performance than firms that do not share the same external environment. This is due to shared agglomeration externalities. In this fashion, we can assess the extent to which variance in the survival rates of new establishments can be attributed to between-firm variance, between-area variance, or between-sector variance (McGahan and Porter 1997). Hence, we are able to assign variability to the appropriate context (Bullen et al. 1997).

Second, multi-level analysis allows us to incorporate unobserved heterogeneity into the model by including random intercepts and allowing relationships to vary across contexts through the inclusion of random coefficients. Whereas "standard" regression models are geared at modeling the mean, multi-level analysis focuses on modeling variances explicitly. For example, the effect of urbanization externalities may vary across small and large firms or across sectors. This kind of complexity can be captured in a multi-level framework through the inclusion of random coefficients (Snijders and Bosker 1999).

4.2 A Mixed Hierarchical and Cross-classified Model

Multi-level analysis has been concerned with modeling hierarchically nested structures (e.g., firms located in the same region are also located in the same country due to the nesting of the two levels). However, the external environment of the establishment may consist of several components that have a non-hierarchical nesting structure, as they are grouped along more than one dimension, or cut across hierarchies (Goldstein 2003). For example, sectors are not nested in regions and vice versa. Hence, establishments can be in the same sector but located in different regions. These different facets of the external environment may explain variation in establishment performance.

In our model, we distinguish between the following four classifications: (1) regions [40 NUTS-3 regions], (2) sectors [19 sectors in the advanced producer services], (3) sectors-by-regions [$40 \times 19 = 781$ clubs], (4) establishments [46,038 new establishments]. To begin with, establishments may be affected by the region in which they are located. As indicated earlier, these location factors may be general (to which all establishments in a given location are exposed) or sector-specific (restricted to a subset of establishments nested within a given sector in that location). However, establishments may also be affected by external factors that are not location-specific, but are common to all establishments within a given sector. As the classification of clubs is intersected by the classification of the sectors and the classification of the locations, it cannot be independently estimated. Hence, we should disentangle the general location factors and the nation-wide sector-specific factors from the sector-specific factors that are spatially bounded. The sector-specific effects that are location-specific are assessed at the appropriate sectors-by-regions or "club" level (Gordon and McCann 2000). To illustrate this, we use a *mixed hierarchical and cross-classified model* (presented in Fig. 2). We have a three-level model (with four classifications) with a random intercept for establishments at the lowest level and random intercepts for regions ($k1$), sectors-by-regions (j), and sectors ($k2$) at the higher levels. More formally, we estimate the following base probit model for the probability of survival $Y_{ij(k1,k2)}$ of new establishments in the advanced producer services:

$$
\begin{aligned}
&y_{ij(k1,k2)} = Binomial(n_{ij(k1,k2)}, \mu_{ij(k1,k2)}) \\
&probit(\mu_{ij(k1,k2)}) = X_{ij(k1,k2)}\beta_0 + u_{0j(k1,k2)} + v_{0k1} + v_{0k2} \\
&\text{where } u_{0j(k1,k2)} \sim N(0, \sigma^2_{u0j(k1,k2)}),\ v_{0k1} \sim N(0, \sigma^2_{v0k1}),\ v_{0k2} \sim N(0, \sigma^2_{v0k2}) \qquad (1)
\end{aligned}
$$

in which the probability of survival of new establishments $\mu_{ij(k1,k2)}$ is explained by the single fixed intercept term $X_{ij(k1,k2)}\beta_0$, which is the average survival rate of new establishments in the advanced producer services. The three separate random terms $u_{0j(k1,k2)} + v_{0k1} + v_{0k2}$ are related to the intercept and mirror, the remaining residual variation at the higher levels. This differs from a typical regression model in that we assume that each sector-by-region j, region $k1$ and sector $k2$ has a different intercept. Note the mixed-hierarchical and cross-classified structure here: the indexing structure $\mu_{ij(k1,k2)}$ refers to the ith establishment in the jth club, which is nested in region $k1$ and sector $k2$. This null model allows us to understand how to

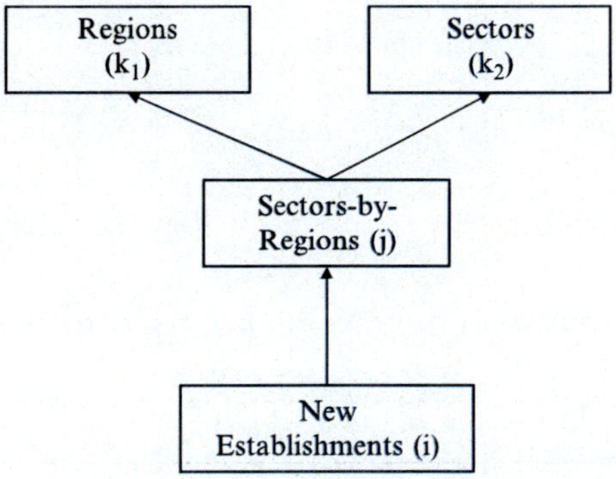

Fig. 2 A mixed hierarchical and cross-classified model of the external environment of new establishments

attribute variation in the probability of new establishment survival to various contexts.

The variance partition coefficient (VPC) can measure the extent to which the probability of survival of new establishments in the same club/region/sector resemble one other relative to those from new establishments in different clubs/regions/sectors. This figure may also be interpreted as the proportion of the total residual variation in survival that is due to differences between clubs, regions, or sectors. For example, the VPC for regions represents the percent of variation explained by the region level differences for firm i in club j and sector $k2$ (2).

$$VPC_{k1} = \frac{\sigma^2_{v0k1}}{\sigma^2_{u0j(k1,k2)} + \sigma^2_{v0k1} + \sigma^2_{v0k2}} \tag{2}$$

In (2), the term $\sigma^2_{u0j(k1,k2)}$ is the between-club variance, σ^2_{v0k1} is the between-region variance, and σ^2_{v0k2} is the between-sector variance. We assume that the probit distribution for the establishment-level residual implies a variance of 1 (Goldstein 2003).

4.3 Adding Predictor Variables and Cross-level Interactions

So far, we have only partitioned the variability in the probability of survival of new establishments over regions, sectors-by-regions, sectors, and establishments. However, in order to see to what extent they explain the partitioned variability, we can add predictor variables to these classifications. More specifically, the predictors (or fixed parameters) we add here contain measures related to establishment characteristics, sector-by-region characteristics and region characteristics. Since we are mainly interested in the effects of regional and sector-by-region characteristics on firm performance, we include sector fixed effects (δ_{k2}) by including sector dummies (based on the classification in Table 1) at the sector level. More formally,

$$y_{ij(k1,k2)} = Binomial\left(n_{ij(k1,k2)}, \mu_{ij(k1,k2)}\right)$$

$$probit\left(\mu_{ij(k1,k2)}\right) = X_{ij(k1,k2)}\beta_0 + \beta_{10}X_{1ij(k1,k2)} + \sum_{j=1}^{q}\beta_{qo}X_{qj(k1,k2)} + \sum_{k1=1}^{r}\beta_{r0}X_{rk1}$$

$$+ \delta_{k2} + u_{oj(k1,k2)} + v_{0k1}$$

$$\text{where } u_{oj(k1,k2)} \sim N\left(0, \sigma^2_{u0j(k1,k2)}\right), v_{0k1} \sim N\left(0, \sigma^2_{v0k1}\right) \tag{3}$$

In (3), the segment $\beta_{10}X_{1ij(k1,k2)} + \sum_{j=1}^{q}\beta_{q0}X_{qj(k1,k2)} + \sum_{k1=1}^{r}\beta_{ro}X_{rk1}$ contains the predictor variables X at the firm, club and region levels that enter the analysis.

The subscripts q and r indicate the number of predictor variables included at the club and regional levels, respectively (please note that with respect to the establishment level, we only include initial establishment size, X_1). The βs refer to the associated regression slope terms.

Equation (3) is a random intercept model. Only the intercept varies across clubs and regions. However, parameter estimates may also vary across different subpopulations. For example, the effects of localization and urbanization economies may vary over small and large firms. This can be modeled using a cross-level interaction between firm size (X_1) and the respective agglomeration economies. Including firm size as predictor variable at the firm level, we obtain (4).

$$
\begin{aligned}
y_{ij(k1,k2)} &= Binomial\left(n_{ij(k1,k2)},\ \mu_{ij(k1,k2)}\right) \\
probit\left(\mu_{ij(k1,k2)}\right) &= X_{ij(k1,k2)}\beta_0 + \beta_{10}X_{1ij(k1,k2)} + \sum_{j=1}^{q}\beta_{qo}X_{qj(k1,k2)} + \sum_{k1=1}^{r}\beta_{r0}X_{rk1} + \delta_{k2} \\
&\quad + \sum_{j=1}^{q}\beta_{q10}X_{1ij(k1,k2)}X_{qj(k1,k2)} + \sum_{k1=1}^{r}\beta_{r10}X_{1ij(k1,k2)}X_{rk1} \\
&\quad + u_{1j(k1,k2)}X_{1ij(k1,k2)} + v_{1k1}X_{1ij(k1,k2)} + u_{0j(k1,k2)} + v_{0k1}
\end{aligned}
\tag{4}
$$

Where $u_{oj(k1,k2)} \sim N\left(0,\ \sigma^2_{u0j(k1,k2)}\right)$, $v_{ok1} \sim N\left(0,\ \sigma^2_{v0k1}\right)$

In (4), $\sum_{j=1}^{q}\beta_{q10}X_{1ij(k1,k2)}X_{qj(k1,k2)} + \sum_{k1=1}^{r}\beta_{r10}X_{1ij(k1,k2)}X_{rk1}$ now represent the cross-level interactions between establishment size and the club-level variables and between establishment size and the region-level variables, respectively, while $u_{1j(k1,k2)}X_{1ij(k1,k2)} + v_{1k1}X_{1ij(k1,k2)} + u_{oj(k1,k2)} + v_{0k1}$ represents the random part of the model. The expressions $u_{1j(k1,k2)} + v_{1k1}$ are the random slope parameters that make the effect of firm size on the probability of survival dependent on the club and region in which the firm is embedded. The cross-level interactions that aim to explain the random slopes can be interpreted as the variation of the effect of the club and region variables across small and larger firms. In the remainder of the chapter, we focus in particular on the interaction between firm size and the different agglomeration economies. However, it should be noted that the range of possible interactions is not limited to these variables.

5 Data and Methodology

5.1 Data and Variables

Data on employment at the establishment level was obtained from the LISA (*Landelijk Informatie Systeem Arbeidsplaatsen – National Information System of Employment*) database, an employment register that covers all establishments in the

Netherlands for the period 1996–2006. For each firm, we retrieved detailed information about the number of employees, economic activity (sector) and geographic position. Our dependent variable, *SURVIVAL* (2000–2006) is a Boolean dummy variable measured at the level of the establishment, which takes the value 1 if a new establishment in 2000 or 2001 survived the first five years of its existence.

As indicated in our theoretical framework, we focus on two types of agglomeration economies: localization economies and urbanization economies. *LOCALIZATION ECONOMIES*, or sector-specific scale economies, are defined at the sector-by-region level and measured as the concentration of own-sector employment in the region under observation. *URBANIZATION ECONOMIES*, or economies available to all firms in a region irrespective of sector they are in, is defined at the region level and measured by the concentration of total employment, which arises from urban size and density.

Besides indicators for localization and urbanization economies, we introduce control variables related to the firm, sector-by-region, and region. At the establishment level, we take *INITIAL FIRM SIZE* into account, measured as the natural logarithm of the number of employees in the year the establishment was founded. Size represents the economies of scale available to a new establishment. Economies of scale, internal to the establishment, refer to the fact that the unit costs of production are a decreasing function of output. By explicitly differentiating between internal and external economies of scale, we try to account for compositional effects.

With respect to the sector-by-region (club) level, the variable *COMPETITION* is introduced to control for market structure. *COMPETITION* is measured as the natural logarithm of the number of entries and exits in the regional sector between 2000 and 2006 divided by the number of firms in 2000. Finally, at the regional level, we have chosen to include R&D expenditures and human capital stock as the main control variables. Regional *R&D EXPENDITURES* are measured as the natural logarithm of the R&D expenditures of firms, research institutes and government agencies in 2000. The *HUMAN CAPITAL* stock in a region is measured as the natural logarithm of the percentage of the workforce that is highly educated (ISCED 5–6) in 2000. Both the *R&D* and the *HUMAN CAPITAL* indicators stem from the research of Broersma and Oosterhaven (2004).

5.2 Estimation Strategy

The mixed hierarchical and cross-classified models specified in the previous section are estimated using the MLWIN 2.10 software (Rashbash et al. 2008). More specifically, we estimate six models. First, we estimate a random-intercept probit model (1) for survival without including any predictor variables. The VPCs are derived from these models (2), which serve as a tool to indicate to what extent location matters by explicitly disentangling the between-location variance from the between-firm and -sector variance. Second, we estimate a random intercept probit model (3) to assess the importance of the different types of agglomeration economies in new establishment survival. Third, we estimate random coefficient models to assess whether the effect of agglomeration economies varies across firms of different sizes (4).

In (3) and (4), we assume that the establishment-level predictor variables are uncorrelated with the club- and regional-level error terms and that the club-level predictor variables are uncorrelated with the regional-level error terms. However, both theoretically and empirically, such an assumption is difficult to meet. Not correcting for this would lead to inconsistent parameter estimates. However, as shown by Snijders and Berkhof (2007), the correlation between the lower level predictor variables and higher level error terms can be easily removed by including club- or region-level means of the lower-level predictor variables in the regression model, a procedure known as the Mundlak (1978) correction. Hence, our multi-level probit models are augmented with this correction.

6 Empirical Results

6.1 Partitioning the Variance

As indicated in the previous paragraph, one advantage of multi-level modeling is the decomposition of the variance. This has three higher-level classifications, in our case: (1) regions; (2) sectors; (3) sectors-by-regions (clubs). The VPC measures the extent to which the probability of survival of new establishments in the same club/region/sector resemble one other relative to those from new establishments in different clubs/regions/sector. Although the VPC is mainly a descriptive tool, it provides insight into the extent to which the region or sector matters for the performance of firms compared to firm characteristics. In empirical research on firm performance, the use of variance decomposition analysis dates back to the work of Schmalensee (1985), who disentangled sector and corporate effects from business unit effects. Today, this has resulted in a large empirical literature in industrial organization, which focuses on whether the appropriate unit of analysis is the firm or the industry (see Rumelt 1991 and McGahan and Porter 1997). However, location remains an under-studied factor in this type of analysis.

Table 2 shows the proportion of the total residual variation in new establishment survival in the advanced producer services sector that is due to differences between clubs, regions, or (sub)sectors. We see that firm survival (survival and growth) is mainly affected by internal (establishment) characteristics. More than 90% of the total variance is between establishment variance. The between-region variance is 3.3%, while the between-club variance is 1.3%. Although the external environment explains only a marginal amount of the variation in the probability of new establishment survival, we argue here that the region contributes to firm performance given the enormous diversity of firms. As we defined agglomeration economies as both region (urbanization economies) and club (localization economies) related, we argue that those externalities “explain” about 3.3–4.6% of the variance in the

Table 2 Variance partition coefficients (VPCs) for the survival of new establishments

	Model 1 survival
VPC (firm) – between firm variance	90.9%
VPC (club) – between club variance	1.3%
VPC (region) – between region variance	3.3%
VPC (sector) – between sector variance	4.5%
N	46,038

probability of new establishment survival. In the next paragraph, we model the contributions of these agglomeration variables to new establishment survival.

6.2 Agglomeration Economies and New Establishment Survival

Table 2 shows the results of our model estimates. With respect to establishment size and survival opportunities (due to downscaling possibilities), we find a small positive and significant effect, with a marginal effect at the mean of 0.005. This effect may be small due to the fact that our "sample" of new establishments mainly consists of smaller firms and that the heterogeneity of size in relation to the probability of survival is relatively low. This is in contrast with other studies, which find a much larger positive relationship between size and survival for incumbent firms (e.g., see Audretsch and Dohse 2007; Raspe and Van Oort 2011).

We now turn to the effect of agglomeration economies on the probability of new establishment survival in the advanced producer services sector. From the previous section, we already noticed a "solitary spatial effect." But looking into this deeper, we conclude that the concentration of own-sector employment (localization economies) has a small, positive effect on new establishment survival, with a marginal effect at the mean of 0.036. The urban density effect, stemming from urbanization economies, has a much higher impact on new establishment performance in the advanced producer services. New establishments located in dense urban regions experience higher survival rates, where the marginal effect at the mean is 0.108. Hence, we conclude that new establishments in the advanced producer services sector have fewer difficulties surviving in cities.

But the relationship between agglomeration and new establishment survival might not be a fixed relationship over all establishments. On the contrary, we argue that some establishments (based on establishment-specific characteristics) profit more than others, or that externalities only appear for some types of firms. In this section, we test for so-called "cross-level interaction effects," interactions between variables measured at hierarchically structured data on different levels (Hox 2002). We focus on initial firm size, analyzing the possibility that agglomeration economies are mainly effective for the larger start-ups.

It appears that initial establishment size has a significant slope variance (the basic underlying condition for the existence of cross-level interaction effects). Table 3 shows the results of the random coefficient models, where we allowed

Table 3 Multi-level probit on new establishment survival

	Model 3 – *PROBIT* survival	Model 4 – *PROBIT* survival
Intercept	0.153 (0.659)	−0.041 (0.636)
Initial establishment size (ln)	0.013 (0.007)*	−0.138 (0.131)
Localization economies (ln)	0.094 (0.053)*	0.083 (0.054)
Competition (ln)	−0.085 (0.054)	−0.087 (0.054)
Urbanization economies (ln)	0.277 (0.102)***	0.254 (0.098)***
Human capital (ln)	−0.129 (0.146)	−0.181 (0.138)
R&D expenditures (ln)	−0.064 (0.035)*	−0.064 (0.033)*
Est. Size * Localization Economies		0.022 (0.013)*
Est. Size * Urbanization Economies		0.051 (0.021)**
Sector fixed effects	Yes	Yes
Mundlak correction	Yes	Yes
Random part		
u_{0jkl}	0.013 (0.002)	0.015 (0.003)
u_{1jkl}	0.009 (0.002)	0.009 (0.002)
v_{0kl}	0.023 (0.006)	0.033 (0.008)
v_{1kl}	0.008 (0.003)	0.006 (0.002)
Observations		
Regions	40	40
Sectors-by-regions	781	781
Establishments	46,038	46,038

Standard errors between parentheses; covariance between the region's intercept and slope in the random part not displayed
$^{*}p < 0.10$, $^{**}p < 0.05$, $^{***}p < 0.01$

for the possibility that the effect of initial firm size can vary from region to region (regions have different slopes), including an interaction effect on size and localization and urbanization economies. The random part in Table 3 shows that the covariance between the region's intercept and slope is significant and *positive*. This positive covariance suggests that a higher intercept is associated with a higher slope. In other words, either larger firms perform better in some regions or their smaller counterparts perform less well in some regions. The question is whether the different agglomeration economies influence this relationship. To test this, we account for cross-level interaction effects in Model 4. We find that the interaction effects between initial establishment size and localization and urbanization economies are significant and positive. This means that *larger* start-ups profit more from own industry and urban density.

7 Conclusions and Discussion

A large empirical literature examines whether spatial circumstances give rise to agglomeration economies – external economies from which firms can benefit through co-location – that endogenously induce localized economic growth. Many existing empirical studies show that agglomeration economies may be one source of the uneven distribution of economic activities and economic growth across cities and regions. At the same time, relatively little is known about the importance of agglomeration economies to the performance of firms. This absence is remarkable because the theories that underlie agglomeration economies are microeconomic in nature. Agglomeration economies do not directly foster regional economic growth, but do so indirectly through their effect on firm performance.

We analyzed this relationship by focusing on the determinants of the survival of new establishments in the advanced producer services sector in the Netherlands. Employing a mixed hierarchical and cross-classified logistic regression, we introduced a model of firm survival specific to characteristics of the internal and external environment of the firm. The external environment consists of several components, such as its region, sector or club (sector-by-region). Controlling for firm and sector characteristics, we find that location accounts for 3.3–4.6% of the variance in new establishment survival. Although this spatial effect appears to be small at first sight – it was obtained that over 90% of the variance in the probability of new establishment survival is due to differences between establishments and the effect of location seems to be rather small at first hand – we still argue that "space matters significantly," especially given the enormous establishment heterogeneity that exists.

We do not find evidence for the widespread hypothesis that larger firms have more survival opportunities. Agglomeration externalities, defined on the regional and sector-by-region levels, perform unevenly in relation to survival and (subsequent) growth of new producer service firms. Localization and urbanization economies do have a positive effect on new establishment survival. However, urbanization economies appear to be more conducive to new establishment survival in the advanced producer services industry than localization economies. Introducing cross-level interaction effects in our models, we find that larger start-ups profit from proximity to a concentration of own-sector employment and urban density with respect to survival opportunities. These outcomes have implications for policymakers. Region and sector conditions have a significant but relatively limited impact on business service firms' survival and growth prospects. Common and popular policies aiming at stimulating spatial producer service clusters (defined as clubs by sector-specific concentrations) may increase the survival chances of only the largest start-ups. In addition, larger urban areas exhibiting potentially larger urbanization economies have a more robust and distinct impact on new establishments. This suggests that localized policy measures should be limited to areas outside these largest economic agglomerations.

In this chapter, we have shown that multi-level analysis provides an analytical tool to assess and magnify the link between the macro-level and micro-level. Yet, there are some limitations to the use of multi-level analysis in spatial research. Multi-level analysis does not fully account for the spatial dependence present in data, in that it does not allow for the effect of neighboring regions on the performance of a firm. Spatial spillover effects between regions may notwithstanding be highly relevant and not accounting for this may underestimate the importance of "space" in explaining the performance of firms. For example, R&D and human capital are well known for spatial spillover effects. Viable solutions here would be to include spatially weighted independent variables in the model (e.g., Florax and Folmer 1992), use a conditional autoregressive multi-level model (e.g., Breslow and Clayton 1993) or employ a spatial multiple membership model (e.g., Browne et al. 2001). Combining such empirical strategies with a micro-macro framework will put the literature on agglomeration economics a step forward in answering the question to what extent the environment of firms is important for the performance of firms.

References

Acs Z, Armington C (2004) The impact of geographic differences in human capital on service firm formation rates. J Urban Econ 56:244–278

Alker HR Jr (1969) A typology of ecological fallacies. In: Dogan M, Rokan S (eds) Quantitative ecological analysis. MIT, Cambridge, MA, pp 69–86

Audretsch DB, Dohse D (2007) Location: a neglected determinant of firm growth. Rev World Econ 143:79–107

Audretsch DB, Mata J (1995) The post-entry performance of firms: introduction. Int J Ind Organ 14:413–419

Audretsch DB, Keilbach M, Lehmann EE (2006) Mansfield's missing link: the impact of knowledge spillovers on firm growth. J Technol Transf 30:207–210

Baldwin R, Okubu T (2006) Heterogeneous firms, agglomeration economies and economic geography. J Econ Geogr 6:323–346

Baldwin R, Brown WM, Rigby DL (2010) Agglomeration economies: microdata panel estimates from Canadian manufacturing. J Reg Sci 50:915–934

Bennett R, Smith C (2002) The influence of location and distance on the supply of business advice. Environ Plann A 34:251–270

Breslow NE, Clayton DG (1993) Approximate inference in generalised linear models. J Am Stat Assoc 88:9–25

Broersma L, Oosterhaven J (2004) Regionale arbeidsproductiviteit: niveau, groei en verklaring. Ministry of Economic Affairs, The Hague

Browne WJ, Goldstein H, Rasbash J (2001) Multiple membership multiple classification (MMMC) models. Stat Model 1:103–124

Brülhart M, Mathys NA (2008) Sectoral agglomeration economies in a panel of European regions. Reg Sci Urban Econ 38:348–362

Brülhart M, Traeger R (2005) An account of geographic concentration patterns in Europe. Reg Sci Urban Econ 35:597–624

Bullen N, Jones K, Duncan C (1997) Modelling complexity: analysing between-individual and between-place variation: a multilevel tutorial. Environ Plann A 29:585–609

Burger MJ, Van Oort FG, Van der Knaap B (2010) Trattato sulla scala geografica delle esternalità di agglomerazione ed il MAUP (a treatise on the geographical scale of agglomeration externalities and the modifiable areal unit problem). Sci Reg 4:19–39

Castells M (1989) The informational city. Information technology, economic restructuring and the urban-regional process. Blackwell, Oxford

Coleman JS (1990) Foundations of social theory. Harvard University Press, Cambridge, MA

Combes PP (2000) Economic structure and local growth: France 1984–1993. J Urban Econ 47:329–355

Combes PP, Duranton G, Gobillon L (2008) Spatial wage disparities: sorting matters. J Urban Econ 63:723–742

De Groot HFL, Poot J, Smit MJ (2009) Agglomeration externalities, innovation and regional growth: theoretical reflections and meta-analysis. In: Capello R, Nijkamp P (eds) Handbook of regional growth and development theories. Edward Elgar, Cheltenham, UK, pp 256–281
Duranton G, Puga D (2000) Diversity and specialisation in cities: why, where and when does it matter? Urban Stud 37:533–555
Florax R, Folmer H (1992) Specification and estimation of spatial linear panel models: Monte Carlo evaluation of pre-test estimators. Reg Sci Urban Econ 22:402–432
Florida R (2002) The rise of the creative class. Basic Books, New York
Geroski PA (1995) What do we know about entry? Int J Ind Organ 13:421–440
Glaeser EL (2000) The new economics of urban and regional growth. In: Clark GL, Feldman MP, Gertler MS (eds) The Oxford handbook of economic geography. Oxford University Press, Oxford, pp 83–98
Glaeser EL, Kallal HD, Scheinkman A, Shleifer A (1992) Growth in cities. J Political Econ 100:1126–1152
Goldstein H (2003) Multilevel statistical models, 3rd edn. Edward Arnold, London
Gordon IR, McCann P (2000) Industrial clusters: complexes, agglomeration, and/or social networks? Urban Stud 37:513–532
Grabher G (1993) Rediscovering the scoial in the economics of interfirm relations. In: Grabher G (ed) The embedded firm. On the socioeconomics of interfirm relations. Routledge, London, pp 1–33
Granovetter MS (1985) Economic action and social structure: the problem of embeddedness. Am J Sociol 81:481–510
Harrison B, Kelley MR, Gant J (1997) Innovative firm behavior and local milieu: exploring the intersection of agglomeration, firm effects, and technological change. Econ Geogr 72: 233–258
Henderson JV (2003) Marshall's scale economies. J Urban Econ 53:1–28
Henderson JV, Kuncoro A, Turner M (1995) Industrial development in cities. J Political Econ 103: 1067–1085
Hoover EM (1948) The location of economic activity. McGraw Hill, New York
Hox JJ (2002) Multilevel analysis: techniques and applications. Erlbaum, Mahwah, NJ
Illeris S (1996) The service economy: a geographical approach. Wiley, Chichester
Isard W (1956) Location and space-economy. MIT, Cambridge, MA
Jones K (2004) An introduction to statistical modelling. In: Somekh B, Lewin C (eds) Research methods in the social sciences. Sage, London, pp 236–251
Lambooy JG (1993) The organisation of firms and spatial agglomeration. Department of Economics, University of Amsterdam, Amsterdam, Research Memorandum, No. 1993–7
Lucas RE (1988) On the mechanics of economic development. J Monetary Econ 22:3–42
Macintyre S, MacIver S, Sooman A (1993) Area, class and health: should we be focusing on places or people? J Soc Policy 22:213–234
Marshall A (1890) Principles of economics. MacMillian, London
Martin P, Ottaviano JP (2001) Growth and agglomeration. Int Econ Rev 42:947–968
Martin P, Mayer T, Mayneris F (2011) Spatial concentration and firm-level productivity in France. J Urban Ec 69:182–195
McCann P, Van Oort FG (2009) Theories of agglomeration and regional economic growth: a historical review. In: Capello R, Nijkamp P (eds) Handbook of regional growth and development theories. UK, Edward Elgar, Cheltenham, pp 19–32
McGahan AM, Porter ME (1997) How much does industry matter really? Strateg Manag J 18: 15–30 (Summer Special Issue)
Melo PC, Graham DJ, Noland RB (2009) A meta-analysis of estimates of agglomeration economies. Reg Sci Urban Econ 39:332–342
Miles I, Kastrinos N, Flanagan K, Bilderbeek R, Den Hertog P, Huntink W, Bouman M (1995) Knowledge intensive business services: their roles as users, carriers and sources of innovation. Report to DG13 SPRINT-EIMS, PREST, Manchester, March 1995

Mion G, Naticchioni P (2009) The spatial sorting and matching of skills and firms. Can J Econ 42: 28–55

Moon G, Subramanian SV, Jones K, Duncan C, Twigg L (2005) Area-based studies and the evaluation of multilevel influences on health outcomes. In: Bowling A, Ebrahim S (eds) Handbook of health services research: investigation, measurement and analysis. Open University Press, Berkshire, pp 262–296

Muller E, Zenker A (2001) Business services as actors of knowledge transformation: the role of KIBS in regional and national innovation systems. Res Policy 30:1501–1516

Mundlak Y (1978) On the pooling of time series and cross-section data. Econometrica 46:69–85

Neffke FMH (2009) Productive places. The influence of technological change and relatedness on agglomeration externalities. PhD Thesis, Utrecht University, Utrecht

Ohlin B (1933) Interregional and international trade. Harvard University Press, Cambridge, MA

Quigley JM (1998) Urban diversity and economic growth. J Econ Perspect 12:127–138

Rashbash J, Browne W, Healy M, Cameron B, Charlton C (2008) MlWin, Version 2.10. Centre of Multilevel Modelling, University of Bristol, Bristol

Raspe O, Van Oort FG (2011) Growth of new firms and spatially bounded knowledge externalities. Ann Reg Sci 46:495–518

Robinson WS (1950) Ecological correlations and the behavior of individuals. Sociol Rev 15: 351–357

Romer PM (1986) Increasing returns and long-run growth. J Political Econ 94:1002–1037

Rosenthal SS, Strange WC (2003) Geography, industrial organization, and agglomeration. Rev Econ Stat 85:377–393

Rosenthal SS, Strange WC (2004) Evidence on the nature and sources of agglomeration economies. In: Henderson JV, Thisse JF (eds) Handbook of regional and urban economics: cities and geography, vol 4. North Holland, Amsterdam, pp 2119–2179

Rumelt RP (1991) How much does industry matter? Strateg Manag J 12:167–185

Santarelli E, Vivarelli M (2007) Entrepreneurship and the process of firm's entry, survival and growth. Ind Corp Change 16:455–488

Schmalensee RE (1985) Do markets differ much? Am Econ Rev 75:341–351

Snijders TAB, Berkhof J (2007) Diagnostic checks for multilevel models. In: De Leeuw J, Meijer E (eds) Handbook of multilevel analysis. Springer, New York, pp 139–173

Snijders TAB, Bosker RJ (1999) Multilevel analysis: an introduction to basic and advanced multilevel modelling. Sage, London

Stam E (2005) The geography of gazelles in the Netherlands. J Econ Soc Geogr 96:121–127

Van Oort FG (2004) Urban growth and innovation: spatially bounded externalities in the Netherlands. Ashgate, Aldershot

Van Oort FG (2007) Spatial and sectoral composition effects of agglomeration economies. Pap Reg Sci 86:5–30

Social Capital in Australia: Understanding the Socio-Economic and Regional Characteristics

Scott Baum

1 Introduction

Understanding the factors associated with broad regional economic outcomes and processes has become increasingly important in recent years. While traditional inputs such as capital and labour have, for a long time, driven debates surrounding regional economic processes, a range of new inputs have also begun to appear as important drivers within the literature. Among these new inputs has been the role of social capital, with several researchers including indicators of social capital in their research and modelling (Putnam et al. 1993; Iyer et al. 2005; Leonardi 1995; Flora et al. 1997; Beugelsdijk and van Schaik 2005a; Bartolini and Bonatti 2008; Hauser et al. 2007; Neira et al. 2009).

The broad focus of these papers have been on addressing questions on how social capital fits into discussions about and contributes to regional development. The most noteworthy contribution to the social capital – regional development literature was the publication of "Making Democracy Work" by Putnam et al. (1993). In what is seen to be the seminal work on this topic, the authors studied Italian regions to find that social capital matters in explaining regional economic development. The definition used by Putnam and his colleagues considers social capital as "features of social organisation, such as trust, norms and networks that can improve the efficiency of society by facilitating co-ordinated actions" (p. 167). Significantly they find that the most successful regional governments and regional economies were those possessing high levels of participation in associational life and those is which citizens displayed high levels of trust in social and political institutions.

La Porta et al. (1997) found a similar link between regions with higher levels of trust and civic engagement having better quality government, while Knack and Keefer (1997) find that social capital matters for measurable economic performance with trust and civic norms being stronger in nations with higher and more equal

S. Baum
Urban Research Program, Griffith University, Nathan, Brisbane, QLD 4111, Australia
e-mail: s.baum@griffith.edu.au

K. Kourtit et al. (eds.), *Drivers of Innovation, Entrepreneurship and Regional Dynamics*, Advances in Spatial Science, DOI 10.1007/978-3-642-17940-2_4,

incomes, with institutions that restrain predatory actions of chief executives, and with better-educated and ethnically homogeneous populations. Hauser et al. (2007) using the European Values Study found that a factor accounting for associational activity had a positive impact on innovation, while other indicators such as trust had no association. In Australia, Woodhouse (2006) tested the hypothesis that differential social capital was associated with economic growth outcomes in two rural towns and found that in the face of similar economic challenges, regions with high levels of social capital were likely to fare better than regions with lower social capital.

Taken as a whole this type of research is considered important because the increased interest in understanding the "new" inputs into regional economic outcomes has required the development of a strong evidence-base. This is especially the case if policy makers and practitioners are to make sound decisions regarding regional economic goals and prospects.

Besides this research that takes a direct approach to understanding social capital's impact of regional outcomes, an equally important strand of research has developed which asks what factors contribute to strong or weak social capital. The factors include the general socio-economic characteristics of individuals, but importantly for strengthening the regional development evidence-base, also consider the ways in which space and regional location are important. The questions these studies ask relate to the extent to which spatial location is important to understanding differences in social capital over and above other characteristics.

Iyer et al. (2005) using data from the Social Capital Benchmark Survey show the important geographical diversity that is apparent across regions in the United States. Identifying various individual factors of social capital the research by Iyer et al. (2005) illustrate the complex multi-dimensional nature of social capital and show that spatial differences between urban and rural areas hold even given the impact of other factors. Considering the European picture Beugelsdijk and van Schaik (2005b) have shown that there are significant regional differences in the measures of social capital they developed. For instance when indicators accounting for trust and active group membership are considered the authors find that regions in the Netherlands appear to be largely homogeneous, while within Italy there is significant regional variation across both indicators. In Australia the small scale research by Onyx and Bullen (2000) together with the larger Australian wide analysis by the Commonwealth Bureau of Transport and Regional Economics (BTRE) (2005) have found differences in the social capital responses of those respondents living in urban areas and those living in regional or rural areas and across broader levels of regional differentiation.

This chapter contributes to this growing evidence base by following this second line of empirical endeavour. It considers the factors associated with differences in social capital and specifically includes the importance of spatial factors. In what follows the data and methodological approach is set out, prior to presenting the findings of the analysis. Following this the chapter concludes by considering the issues and implications associated with the analysis.

2 Methods

The analysis set out in this paper uses data from the Household Income and Labour Dynamics Australia (HILDA) survey to develop indicators of social capital and following the methodology develops a multi-dimensional analysis of the determinants of social capital that include the potential geography variations.

2.1 Defining Social Capital

There has been a large body of literature which provides background to the definition and measurement of social capital (Woolcock 1998; Woolcock and Narayan 2000; Putnam 2000; Putnam et al. 1993; Durlauf 1999; Granovetter 1973; Portes 1998). "The basic idea of social capital is that one's family, friends and associated constitute an important assets, one that can be called upon in a crisis, enjoyed for its own sake, and/or leveraged for material gain" (Woolcock 2001, p. 3). Social Capital is a resource that can be used to achieve a variety of ends and as a resource it is generated by individuals or groups of individuals through deliberate processes of accumulation involving interaction with other people. It can be considered, like other forms of capital, as a stock from which future (positive) benefits may flow (Krishna 2000).

Two aspects to social capital can be identified in the research literature: social structures or social networks (informal and formal networks or structures); and the norms governing behaviour in these social structures or social networks. In discussing social networks, Putnam argues that informal networks include relationships people have with their families, partners, friends and neighbours; whereas formal networks include relationships at work, within community groups and churches and with formal bodies such as business and governments. Norms governing network behaviour relate to trust, unity and reciprocity. Within formal networks, individuals have what is called particularised trust; a trust that is specific to the individual a person knows (Uslaner 1999; Cox and Caldwell 2000). This is different to the trust people have for strangers since the probity of a stranger cannot be predicted with the same certainty as it can for a person known to the individual. The trust afforded to strangers is by its very nature generalised and is termed generalised trust (Dasgupta 1998; Uslaner 1999). Trust in formal networks, which is referred to as institutional trust, is similarly general because it is not aimed at individuals but rather at institutions and relates to, for example, trust of "the government", of the police or of the church (Giddens 1990; Black and Hughes 2001; Stone 2001).

A combination of network types and norms produces differing levels of social capital. A high level of social capital is seen in situations where there are cohesive

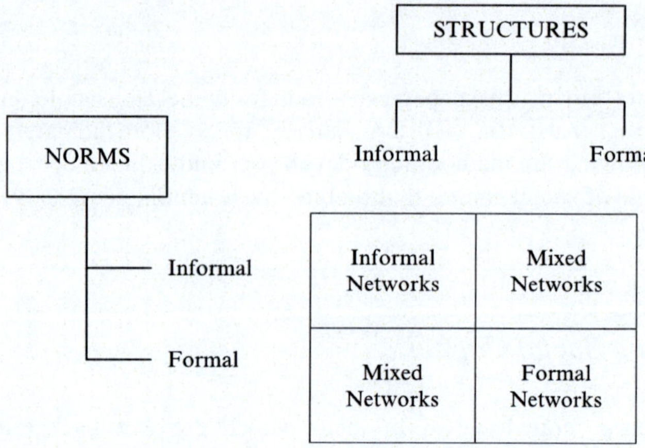

Fig. 1 Structures and norms in social capital (*Source*: Western et al. 2005)

networks of considerable density and where interactions are governed by norms of trust and reciprocity. More particularly, as seen in Fig. 1, it is possible to conceptualise social capital in terms of a four-fold classification of norms and structures with combinations of structures and norms presenting differing types of networks.

2.2 *Developing the Social Capital Indicators*

The empirical measurement of social capital has, like the definition of the concept, been witness to a broad and growing body of literature (see, e.g., Stone 2001; Narayan and Cassidy 2001; Onyx and Bullen 2000; Grootaert et al. 2004; Krishner and Shrader 1999). The development of the measures used in this chapter has taken some of this literature as a guide. The development of the indicators of social capital was undertaken using a range of scales from a social survey and conducting a data reduction exercise.

The survey data used comes from the fifth wave of the Household Income and Labour Dynamics Australia (HILDA) survey. The HILDA survey is a broad social and economic survey conducted annually which contains information on employment, individual socio-economic characteristics and household/family characteristics. It also contains identifiers to allow broad spatial characteristics (such as labour market or local area available from census data and labour force surveys) to be considered. The wave five survey file contains a total of around 17,000 respondents. After data cleaning to remove non-responses and missing variables a workable data set containing around 10,000 respondents was produced.

The wave five HILDA survey contains several possible questions which can be considered as indicators of the different components of social capital. In this analysis we used nine individual variables. These were:

1. Has your physical or emotional health interfered with normal social activities? (1 not at all, 5 extremely)
2. Are you currently an active member of a sporting, hobby or community based association (yes/no)
3. How often get together socially with friends or relatives not living with you? (every day, several times a week, about once a week, two or three times a month, about once a month, once or twice every 3 months, less often)
4. I often need help from other people but can't get it (1 strongly disagree to 7 strongly agree)
5. I have no one to lean on in times of trouble (1 strongly disagree to 7 strongly agree)
6. I often feel very lonely (1 strongly disagree to 7 strongly agree)
7. How many hours per week do you spend doing volunteer or charity work? (hours per week)
8. Generally speaking, most people can be trusted (1 strongly disagree to 7 strongly agree)
9. Most people would try to take advantage of you (1 strongly disagree to 7 strongly agree)

A priori it can be expected that these individual factors will produce several different factors associated with the conceptualisation presented above – that is measures of informal and formal structures and norms.[1]

Given the structure of the individual variables, an optimal scaling exercise was conducted using Categorical Principal Components (CATPCA) routine in SPSS. CATPCA simultaneously quantifies categorical variables while reducing the dimensionality of the data. As with all data reduction exercises CATPCA reduces an original set of variables into a smaller set of uncorrelated components that represent most of the information found in the original variables. The technique is most useful when a large number of variables prohibit effective interpretation of the relationships between objects (subjects and units). By reducing the dimensionality, you interpret a few components rather than a large number of variables. Standard principal components analysis assumes linear relationships between numeric variables. On the other hand, the optimal-scaling approach allows variables to be scaled at different levels. Categorical variables are optimally quantified in the specified dimensionality. As a result, nonlinear relationships between variables can be modelled.

From the nine individual variables three components were apparent from the analysis (Table 1). The first component accounts for the largest share of the variance and is labelled "general support" and accounts for the extent to which individuals are able to access support from friends and family within the community (informal

[1]In developing the indicators of social capital the focus has been on the outer categorization of Fig. 1, rather than on the contents of the cells.

Table 1 Components loadings, CATPCA

	General support	Trust	Participation
I often feel very lonely	0.790		0.110
I often need help from other people but can't get it	0.774		
I have no one to lean on in times of trouble	0.747		0.221
Has your physical or emotional health interfered with normal social activities?	0.624		−0.222
Most people would try to take advantage of you	0.103	0.859	
Generally speaking, most people can be trusted		−0.858	
How many hours per week do you spend doing volunteer or charity work?	0.138	0.202	−0.652
How often get together socially with friends or relatives not living with you?	0.281		0.578
Currently an active member of a sporting, hobby or community based association		0.138	0.491

structures). The second component is labelled "trust" and is simply accounting for the extent to which individuals trust others. The final component is labelled "participation" and measures the extent to which individuals participate in their local communities (formal structures). Each of these components might be thought of as measuring different characteristics of social capital and reflect the types of indicators developed elsewhere (Western et al. 2006; BTRE 2005). Following Iyer et al. (2005) each of the components was re-scaled into a categorical variable – high, medium-high, medium-low, low.

3 Analysis

In order to begin considering the potential for regional differences in social capital to be discerned we run a series of ordered probit models on our three separate indices, with the original index re-scaled into four categories-high, medium-high, medium-low and low. The independent variables included in the models are developed with reference to the existing social capital literature and the availability of indicators within the HILDA survey. A range of variables accounting for the social and demographic characteristics of individuals are included:

- Age (age at last birthday, continuous variable, also age squared)
- Gender (1 = male, 0 = female)
- Income (annual household income, continuous variable, also income squared)
- Education (1 = higher degree, 0 = other)
- Home ownership (1 = home owner, 0 = other tenure)
- Years of residence (years at current address, continuous variable)

- Socio-economic status of the local community (SEIFA index of disadvantage, continuous variable[2])
- Labour force participation (1 = employed, 0 = not working)

Each of these factors has been shown to be associated with individual measures of social capital (Iyer et al. 2005).

The geographical context of social capital is accounted for by including an indicator of the place of residence of each respondent who has a score on a social capital indicator. The HILDA data does, in theory, have a wide range of potential spatial or geographic reference points. Researchers are however largely restricted to the scale of unit used and in this analysis a four level differentiation – Major Urban, Other urban, bounded locality and rural – are used.

As our social capital indicators are represented as categorical variables we fit a series of ordered probit models, one for each of the social capital indicators. The results are discussed below.

3.1 General Support

The results in Table 2 are for the general support indicator. Being male reduced the likelihood of having high general support and increased the likelihood of low general support. Older age was generally related to having lower general support (reduced likelihood of high general support), but the very old have a higher likelihood of having high general support. The socio-economic status of the local community is associated with an increased likelihood of having high general support as is being a home owner, being highly educated and being employed. The significant income variables suggest that having higher incomes is associated with a lower likelihood of having high general support, although for the highest incomes there was a small likelihood of high general support. In the case of general support there were no significant associations with the region of residence.

3.2 Trust

The results for the indicator of trust are presented in Table 3. The significant age_sq variable suggests that extremely older persons have a greater likelihood of having

[2]Socio-Economic Index For Areas (SEIFA) consists of several indexes developed by the Australian Bureau of Statistics. Each index summarises a different aspect of the socio-economic conditions of the Australian population using a combination of variables from the Census of Population and Housing. The Index of Relative Socio-Economic Disadvantage (2001) was used for this analysis and includes variables that reflect or measure relative disadvantage. The indicator is for a local community rather than an individual.

Table 2 Ordered probit results, community involvement indicator

Community involvement	Low			Medium-low			Medium-high			High		
	dy/dx	Std. Err.	z	dy/dx	Std. Err.	z	dy/dx	Std. Err.	z	dy/dx	Std. Err.	z
Gender	−0.00211	0.00238	−0.89	−0.00531	0.00598	−0.89	0.00409	0.00462	0.89	0.003329	0.00375	0.89
Age	0.00023	0.00042	0.55	0.00058	0.00105	0.55	−0.00044	0.00081	−0.55	−0.00036	0.00066	−0.55
Age_sq	0.00000	0.00000	−0.92	−0.00001	0.00001	−0.92	7.39E-06	0.00001	0.92	6.02E-06	0.00001	0.92
SES	−0.00010	0.00001	−6.95	−0.00024	0.00003	−7.00	0.000186	0.00003	6.99	0.000151	0.00002	7
Years	−0.00036	0.00012	−2.89	−0.00090	0.00031	−2.89	0.000694	0.00024	2.89	0.000566	0.0002	2.89
Owner	−0.02206	0.00346	−6.37	−0.04962	0.00690	−7.19	0.040792	0.00606	6.73	0.030883	0.00428	7.22
Employed	0.01136	0.00293	3.88	0.02984	0.00797	3.74	−0.02235	0.00582	−3.84	−0.01884	0.00507	−3.72
Education	−0.01718	0.00272	−6.32	−0.04877	0.00856	−5.70	0.034734	0.00566	6.13	0.031221	0.0056	5.57
Income	0.00015	0.00005	3.02	0.00038	0.00013	3.02	−0.00029	0.0001	−3.02	−0.00024	0.00008	−3.02
Income_sq	0.00000	0.00000	−3.25	0.00000	0.00000	−3.26	6.02E-07	0	3.26	4.91E-07	0	3.26
Urban	0.00778	0.00364	2.14	0.01998	0.00949	2.11	−0.01519	0.00714	−2.13	−0.01257	0.00599	−2.1
Other urban	−0.00685	0.00390	−1.76	−0.01802	0.01069	−1.69	0.013501	0.0078	1.73	0.011371	0.00679	1.68
Bounded locality	0.000739	0.00756	0.1	0.001849	0.01876	0.1	−0.00143	0.01458	−0.1	−0.00116	0.01174	−0.1

Number of observations = 10,715
Log likelihood = −11,572.564

Table 3 Ordered probit results, general support indicator

General support	Low			Medium-low			Medium-high			High		
	dy/dx	Std. Err.	Z	dy/dx	Std. Err.	Z	dy/dx	Std. Err.	Z	dy/dx	Std. Err.	Z
Gender	0.01791	0.00591	3.03	0.01	0.00243	3	−0.00647	0.00214	−3.03	−0.01873	0.00621	−3.02
Age	0.00708	0.00104	6.78	0.00	0.00043	6.63	−0.00257	0.00039	−6.63	−0.00737	0.00109	−6.78
Age_sq	−0.00007	0.00001	−6.53	0.00	0	−6.39	2.46E-05	0	6.39	7.07E-05	0.00001	6.53
SES	−0.00018	0.00003	−5.32	0.00	1.00E-05	−5.25	6.53E-05	0.00001	5.25	1.88E-04	0.00004	5.32
Years	−0.00010	0.00031	−0.32	0.00	0.00013	−0.32	3.58E-05	0.00011	0.32	0.000103	0.00032	0.32
Owner	−0.02267	0.00765	−2.96	−0.01	0.00278	−3.12	0.008382	0.00289	2.9	0.022966	0.00753	3.05
Employed	−0.02001	0.00778	−2.57	−0.01	0.00294	−2.66	0.007348	0.0029	2.53	0.020483	0.00783	2.62
Education	−0.01716	0.00759	−2.26	−0.01	0.00341	−2.14	0.006082	0.00263	2.31	0.018393	0.00837	2.2
Income	0.00032	0.00012	2.64	0.00	5.00E-05	2.63	−1.17E-04	0.00004	−2.63	−3.37E-04	1.30E-04	−2.64
Income_sq	0.00000	0.00000	−2.41	0.00	0	−2.4	1.96E-07	0.00E+00	2.4	5.63E-07	0.00E+00	2.41
Urban	0.00717	0.00923	0.78	0.00	0.00379	0.77	−0.00259	0.00333	−0.78	−0.0075	0.0097	−0.77
Other urban	0.002808	0.01032	0.27	0.001127	0.00411	0.27	−0.00102	0.00377	−0.27	−0.00291	0.01066	−0.27
Bounded locality	−0.02083	0.01762	−1.18	−0.00934	0.00872	−1.07	0.007206	0.00577	1.25	0.022974	0.02057	1.12

Number of observations = 10,715
Log likelihood = −14,652.523

greater trust and a lower likelihood of having low trust. The socio-economic status of the local community is associated with higher trust with a greater likelihood of high than low trust. As an individual lives longer in their particular community trust increases, an outcome also reflected in individuals who own their own homes. In both cases the likelihood of high trust increases as years of residence increases or were the individual is a homeowner. Being employed rather than non-employed increases the likelihood of having high trust rather than low trust and those with higher education also have greater likelihood of having high trust. For income, as the annual income received increases so does the likelihood that an individual will have high trust. Finally, for those living in urban areas, the likelihood of having high trust, net of other factors is lower compared to those living in rural areas.

3.3 Community Participation

The final social capital indicator is community participation and the results of the probit model are presented in Table 4. Living in a higher socio-economic status community is associated with an increased likelihood of having high community participation as does living at the same address for a longer period. Those who were home owners had a higher likelihood of high community involvement. For individuals who were employed (compared to those not working) the likelihood of high community participation was lower. Higher education was associated with a higher likelihood of high community participation. For the income variables, while higher incomes were associated with a lower likelihood of high community participation, extremely high levels of income were also associated with high community participation. For the geographic variables, when compared with individuals living in rural areas, those in urban areas had a significantly lower likelihood of having high community participation (and conversely a higher likelihood of low community participation). The variable "other urban" was only weakly significant but suggested that when compared to individuals in rural areas, those living in other urban areas (smaller cities) had lower likelihood of having high community participation.

4 Discussion

This paper has presented the findings of an analysis of social capital within Australia. The multivariate analysis draws on the empirical analysis undertaken in the US by Iyer et al. (2005) and expands on earlier Australian research (BTRE 2005; Western et al. 2005). The analysis proceeds by developing three measures or indicators of social capital – general support (informal structures), trust and community participation (formal structures) – using data from an Australian social survey and then undertakes analysis of the general determinants of social capital.

Table 4 Ordered probit results, trust indicator

Trust	Low			Medium-low			Medium-high			High		
	dy/dx	Std. Err.	z	dy/dx	Std. Err.	z	dy/dx	Std. Err.	z	dy/dx	Std. Err.	z
Gender	−0.00119	0.00449	−0.26	−0.00100			0.00097	0.00366	0.26	0.00122	0.00464	0.26
Age	0.00101	0.00079	1.28	0.00086	0.00067	1.28	−0.00082	0.00064	−1.28	−0.00105	0.00082	−1.28
Age_sq	−0.00003	0.00001	−4.04	−0.00003	0.00001	−4.02	0.00003	0.00001	4.02	0.00003	0.00001	4.04
SES	−0.00029	0.00003	−11.24	−0.00025	0.00002	−10.97	0.00024	0.00002	10.87	0.00030	0.00003	11.23
Years	−0.00078	0.00023	−3.31	−0.00066	0.00020	−3.30	0.00063	0.00019	3.30	0.00080	0.00024	3.31
Owner	−0.03805	0.00611	−6.22	−0.02957	0.00438	−6.75	0.03146	0.00514	6.12	0.03616	0.00535	6.75
Employed	−0.01752	0.00596	−2.94	−0.01443	0.00478	−3.02	0.01436	0.00492	2.92	0.01759	0.00582	3.02
Education	−0.06475	0.00482	−13.45	−0.06612	0.00587	−11.26	0.04817	0.00335	14.37	0.08269	0.00749	11.05
Income	−0.00033	0.00009	−3.51	−0.00028	0.00008	−3.50	0.00027	0.00008	3.50	0.00034	0.00010	3.51
Income_sq	0.00000	0.00000	0.93	0.00000	0.00000	0.93	0.00000	0.00000	−0.93	0.00000	0.00000	−0.93
Urban	0.03961	0.00675	5.87	0.03488	0.00618	5.64	−0.03177	0.00536	−5.93	−0.04272	0.00759	−5.63
Other urban	0.01423	0.00808	1.76	0.01161	0.00635	1.83	−0.01169	0.00669	−1.75	−0.01415	0.00774	−1.83
Bounded locality	0.02141	0.01551	1.38	0.01643	0.01073	1.53	−0.01779	0.01308	−1.36	−0.02005	0.01317	−1.52

Number of observations = 10,715
Log likelihood = −13,422.453

The findings reported in this paper illustrate several regularities across the three types of social capital – general support, trust and community participation. Importantly reflect a range of existing literature suggesting some consistency both between small scale Australian studies and this larger one, and across studies in different countries.

Socio-economic markers such as high education, high income and employment were significantly associated with higher social capital. While education was always positively associated with high social capital the outcomes for income and employment differed across the different types of social capital indicator. Being a home owner rather than living in some other tenure was consistently associated with higher social capital. Finally the socio-economic status of the local community is positively associated with high social capital across all three models, suggesting that a local neighbourhood effect may be important in determining social capital outcomes. Other individual socio-economic/demographic factors are associated with only some of the social capital measures. Being a male reduces the likelihood of having high general support and age was associated with general support and trust as was the period of residence at a particular address.

With reference to the spatial characteristics of social capital the analysis found that across both the measure of community participation and trust, urban areas seemed to display less social capital than rural areas, especially when the distinction was between major urban areas and rural areas. This finding is important as it suggests that even when account is taken of a range of other potential important variables, regional factors remain important indicators of social capital outcomes, at least for the regional scale adopted here.

What does the analysis suggest for broad regional processes and policies? As was suggested in the introduction, social capital is being thought of as part of a "new" set of inputs in regional processes that sits alongside of the more traditional measures of labour and physical capital. But as the analysis has shown, the broad characteristics associated with social capital are complex and this means that any attempt to consider social capital as an input into the regional development processes will need to contend with this complexity.

If there is an assumption that social capital is a good thing for regional development then were should governments begin? It would appear that education might be an important factor, as is a strong local labour market that encourages strong employment. But if declining regions are looking to social capital to help boost local development then the employment link may be problematic in that these regions may have poorly performing labour markets in the first place. Even if policy makers were sure about what drivers to focus on there is also the issue of how policy should be targeted. There has been a long running policy discussion in regional science about the people versus place policy dilemma – that is should governments focus policy at the people level regardless of where they live or should they focus policy on specific places. Much of the analysis presented here suggests that a broad people based approach might be appropriate. However, the crude spatial analysis that has been included suggests that maybe a place based approach might also be appropriate. In reality it is likely to be a mix of people based and place based approaches that will needed.

These are some of the broad policy issues that might be suggested by the analysis presented in this chapter. However, the analysis also leaves some gaps in the knowledge which need to be considered. What the analysis has not identified here is the factors that might explain why there is a regional difference in social capital measures and therefore how regional differences in social capital might help explain differences in regional development and performance. This large scale analysis therefore needs to be backed up by more in-depth analysis of the associations between regional development/performance and social capital. However, such analysis need to recognise that the impact of social capital, if it is to be correctly conceptualised and identified, must be set within a broad multi-factor framework whereby social capital sits alongside other forms of capital. Furthermore, while broad scale analysis may provide some indicators of potential patterns research will also need to focus on case study material in order to obtain a truly robust account of the potential of social capital to impact on regional development.

References

Bartolini S, Bonatti L (2008) Endogenous growth, decline in social capital and expansion of market activities. J Econ Behav Organ 67(3–4):917–926

Beugelsdijk S, van Schaik T (2005a) Social capital and growth in European regions: an empirical test. Eur J Polit Econ 21(2):301–324

Beugelsdijk S, van Schaik T (2005b) Differences in social capital between 54 Western European regions. Reg Stud 39(8):1053–1064

Black A, Hughes P (2001) The identification and analysis of indicators of community strength and outcomes. Occasional Paper No. 3, Department of Family and Community Services, Canberra

BTRE (2005) Focus on regions, No. 4, Social Capital, Department of Transport and Regional Services, Canberra

Cox E, Caldwell P (2000) Making social policy. In: Winter I (ed) Social capital and public policy in Australia. Australian Institute of Family Studies, Melbourne, pp 43–73

Dasgupta P (1998) Trust as a commodity. In: Gambetta D (ed) Trust. Basil Blackford, Oxford, pp 49–72

Durlauf SN (1999) The case 'against' social capital. Focus 20:1–4

Flora JL, Sharp J, Flora C, Newton B (1997) Entrepreneurial social infrastructure and locally initiated economic development in the non-metropolitan United States. Sociol Q 38 (4):623–645

Giddens A (1990) The consequences of modernity. Stanford University Press, Stanford

Granovetter M (1973) The strength of weak ties. Am J Sociol 78:1360–1380

Grootaert C, Narayan D, Jones V, Woolcock M (2004) Measuring social capital: an integrated questionnaire. The World Bank, Washington

Hauser C, Tappeiner G, Walde J (2007) The learning region: the impact of social capital and weak ties on innovation. Reg Stud 41(1):75–88

Iyer S, Kitson M, Toh B (2005) Social capital, economic growth and regional development. Reg Stud 39(8):1015–1040

Knack S, Keefer P (1997) Does social capital have an economic payoff? A cross country investigation. Q J Econ 112:1251–1288

Krishna A (2000) Creating and harnessing social capital. In: Dasgupta P, Seragenldin I (eds) Social capital: a multi-faceted perspective. World Bank, Washington

Krishner A, Shrader E (1999) Social capital assessment tool. Paper presented at the conference on social capital and poverty reduction, World Bank, Washington, DC

La Porta R, Lopez-de-Silianes F, Schleifer A, Vishny R (1997) Trust in large organisations. Am Econ Rev 87:333–338

Leonardi R (1995) Regional development in Italy: social capital and the Mezzogiorno. Oxf Rev Econ Policy 11(2):165–179

Narayan D, Cassidy M (2001) A dimensional approach to measuring social capital: development and validation of a social capital inventory. Curr Sociol 49(2):59–102

Neira I, Vazquez E, Portella M (2009) An empirical analysis of social capital and economic growth in Europe (1980–2000). Reg Stud 92(1):111–129

Onyx J, Bullen P (2000) Sources of social capital. In: Winter I (ed) Social capital and public policy in Australia. Australian Institute of Family Studies, Melbourne, pp 105–135

Portes A (1998) Social capital: its origins and application in modern sociology. Annu Rev Sociol 24:1–24

Putnam R (2000) Bowling alone: the collapse and revival of American community. Simon and Schuster, New York

Putnam R, Leonardi R, Nanetti R (1993) Making democracy work: civic traditions in modern Italy. Princeton University Press, Princeton

Stone W (2001) Measuring social capital: towards a theoretically informed measurement framework for researching social capital in family and community life. Research Paper No. 24, Australian Institute of Family Studies, Canberra

Uslaner EM (1999) Trust but verify: social capital and moral behaviour. Soc Sci Inf 38:29–56

Western J, Stimson R, Baum S, Van Gellecum Y (2005) Measuring community strength and social capital. Reg Stud 39(8):1095–1109

Woodhouse A (2006) Social capital and economic development in regional Australia: a case study. J Rural Stud 22:83–94

Woolcock M (1998) Social capital and economic development: towards a theoretical synthesis and policy framework. Theory Soc 27:151–208

Woolcock M (2001) The place of social capital in understanding social and economic outcomes. Can J Policy Res 2(1):12–22

Woolcock M, Narayan D (2000) Social capital: implications for development theory, research and policy. World Bank Res Obs 15:225–249

Part II
Evidence-Based Analysis: European Studies

Entrepreneurship, Innovation and Regional Development: A Southern European Perspective

George Petrakos, Pantoleon Skayannis, Apostolos Papadoulis, and George Anastasiou

1 Introduction

Knowledge, innovation and entrepreneurship are considered to be of utmost importance for regional growth, employment and social cohesion in the EU (Asheim et al. 2006). Most regions are meant to adjust their development policies or design new ones in order to incorporate the basic guidelines of the European Lisbon Strategy. The critical question arising is whether the institutional environment and the characteristics of the productive base in the European periphery of the South are suitable for the effective implementation of such policies.

This paper examines the potential of structurally weak regions in the southern European periphery to take advantage of the new policy environment, based on knowledge and innovation, in order to grow and converge. The analysis is based on the examination of the policy context for R&D and innovation in Greece, both at a national and regional level, using as a reference example the region of Thessaly.

It mainly focuses on the key characteristics of the manufacturing industry productive environment in the region of Thessaly, in Greece where a business survey was conducted. Thessaly is by and large a traditional agricultural economy having lately developed the tertiary sector (esp. in tourism and administration). Nevertheless, it always used to have a certain level of manufacturing industry production (initially linked to the primary sector) which over the last years has shown a decreasing trend. Hence, the necessity to secure the level of manufacturing industry was made obvious. In contemporary world, this inevitably had to follow the path of innovation and entrepreneurship. The Lisbon strategy, aiming to activate

G. Petrakos (✉)
South and East European Development Center, Department of Planning and Regional Development, University of Thessaly, 38334 Volos, Greece
e-mail: petrakos@uth.gr

P. Skayannis, A. Papadoulis, and G. Anastasiou
Department of Planning and Regional Development, School of Engineering, University of Thessaly, Pedion Areos, 38334 Volos, Greece

K. Kourtit et al. (eds.), *Drivers of Innovation, Entrepreneurship and Regional Dynamics*, Advances in Spatial Science, DOI 10.1007/978-3-642-17940-2_5,

innovative policy selections seems from a first glance to be a useful strategy for manufacturing industry in regions such as Thessaly.

In this sense, the survey inquired into the patterns of change in the innovative activity of manufacturing industry firms. It focused on the role of both the internal and the external environment, specialisation, human resources and inter-firm relations. It also focused on the ability of firms to innovate and compete in an increasingly open and demanding environment.

This survey, focusing on endogenous entrepreneurial activity does not deal with Multi National Corporations which in any case are not very much present in Thessalian manufacturing industry. In any case, the recent economic crisis has revealed that activating endogenous capacity is a crucial development factor.

The paper is organised as follows: Sect. 2 presents a brief literature review on entrepreneurship, innovation and regional development. Section 3 analyses to experience of Greece, putting emphasis on the evolution and the structural characteristics of industry, the regional structure of the economy and the national and regional innovation systems. Section 4 focuses on the region of Thessaly and presents the main findings of the business survey. Finally, Sect. 5 reports the main findings and the conclusions of the paper.

2 Entrepreneurship, Innovation and Regional Development: A Review of the Literature

During the last few years, the regions have been the focus of attention for literature on innovation policy (Koschatzky 2006). Although in the works of Lundvall (1992) and Nelson (1993) about "National Innovation Systems (NIS)", the regional dimension was absent, this shortcoming was soon overcome, once the importance of spatial and relational proximity in innovation processes was recognised.

The regional dimension was explicitly considered later on, in the approaches of "Regional Innovation Systems (RIS)" (Cooke et al. 2004). The notion of RIS has emerged as a territorially focused perspective of analysis, derived from the broader concept of NIS (Iammarino 2005). Whilst not denying that the national (as well as the international) dimension are important, it assigns to the national level the subsidiary role of assisting the regions to overcome their deficiencies. Innovation and technological progress are the result of a series of complex relationships that exist between private enterprises, universities and public research institutes and the people within them. There are knowledge flows among these actors through channels of interaction, diffusion and personnel mobility (OECD 1997). The European Commission (CEC 2007a), seems to adopt the notion that innovation is most effectively addressed at the regional level, as physical proximity fosters partnerships between actors in both the public and private sectors. The formation of regional clusters is often the key to the successful promotion of research, technological development and innovation. The capacity of regional decision makers and entrepreneurs to turn knowledge, skills and competencies into sustainable competitive advantage is crucial to regions' economic performance.

However, European regions vary considerably in their capacity to absorb and develop knowledge and technology. This impedes their growth prospects and is likely to reinforce the considerable disparities in prosperity across the EU. As Fritsch and Stephan (2005) point out, innovation processes are not spread evenly across space. In the EU context, regional differences exist, regarding the amount and share of innovation between the core regions and those located in the periphery. Overall, peripherality, apart from the geographical distance, can be attributed to rather weak financial capabilities of firms and their dependence on important knowledge sources from outside the region, through non-localized forms of interaction (Lagendijk and Lorentzen 2007). The main problems peripheral regions face, especially those of the European South, are polarisation, insufficient infrastructure, inadequate human and social capital, a low level of R&D and innovation due to a predominance of SMEs in traditional industries, weakly-developed firm clusters, few knowledge providers and a weak endowment of innovation support institutions.

The low level of R&D does not only hamper the internal innovation activity in the region, it also leads to a low absorption capacity on the part of the regional firms. As a consequence, interregional knowledge spillovers as well as public innovation funds cannot be absorbed to a sufficient extent in such regions (Tödtling and Trippl 2005). This is also referred as the "Regional Innovation Paradox".[1] Furthermore, a supply-oriented approach in technology transfer can often be found, which reaches larger firms better than the smaller ones. The demand of SMEs is often not well met and interactive learning is rarely achieved (Asheim and Isaksen 2003). Iammarino (2005) adds that the need for technology in lagging regions is "satisfied mainly by mere adaptation of imported innovation". Additionally they have limited or no capacity to recombine and integrate old and new pieces of knowledge. Another condition that may explain why an RIS does not develop easily in peripheral regions is the absence of innovation and cluster dynamics, because there is neither a critical mass of actors nor the support infrastructure necessary for the emergence of technological innovation (Doloreux and Dionne 2008). The role of historical evolution is also important because it often acts as a filter for assessing new growth opportunities and policy options (Iammarino 2005; Asheim et al. 2006). This applies especially in the Southern European productive system, which is a distinct model of growth based on traditional economic activities throughout the post-war period, small family-owned firms and substantial informal economic activity (Zambarloukou 2007).

According to Tödtling and Trippl (2005), the main policy agenda for peripheral regions is the strengthening and upgrading of the regional economy, giving

[1]The regional innovation paradox refers to the apparent contradiction between the comparatively greater need to spend on innovation in lagging regions and their relatively lower capacity to absorb public funds earmarked for the promotion of innovation and to invest in innovation related activities compared to more advanced regions (Oughton et al. 2002).

priority to organisational and technological "catching-up learning", targeting firms (especially SMEs) and their innovation weaknesses, attracting new firms to the region and strengthening potential clusters. This should be accompanied by behavioural changes, improving the attitude of firms towards innovation and cooperation (Asheim et al. 2006). Of equal importance is the linking of firms to knowledge sources inside and outside the region, encouraging collaboration with the research base and enabling them to benefit from major technological developments, and R&D cost sharing (Garcia-Aracil and Fernandez De Lucio 2008). Uyarra (2007) argues that instead of the prevalent attitude to favour high-tech industries, increased attention should also be given to the "traditional sectors". Hospers (2005) finally adds that when supporting traditional sectors does not seem a viable solution maybe a recombination of the "old" with the "new" could create a more appropriate direction for policies.

3 Innovation, Competitiveness and Development in the European South: The Case of Greece

The less advanced and peripheral EU countries and regions often have a limited ability to adjust to the conditions and demands of the newly emerging European economic space (Davis and Weinstein 1999; Overman et al. 2001). As a result, spatial imbalances continue to exist and in many cases they become wider (Brülhart et al. 2004; Petrakos et al. 2004; Petrakos 2008).

When Greece joined the EU, in 1981, it was the tenth and least-developed member of the Union. Membership was initially received as a shock by the unprepared to join a competitive market Greek economy. Indeed, membership was followed by a divergence of Greece from the EU average in terms of GDP per capita in the period 1981–1994 (Petrakos and Pitelis 2001). Although convergence resumed after 1995 and has continued uninterrupted until the present time, this should be conceived more as the outcome of European cohesion policies rather than the competitiveness of the economy in the integrated market. It should also be kept in mind that, despite convergence in terms of income, the country has maintained its structural deficiencies, especially the ones related to its industrial sector (Petrakos and Kallioras 2006; Petrakos et al. 2008).

3.1 The Industrial Structure of Greece

The evolution and structure of Greek industry is characterised by a number of deficiencies and shortcomings that do not encourage innovative activity. The first one is related to the size of the industrial sector, which, as a share of GDP, is the smallest in the EU-15, and has been continuously declining. In 2005, the share of industry as part of the GDP in Greece was 20.7%, while the share of manufacturing was 9.5%. Over half of the industrial activity in Greece is accounted for by construction. These figures are well below those for the EU-15 as a whole (26.5 and 18.1%, respectively) and indicate limited competitiveness in the integrated EU market (World Bank 2008).

The second unfavourable characteristic of Greek industry is structural in nature and is related to its sectoral composition. Actually, one of the factors behind the weak performance of industry is considered to be its sectoral orientation, which is characterised by the dominance of labour-intensive or low-tech sectors and the limited presence of capital-intensive or high-tech sectors[2,3] in employment terms (Table 1). Note that almost half (48%) of Greek industrial employment in the year 2005 was in low-tech sectors (see footnote 3), while the high-tech sectors account for a very low share (14%). This structure differs significantly from the structure of EU industry as a whole, and remained virtually unchanged throughout the 1995–2005 period. The strong presence of low-tech sectors is an indication that Greek manufacturing industry is traditionally dependent on domestic demand, while the limited presence of capital-intensive sectors is a long-term structural weakness that does not allow for significant innovative activity to take place (Petrakos et al. 2008).

The third unfavourable characteristic of the Greek manufacturing industry is also structural and it is related to the size of firms. As Table 2 shows, the average Greek industrial firm is very small compared to the average size of EU-27 countries. With five employees on average, Greek firms have little room to benefit from internal economies of scale or to develop R&D activities.

Overall, Greek manufacturing industry is relatively small in size and declining over time, concentrated in low-tech sectors and small firms. These characteristics are partly the outcome of historical processes and the geographic coordinates of the country, and partly the outcome of the recent integration experience of the European South (Petrakos et al. 2008). They are at variance with the characteristics of the average EU industry and do not seem to encourage innovation and entrepreneurship.

Table 1 Sectoral composition of industrial employment of the EU-15 and Greece, 1995 and 2005

	EU-15		Greece	
	1995	2005	1995	2005
Low-tech sectors	23	25	48	45
Resource or scale intensive sectors	43	43	39	41
High-tech sectors	34	32	14	14
Total	100	100	100	100

Source: Authors' estimation from Eurostat (2008b)

[2]According to Eurostat (2008a), "high-technology or 'high-tech' sectors are key drivers of economic growth, productivity and social protection, and are generally a source of high value added and well-paid employment".

[3]High Tech and Low Tech sector definitions according to NACE Rev 1.1 and NACE Rev 2. Available at: http://epp.eurostat.ec.europa.eu/cache/ITY_SDDS/Annexes/htec_esms_an2.pdf and http://epp.eurostat.ec.europa.eu/cache/ITY_SDDS/Annexes/htec_esms_an3.pdf

3.2 The Regional Structure of Greece

The structural difficulties faced by the Greek economy are even greater in most regions outside Attiki (the Athens region). Greece has traditionally been a polarised economy where a significant share of population and activities are concentrated primarily in the metropolitan region of Attiki. As Table 3 shows, Athens has a concentration of nearly 40% of the population and 50% of the GDP of the country. It is also by far the most advanced region of Greece, with a GDP per capita higher that the EU-27 average and more than double the level of the least advanced Greek region. Athens contains nearly 50% of the industrial activity of the country, the majority of academic and research institutes and the great majority of the most advanced human resources. Thessaloniki, the second major urban agglomeration

Table 2 The average size of industrial enterprises in the EU-27 countries, 2003

Countries		Countries	
Ireland	64	Finland	16
Slovakia	60	Belgium	16
Luxembourg	38	France	15
Germany	35	Netherlands	15
Romania	34	Sweden	13
Lithuania	29	Spain	11
Latvia	27	Portugal	11
Estonia	26	Hungary	8
UK	21	Poland	7
Bulgaria	20	Italy	7
Austria	20	Czech	7
Denmark	20	Greece	5
		Slovenia	4

Source: UNIDO (2007)

Table 3 The regional structure of the Greek economy at the NUTS II level in 2005

	Population		GDP		GDP/cap
	Share	Change (%) 1981–2005	Share	Change (%) 1981–2005	EU-27 = 100
Attiki	35.85	17.92	48.84	126	109
Kentriki Makedonia	17.25	19.25	13.95	48	65
Thessaly	6.66	6.03	5.09	24	61
Dytiki Ellada	6.61	11.76	4.05	12	49
A. Makedonia, Thraki	5.48	5.67	3.60	17	53
Kriti	5.43	19.73	4.62	75	68
Peloponnisos	5.40	3.66	4.71	24	70
Sterea Ellada	5.04	3.81	5.34	23	85
Ipeiros	3.08	5.33	2.20	42	57
Notio Aigaio	2.74	29.80	2.68	98	78
Dytiki Makedonia	2.66	1.88	2.12	29	64
Ionia Nisia	1.99	20.67	1.55	58	62
Voreio Aigaio	1.83	3.79	1.25	48	55
Greece	100	13.78	100	69	80

Source: ESYE (2008) and Eurostat (2008b)

is behind Athens in most indices but still ahead of the rest of the country. It is estimated that nearly 71% of university graduates holding a Ph.D. degree live in Athens and Thessaloniki (GSRT 2008).

With the exception of the Kentriki Makedonia Region (Thessaloniki), that has a critical scale of population and activities, most other Greek regions are sparsely populated with small urban centres unable to generate significant agglomeration economies and act as poles of attraction. Some island regions are relatively advanced by national standards (Kriti, Notio Aigaio), but their economies (especially of the smaller islands) are solely based on tourism. Most other regions have weak economic bases, depend on traditional agriculture, the public sector, or traditional industry and face serious difficulties in modernising their economies. In addition, the latest evidence seems to indicate that since the mid-1990s, regional inequalities have been increasing in Greece (Petrakos and Psycharis 2004).

Overall, the concentration of activities, uneven development levels, weak human resources and the lack of support mechanisms do not encourage the development of innovative activities outside the major metropolitan areas.

3.3 The National and Regional Innovation System

The structure of the Greek innovation system is highly centralised, dominated by the predominant role of the national government, and following the structure of the political system that has been much the same since the establishment of the modern Greek state (Andreou 2006). Greece looked to external influence in the development of its innovation system (Collins and Pontikakis 2006), following a "top-down" approach (Prastacos et al. 2003). However, despite the increased awareness of the importance of innovation policies, the national system remains structurally immature and dependant on EU funding.

According to Skayannis (2002), the weakness of a production structure means (a) less information- and knowledge-intensive industries, (b) branches remotely relevant to high technology, (c) industries in declining branches (d) weaknesses in taking advantage of comparative advantages, (e) weak development of human capital, (f) low technical and social infrastructure endowment, and (g) weak or non-existent financial mechanisms. In addition, institutional lagging behind is very important. This leads less developed regions, among them Greek regions, to an innovation capacity handicap and to weak regional innovation systems.

The performance of Greece is systematically and considerably below the EU average regarding research and innovation, as demonstrated by the main benchmarking indices. In the European Innovation Scoreboard, 2006b, (SII index), Greece occupies the last position among the EU-25 countries (Fig. 1).

The share of the Greek Community Support Framework (2000–2006) that was spent on innovation measures was very low and corresponded to 2.3% of the total funding (EC 2006a, b, c). The low spending on R&D was also evident in recent Eurostat data. In 2006, Greece spent only 0.57% of its GDP on R&D, translating

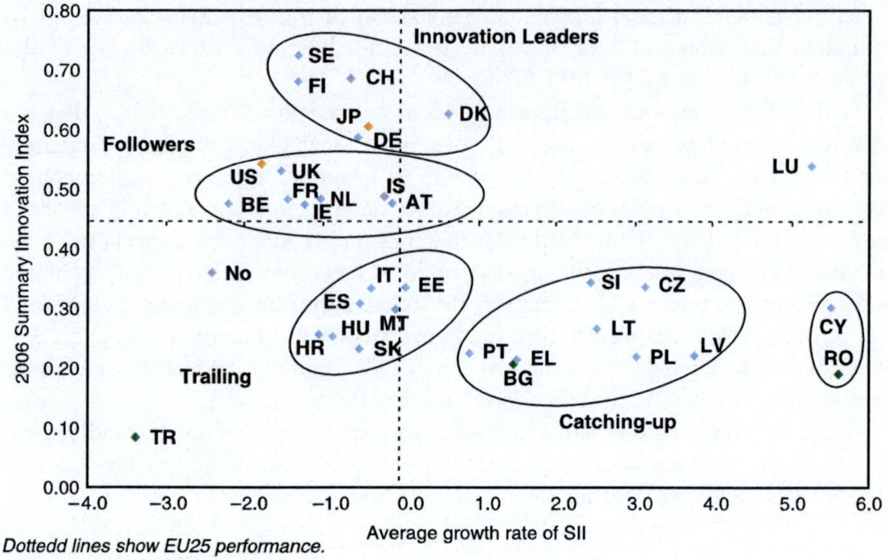

Fig. 1 Summary innovation index of the EU-25 countries, 2006 (*Source*: EC 2006b)

as less than 1/3 of the respective figure for the EU-27 (1.84%), placing them in 21st place (Eurostat 2008c).

Skayannis (1990) has shown that Greece after World War II was rehabilitated and developed under an infrastructure (primarily of the public sector) biased regime of accumulation that practically put manufacturing industry at a second place. Similarly, Collins and Pontikakis (2006) argue that Greece made an early choice favouring public investment in infrastructure in order to reduce regional disparities and enhance growth and competitiveness. Innovation was mostly considered as a risky, high-tech based activity and not a profitable venture.

This mentality has affected spending on education, basic research and life-long learning. The Greek figure on spending per student in higher education is 4,605 euros compared to an average of 5,627 euros in the EU-27 and 6,203 euros in the EU-15 (Eurostat 2008b). The latter ranks Greece in the last place among EU-15 countries on all counts. It seems that education was perceived mostly as welfare expenditure, not as a rewarding investment in human capital with social as well as private returns (Collins and Pontikakis 2006). Indicative of this mentality is the fact that no formal mechanisms have been established for linking tertiary education to industrial needs.

Low levels of innovation and research activities are also explained by the low participation (less than 30%) of the private sector in the Gross Expenditure on Research and Development (GERD) (EC 2006c). Large domestic enterprises that are mainly in low-tech and traditional sectors, have not been at the forefront of investment in new technologies, whereas the fledgling knowledge-based companies in sectors such as health care, software, and communications are typically too small to make a difference in terms of overall R&D investment (Sofouli and Vonortas 2006). Hatzikian (2007) indicates that research activity gives support to the assimilation and the adaptation of existing technology and not to the creation of new knowledge.

At the national, but also the regional level, the deficiency of the innovation governance system is considered to be a serious problem, affecting the quality and effectiveness of innovation policies. According to Andreou (2006), the present spatial setting is characterised by vagueness, uncertainty and asymmetry in the allocation of roles and responsibilities among the administrative levels and the various stakeholders that eventually maintain the centralised character of the system. Instead of empowering local actors and encouraging region-based approaches to economic problems, innovation policy and funds remain under the control of central government. This centralised, complicated and highly bureaucratic administration system reached its limits in the former programming period, increasing delays and difficulties in the implementation of the EU programmes (EC 2006c).

At regional level, the innovation systems are found to be in an embryonic state, especially regarding the elements of production and exploitation of knowledge. The regional innovation base is highly polarised and characterised by serious and increasing disparities in the innovation indices and the R&D infrastructure. Universities, Research Institutions and research employment are concentrated in the metropolitan areas (Alexiadis and Tsagdis 2006; GSRT 2007). In the European Regional Innovation Scoreboard 2006 (RRSII index), the Greek regions account for the last positions. The only notable exception is that of Athens, which has an RRSII index that is close to the EU average.

Table 4 provides a summary account of the performance of the 13 NUTS II regions in the RRSII index in the 2002–2006 period. Attiki (the region of Athens) has the highest score in the innovation index, with a value 53% higher than the national average and 10% lower than the EU-25 average. Second is the region of Kentriki Makedonia (the region of Thessaloniki), with an RRSII value equal to 91% of the national and 53% of the European score. In the next three positions are found regions that combine a relatively large (by Greek standards) peripheral city (Patras, Iraklio, Ioannina) and a relatively old and established university. These three regions (Dytiki Ellada, Kriti, and Ipeiros) maintain an RRSII value that is close to 80% of the national and 45% of the EU-25 score. Then, the RRSII index drops significantly as we move to regions that are either agricultural or peripheral, lack major urban centres, have relatively new universities or lack industrial activities and have an economy which specialises in tourism. In this group of regions, the RRSII index drops to values lower than 50% of the national and 30% of the EU-25 average figures. As can be seen, in the tourist island regions of Notio Aigaio and Ionia Nisia, innovative activity is either very low or completely absent. Overall, innovative activity in the Greek regions is very low by European standards and it is primarily concentrated in the metropolitan region of Athens, which is also the industrial, academic and administrative centre of the country.

Table 4 Revealed Regional Summary Innovation Index (RRSII) in the Greek regions, 2002–2006

Region	Average 2002–2006	GR = 100	EU-25 = 100
Attiki	0.46	153.72	89.70
Kentriki Makedonia	0.27	91.60	53.42
Dytiki Ellada	0.24	78.48	45.81
Kriti	0.23	78.15	45.61
Ipeiros	0.23	76.44	44.55
An.Makedonia-Thraki	0.14	46.67	27.27
Thessaly	0.14	45.12	26.26
Sterea Ellada	0.13	44.07	25.68
Peloponnisos	0.10	31.75	18.49
Dytiki Makedonia	0.08	26.76	15.58
Voreio Aigaio	0.06	21.45	12.50
Notio Aigaio	0.01	1.70	0.99
Ionia Nisia	–	–	–
Greece	0.30	100.00	58.35
EU-25	0.51	171.57	100.00

Source: Authors' estimation from 2006 ERIS (Hollanders 2007)

As argued in this section, the main innovation policy deficiencies found at national level are present and more prominent at regional level. Regional administrations exhibit the same vague attitude towards innovation and the allocation of funds. Despite rhetoric, innovation and knowledge development are not considered a priority and as a result they receive a low share of the Regional Operational Programmes' budget (EC 2006c). The funds allocated to R&D activities in the 2000–2006 period range from 4.4 to 0.7% of the regional budgets. This low budget is often the outcome of pressure from regional lobbies or constituencies, in favour of more tangible projects such as transport or environmental infrastructure. In addition to limited budgets, most regions are also characterised by low operational capacity at the administrative level for R&D and innovation programmes.

In addition to limited demand for R&D, many Greek regions are also faced with a limited supply. This is due to the spatially concentrated character of the research base of the country. More than 70% of research activity in Greece takes place in universities, with the rest taking place in independent research centres or institutes. The great majority of both universities and centres are located in the two metropolitan areas of the country (Athens and Thessaloniki). In most other regions the research base is usually new or very thin.

This is part of the reason why Skayannis (2003, 2005) challenges the Greek infrastructure biased development trajectory stating that innovation and entrepreneurship, especially in the technology sectors (whereby R&D is crucial), should take the lead. Drawing from the Technology Foresight exercise in Greece, he predicts that the prevalent scenarios are not that encouraging for the country and its regions if a major policy change does not happen.

4 Entrepreneurship and Innovation in the Region of Thessaly: A Survey

4.1 The Economic and Innovative Characteristics of Thessaly

Thessaly is a region in central Greece accounting for 6.7% of the population (0.7 million) and 5.1% of the GDP of Greece. Its GDP per capita is below the national average and was equal to 61% of the EU-27 in PPS in 2005. Both population and GDP increased at a slower rate than the national average in the 1981–2005 period (Table 3). Compared to the national average, the region has a higher share of GDP and employment in agriculture (13.7%) and a lower share in services (4.3%) (AllMedia 2007).

The research base of the region primarily consists of the University of Thessaly, the Technological Institute of Larissa and some smaller Institutes. Funded research takes place primarily at the University of Thessaly, which was established in 1988 and currently provides 16 undergraduate and 22 graduate programmes, with 6,500 undergraduate and 1,200 graduate students. The university hosts a tenured or tenure-track faculty of 360 members and a similar number of adjunct or visiting teaching stuff. Like most universities in Greece, the University of Thessaly suffers from serious underfunding in personnel, academic staff and infrastructure. Despite that, it shows a relative dynamism in published academic research and has improved its performance in competitive project funding. Based on Thomson Scientific data, Bontozoglou (2008) has estimated that in 2007 the faculty of the University of Thessaly published nearly 400 papers in refereed journals and had its work cited in nearly 2,400 papers. Also, according to the official report of the University of Thessaly Research Committee, the research budget of the University for the period 2005–2007 exceeded €40 million. Although significant progress has been made in basic and applied research, the interaction of the academic staff with the local economy is still limited.

4.2 The Aim and Methodology of the Survey

In order to examine the innovative characteristics of enterprises in the region, their responses to change and the appropriate policy mix, an industrial business survey was conducted in 2008 by the Regional Innovation Pole of Thessaly (RIP Thessaly 2008). The survey involved 115 industrial firms that responded to a detailed questionnaire of 50 questions divided into six groups. The firms that participated in the survey were not selected randomly from the industrial base of the region. Given that the average industrial firm size in Thessaly is about five employees per firm, it was not meaningful to analyse the behaviour of the representative firm, because this would be too small to be significantly concerned with R&D and innovation. As a result, the firms selected in the survey were in the upper part of the regional distribution, that is, the largest and more established industrial firms in Thessaly. The 115 firms of the survey are very important for the regional industrial base, as they have a total employment size that is equal to 12% (7,326 employees) of the total industrial employment in Thessaly. In terms of employment number, Eurostat and the CIS use a classification of micro-enterprises (<10), small enterprises (10–49), medium-sized (50–250), and large enterprises (>250). However, in the case of Thessaly, these classes would not facilitate useful results, as the (comparatively) very large number of micro firms would not let us draw conclusions for the relatively larger ones for which innovation and the inquiries of this research are more meaningful. It was therefore decided to use a different classification dividing the sample into three size classes (1–20, 21–50, 50+) based on employment (Table 5) in order to detect differences in the performance and behaviour of firms. This classification model, using the terms "small", "medium" and "large" is based on the Greek experience and was derived form the need to differentiate firms so that meaningful results can be yielded. The primary concern in this classification is employment and not its concurrence with economic performance of firms that would lead to a more complex index and to less "legible" results.

As shown in the Table 5, there are significant differences among the three groups. Taking employment as a measure, small firms have an average size of 14 employees, medium-sized firms have an average size of 35 employees and large firms have an average size of 166 employees. Differences are also found in the performance of the three groups. As Fig. 2 shows, average labour productivity

Table 5 Average sales, assets, employment and investment of the sample firms by size group

2008	Total sample	(1–20) 32.7% of total	(21–50) 37.5% of total	(>50) 29.8% of total
Sales (m)	12.8	1.0	5.2	39.1
Assets (m)	17.3	1.8	5.7	51.9
Employees	65	14	35	166
Investment (m)	0.7	0.3	0.2	2.0

Source: RIP Thessaly (2008)

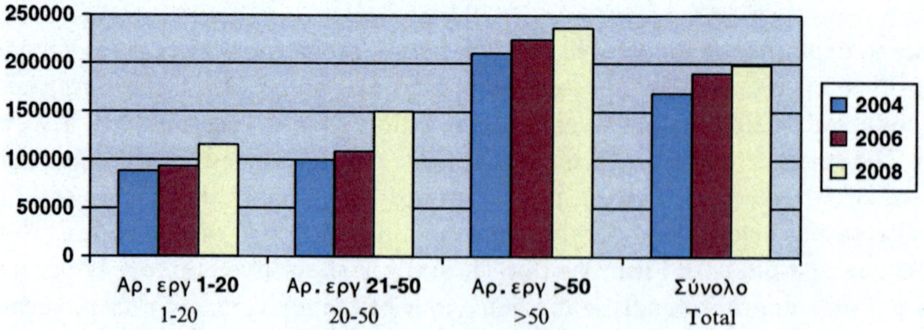

Fig. 2 Labour productivity by size group (current prices) (*Source*: RIP Thessaly 2008)

increases with size, as large firms appear with an average productivity that is twice the level of the small firms.

The analysis below focuses on a selected number of questions, which are important for the aim of this paper. More specifically, the next sections use the results of the survey to analyse the perception of the firms regarding the locational advantages of Thessaly, their competitiveness, their cooperation practices, their R&D and innovative activities, as well as the demand for entrepreneurial development and innovation policies.

4.3 The Locational Advantages of Firms

The location of firms affects their competitiveness through a series of advantages and disadvantages accruing from the characteristics of the host region. In this respect, Table 6 shows the perception of the firms in relation to the strong and weak points of Thessaly as an industrial location.

It indicates that larger firms tend to have a more favourable view of their location than the small ones. It is impressive that this sequence is found in all criteria. In general, however, industrial firms in Thessaly are not satisfied with their location and they give a modest-low overall score (6.65). It seems that the main advantage of the region is its geographic position in the middle of the country (scoring 7.83) and its potential access to national markets. The firms seem to be seriously concerned about the quality of the infrastructure and human resources in the region and the level of financial support for investment. The lowest score is given to business support services that are still in their embryonic stages in the region.

Table 6 The attractiveness of Thessaly for entrepreneurial activity

	Grading in the scale 0–10			
	Total sample	Small firms with 1–20 employees	Medium firms with 21–50 employees	Large firms with more than 50 employees
Geographic position	7.83	7.23	7.87	8.25
Financial services	6.96	6.50	6.71	7.50
Quality of human resources (knowledge, specialisation, experience)	6.34	5.81	6.51	6.65
Transport infrastructure (all modes)	6.28	6.11	6.33	6.32
Level of investment incentives	6.16	5.86	6.03	6.53
Entrepreneurial infrastructure (Industrial areas, technology park)	6.03	5.78	5.68	6.66
Availability of providers and sufficiency of raw material	5.94	5.75	5.83	6.10
Entrepreneurial services (consultancies, marketing, etc.)	5.81	5.44	5.68	6.13
Total grading of Thessaly's attractiveness	6.65	6.29	6.68	6.90

Source: RIP Thessaly (2008)

Table 6 reveals the fundamental problems of peripheral regions that affect negatively entrepreneurship and development. Scores near the middle of the scale suggest a marginally sustainable environment where basic conditions with respect to infrastructure, human resources and support mechanisms are not met. Although large firms seem to be better prepared to deal with these conditions, this is not so with smaller firms. Given that the sample is selected among regional leaders, one can imagine what the opinion of the representative firm in Thessaly with less than five employees would be.

4.4 Markets, Geography and Competitiveness

Table 7 depicts one of the major problems of Greek industry, which is the low level of exporting activity. Overall, 80% of total sales are directed to the regional or the national market and only 11% is exported to the EU. The 28-year long process of integration has not helped Greek firms to expand their market to other EU countries. Moreover, the Thessalian firms have very limited access to the Balkan markets, which are nearby and have lower standards than the advanced European ones. These figures reveal the introvert character and the low level of international competitiveness of the firms. As with the previous Table, size matters. Larger firms seem to be more open, more competitive and less dependent on local demand. On the other hand, the smaller ones depend, almost entirely, on regional and national demand.

4.5 Interactions and Networking in the Local Productive Base

The literature indicates that in regional productive bases dominated by small firms, a successful growth strategy is to develop inter-firm relations and networks at the regional level that will generate external economies of scale and increase the efficiency of the firms. The next two tables provide information for the relations that the firms in Thessaly have developed with each other and the research and support base of the region.

Table 7 The geographical distribution of sales

	Average sales by destination (%)			
	All firms	Small firms	Medium firms	Large firms
Sales to Thessaly	31.08	39.23	30.93	18.31
Sales to rest of Greece	49.38	51.13	51.95	45.47
Sales to the Balkans	3.43	3.26	2.56	4.94
Sales to rest of Europe	11.55	4.87	11.00	21.31
Sales to rest of the world	4.50	1.51	3.56	9.94
Total sales	100.00	100.00	100.00	100.00

Source: RIP Thessaly (2008)

Table 8 depicts the attitude of firms towards cooperation in a number of fields, including production, promotion, design, distribution, supplies, etc., with competing firms (co-operation in competition) or with up-stream and down-stream firms. The first column indicates that the majority of the firms consider that there is no room for cooperation in most fields. This is a shocking position, if one considers the small size of the firms and the multiple problems that reduce their competitiveness. The only areas in which the majority of the firms expect benefits from cooperation are the areas of joint research for new product development and acquisition of know-how.

Despite this negative overall attitude, there is a significant minority of firms that expects benefits from cooperation. On average 9–19% of firms are in favour of cooperation with local competitors, while a smaller group (3–13%) is in favour of cooperation with local upstream and downstream firms. A similar proportion of firms would favour cooperation with firms in other regions, either competing (5–17%), or in related business (4–16%). Cooperation with local competitors is more popular in the fields of product promotion (19.0%), distribution (16.2%) and supplies (14.3%), while cooperation with distant competitors is more popular in the fields of know-how acquisition (17.9%) and production (15.8%).

In general, the spirit of cooperation among the industrial firms in Thessaly is low. The majority of the firms are introvert in character and reluctant to adopt cooperation practices. The analysis by size shows that small firms are less willing to cooperate than large firms (RIP Thessaly 2008). The extent to which this "atomistic turn" to entrepreneurship is the outcome of institutional, cultural, social or other

Table 8 Co-operation between firms in the same or in related activities

Field of co-operation	No (%)	Yes, with competing firms		Yes, with firms related with forward and backward linkages	
		Yes with local firms (%)	Yes, but not with local firms (%)	Yes with local firms %	Yes, but not with local firms %
Co-operation in production	51.8	12.3	15.8	9.6	10.5
Co-operation in promotion	53.3	19.0	8.6	12.4	6.7
Co-operation in product design	57.4	12.0	9.3	9.3	12.0
Co-operation in product distribution	64.8	16.2	4.8	9.5	4.8
Co-operation in supplies	61.1	14.3	9.5	10.5	8.6
Joint research for the development of new products	46.4	13.6	12.7	13.6	13.6
Common use or common purchasing of equipment	79.0	9.5	4.8	2.9	3.8
Co-operation in know-how acquisition	42.5	14.2	17.9	9.4	16.0

Source: RIP Thessaly (2008)

factors is a critical question[4] that needs to be addressed by industrial organisation and perhaps sociological studies. It is interesting to observe that the minority of the firms that are willing to engage in cooperation prefer, on average, to cooperate with competitors rather than upstream or downstream related business. They also have a slight preference for cooperating with local rather than distant partners. These two elements may be an encouraging starting point for the (careful) design of cluster policies in peripheral regions.

The firms in our selected sample were also asked to indicate whether or not they cooperate with the science base of Thessaly, the regional and local administration and the business support organisations. Table 9 reports their responses. In general, cooperation does not seem to be a priority for most firms. At the top of the cooperation list are the local Chambers of Industry and the Regional Industrial Association, with 61 and 53% of the respondents. This is expected, yet is surprisingly low, given that firms are members of these institutional bodies.

Administrative bodies, like the Region of Thessaly, a public business support organisation, the Prefectures and the local Development Agencies come next with shares in the range of 32–45% of the respondents. About 1/3 of the firms declare that they have some sort of cooperation with the University of Thessaly, about 1/4 with the Technical Institute and 1/5 with the Research Centres of the Region. Keeping in mind that the firms in the sample are local leaders and that some of them participate in the Regional Innovation Pole of Thessaly project (RIP Thessaly 2008), the share of firms cooperating with the research and support base of the region is very low.

Table 9 Cooperation of firms with the science and business support base

			If no, intention to co-operate in the future	
	Yes	No	Yes	No
Chambers of industry	60.95	39.05	21.90	6.67
Regional industrial association	53.77	46.23	27.36	8.49
Region of Thessaly	45.19	54.81	26.92	15.38
Centre for entrepreneurship and technology development	36.89	63.11	36.89	11.65
Prefecture	36.54	63.46	31.73	16.35
Development agencies	32.32	67.68	41.41	11.11
Universities	32.08	67.92	35.85	13.21
Centre for professional training	29.81	70.19	33.65	23.08
Technological educational institute	23.81	76.19	48.57	14.29
Research centres	22.00	78.00	50.00	13.00
Municipal enterprises	21.78	78.22	32.67	25.74
Technical chamber of Greece	15.15	84.85	42.42	22.22
Social enterprises	6.32	93.68	33.68	37.89

Source: RIP Thessaly (2008)

[4]The reader should be warned that these are self-reporting answers, so the smaller firms may be biased towards non-cooperation. This is because their conception of clustering makes them perceive it rather as a threat than as an opportunity.

Despite the low shares of cooperation, Table 9 has a positive message for the future. As can be seen in the last two columns, the majority of the firms that have not yet cooperated with the regional research base and support mechanisms are willing to do so in the future. About 35% (50%) of the firms declare that they would like to cooperate in the future with the University (the Research Centres) of the region. Clearly, the firms understand that there are unexplored opportunities associated with their practice and are willing to change.

4.6 R&D and Innovation in the Local Industrial Base

Given the low levels of cooperation with each other and the research and business support base of the region, a critical question is: To what extent are industrial firms internally active in R&D and innovative activities which would allow them to improve their competitiveness? Tables 10 and 11 provide information for the R&D activity and the changes in processes and products initiated by the firms in our sample.

Table 10 indicates that firms which occupy personnel in R&D activity on a steady basis represent a small percentage of the total (24%). Small firms have a lower share (13%) and large firms a higher one (40%). Despite the obvious lack of permanent R&D functions (or because of it), a significant share of firms has a part-time or sporadic engagement, indicating that many firms do actually realise the importance of R&D functions and innovation for their performance. Although the sectoral specialisation of local industry certainly affects the reported figures, we can claim that in general R&D activity is low, even among the leading firms of our sample. Formal R&D activity is mainly concentrated in the larger firms, while the smaller ones are characterised by non-systematic patterns of engagement.

During the last 2 years, the firms in our sample have undertaken some changes in a number of aspects of their activity in order to improve their competitiveness. As Table 11 shows, these changes are modest overall and are characterised by significant variation among different areas of entrepreneurial activity. The most significant major changes (35% of firms) are in equipment, presumably because of the investment subsidies provided. Also, a significant share of the firms have

Table 10 Department or personnel engaged in research activity (R&D activities)

	All firms	Small firms	Medium firms	Large firms
Yes	23.89	13.51	19.51	40.63
No, but some personnel are engaged part time	16.81	21.62	12.20	15.63
No, but some personnel are engaged occasionally, if required	27.43	29.73	29.27	25.00
Nobody	20.35	16.22	26.83	15.63
Nobody in-house, but we cooperate with external laboratories	11.50	18.92	12.20	3.13
Total	100.00	100.00	100.00	100.00

Source: RIP Thessaly (2008)

Table 11 Changes during the last 2 years

	Major changes	Minor changes	No changes
Production equipment	35.85	36.79	27.36
Quality control	26.17	37.38	36.45
Design of product	25.71	41.90	32.38
Range of products	24.77	48.62	26.61
Hygiene and security policy	21.90	38.10	40.00
Packaging	20.19	28.85	50.96
Kind of products	19.09	38.18	42.73
Marketing	11.11	30.30	58.59
Administration	10.68	33.01	56.31
Advertisement	10.20	23.47	66.33
Export policy	9.71	28.16	62.14
Distribution	7.92	14.85	77.23
Stock management	7.14	30.61	62.24
Personnel training	6.80	40.78	52.43
Finance	5.94	19.80	74.26
Supplies policy	5.77	47.12	47.12
Relations with workforce	4.95	30.69	64.36

Source: RIP Thessaly (2008)

introduced major changes in quality control and hygiene policy (a requirement of the law), in product design and in the introduction of new products. Very few firms have introduced major changes in marketing, administration, export policy, personnel training or labour force relations. The great majority of firms that introduced major changes report that these changes have had a positive impact in their business. Also, large firms tend to introduce major changes more often than small ones, although the firms that resist changes the most are those of a medium size (RIP Thessaly 2008).

A significant share of firms, ranging from 30 to 50% of our sample, has introduced minor changes in the areas of entrepreneurial activity of Table 11 during the last 2 years. These changes mostly took place in the domain of production rather that in the softer domains which are, however, the faster changing markets.

Despite significant positive signs of change, it should not escape our attention that in our selected sample of firms comprising many regional leaders, the majority of them have not undertaken any change in most domains. For example, despite their limited competitiveness and poor export performance, the industrial firms of Thessaly have not introduced any change at all in the domains of marketing (58%), administration (56%), advertisement (66%) and exports policy (62%). Overall, in nine out of seventeen domains of entrepreneurial activity, the majority of firms have made no changes during the last 2 years.

4.7 *Innovation Policies*

The last question of the survey asks the firms of the sample to make their suggestions for an effective innovation policy in Thessaly from a list of available measures. Their answers are reported in Table 12. The three most popular policy

Table 12 Suggested innovation policies in Thessaly

	All firms	Small firms	Medium firms	Large firms
Provision of useful information	63.48	66.67	60.98	68.75
Cooperation with the Research base of the Region (UTH, TEI, and RCs)	59.13	66.67	51.22	65.63
Investment subsidies that support clusters	56.52	56.41	48.78	59.38
Best practice transfers from abroad	47.83	53.85	46.34	50.00
Subsidies for innovative activity	46.09	48.72	43.90	50.00
Consultancy services	45.22	35.90	41.46	46.88
Tax incentives that support clusters	44.35	46.15	43.90	50.00
Establishment of an Institute of Entrepreneurship and Innovation in Thessaly	39.13	35.90	41.46	34.38
Forum of knowledge exchange and knowledge diffusion	38.26	33.33	39.02	34.38
Possibility for cooperation with market leaders	22.61	28.21	21.95	25.00
Brainstorming with specialists	21.74	28.21	14.63	18.75
Issuing of a certificate for innovative enterprises	19.13	25.64	12.20	18.75

Source: RIP Thessaly (2008)

measures supported by the majority of the firms are: the provision of useful information (63%), cooperation with universities and research centres (59%) and investment incentives for clusters (56%). Clearly, this table indicates that the firms recognise their weaknesses, which are limited specific knowledge, lack of cooperation with the research base of the region and lack of inter-firm cooperation, and they request regionally based policies that will deal with these factors.

Other policies with significant support from the firms include best practices transfer from abroad (47%), subsidies for innovative activities (46%), better local support mechanisms (45%) and the provision of tax incentives for the development of clusters (44%). It is interesting that all classes of firms rank the requested policies with the same order, regardless of size.

5 Conclusions and Policy Implications

Since the launch of the "Lisbon Strategy" in the year 2000, and especially after the reform of the policy agenda in 2005, boosting regional innovation capacity has been given top priority in National Reform Programmes and the new Cohesion policy (CEC 2007b). However, most of the Southern European regions score below average in the Regional Innovation Scoreboard (Hollanders 2007; CEC 2007b) and still have limited innovative activity. This has raised questions about the ability of the Lisbon strategy to be implemented in all regions, the effectiveness of the funds and the ability of policies to generate convergence among the EU regions (Esposti and Bussoletti 2008). In this respect, studies analysing the innovation environment in Southern Europe can make important contributions towards the alignment of innovation policies.

The findings of this paper suggest that there are a series of conditions that affect the innovative performance of peripheral regions and do not allow them to effectively converge with their more advanced counterparts on the innovation scoreboard.

Firstly, the characteristics of the Greek productive base indicate that many peripheral countries and regions maintain a weak industrial base dominated by traditional, labour- or resource-intensive sectors and small-in-size firms. These two conditions unfortunately imply low levels of entrepreneurial R&D activity. Secondly, the national and regional innovation systems are characterised by low levels of public spending for R&D and a highly centralised and bureaucratic innovation system that further reduces the effectiveness of limited funds. Thirdly, the analysis of entrepreneurial behaviour has revealed low levels of cooperation between firms and low levels of cooperation with the (often promising) research base.

In general, the survey in Thessaly indicates that there are multiple financial, structural, institutional and cultural constraints that generate an unfavourable environment for innovation policies in lagging regions. As a result, the implementation of the Lisbon Strategy in the European regions cannot follow a more or less uniform pattern, but it will be characterised by a great variety in means and results.

Despite difficulties and multiple barriers, there is room for policies that will improve the innovative capacity and performance in lagging regions. First of all, the Greek experience shows that some countries and regions need to re-organise and decentralise their innovation systems, giving more power to regional stakeholders, and to drastically reduce bureaucracy. Nevertheless the decentralisation processes should be exercised with caution, since regions are competing with each other and the fight for resources might lead to sub-optimization from a national perspective.

Secondly, they have to significantly increase R&D funding to levels that overtime come to approach the EU-average figures. Given the budget difficulties that many less advanced countries and regions face, this is a more difficult step. National governments and regional administrations have to convince their constituencies that R&D funding and innovation policies are not a luxury, but a necessary ingredient of a successful growth strategy.

Thirdly, the analysis of entrepreneurial behaviour in Thessaly shows that industrial firms have serious difficulties in cooperating, either because they are competing with each other, or because they think they have deviant interests, or because of cultural reasons. Given the small size of these firms, this practice needs to change where possible, in the spirit of "cooperation in competition". Targeted policies, carefully structured investment incentives, tax breaks and campaigns will be required in order to challenge this deeply embedded entrepreneurial culture of autarky. In this sense, firms, especially the small ones, need incentives and guidance to cooperate, finding common grounds of mutual interest. In the absence of internal economies of scale, they need to seek benefits from external economies of scale in order to improve their collective efficiency. They also need to improve their cooperation with their regional research base, as a potential source of solutions for a wide range of technical or operational problems.

The analysis suggests that a number of characteristics of the productive base and the prevailing institutional environment in lagging regions often require tailor-made policies of innovation. At the same time, it is a tempting "learning from others" practice to look at "success stories" in advanced regions and try to draw policy-related lessons. This exercise needs some caution. Policy makers in less developed regions need to critically assess the EU experience and resist calls to unconditionally adopt and implement policies successfully applied in advanced regions, as those regions have a totally different productive, structural, technological and institutional environment. They are faced with the difficult task of distinguishing truly "successful policies" from policies in "successful regions".

The innovation strategies of the less advanced regions need to be utilising the ingredients of their productive base. To the extent that they have a sectoral focus, this should include traditional or new sectors that have an important participation in local employment. Competing with leading European regions in high-tech sectors that are not available locally may be a strategy of high risk and a possible waste of limited resources. Innovation policies need to provide solutions to pressing problems of the productive base and do not always need to have a high-tech character. Dynamic new sectors in the European South, such as services and tourism can benefit from innovative actions that are not high-tech solutions, but organisational advances that improve efficiency and competitiveness.

References

Alexiadis S, Tsagdis D (2006) Examining the location factors of R&D labor in the regions of Greece. Ann Reg Sci 40:43–54

AllMedia (2007) The economic and social physiognomy of the prefectures and regions of Greece (in Greek). AllMedia, Athens

Andreou G (2006) EU cohesion policy in Greece: patterns of governance and Europeanization. S Eur Soc Polit 11(2):241–259

Asheim BT, Isaksen A (2003) SMEs and the regional dimension of innovation. In: Asheim B, Isaksen A, Nauwelaers C, Tödtling F (eds) Regional innovation policy for small–medium enterprises. Edward Elgar, Cheltenham

Asheim BT et al (2006) Constructing regional advantage: principles-perspectives-policies, report of DG research expert group on constructing regional advantage. European Commission, Brussels

Bontozoglou (2008) Research and innovation in the University of Thessaly' presentation in the conference on research, technology and innovation in Thessaly: 2007-13, Larissa, 20 June 2008

Brülhart M, Crozet M, Koenig-Soubeyran P (2004) Enlargement and the EU periphery: the impact of changing market potential. World Econ 27(6):853–875

CEC (2007a) Competitive European regions through research and innovation: a contribution to more growth and more and better jobs. Communication from the Commission, COM (2007) 474 final, Commission of the European Communities, Brussels, 16 Aug 2007

CEC (2007b) Regions delivering innovation through cohesion policy, SEC (2007) 1547, Commission of the European Communities, Brussels, 14 Nov 2007

Collins P, Pontikakis D (2006) Innovation systems in the European periphery: the policy approaches of Ireland and Greece. Sci Public Policy 33(10):757–769

Cooke P, Heidenreich M, Braczyk HJ (2004) Regional innovation systems: the role of governance in a globalized world. Routledge, London, UK

Davis DR, Weinstein DE (1999) Economic geography and regional production structure: an empirical investigation. Eur Econ Rev 43:379–407

Doloreux D, Dionne S (2008) Is regional innovation system development possible in peripheral regions? Some evidence from the case of La Pocatiere, Canada. Entrepren Reg Dev 20(3): 259–283

EC (2006a) Annual innovation policy trends and appraisal report: Greece 2006, European Trend Chart of Innovation, Directorate General Enterprise and Industry

EC (2006b) European innovation scoreboard 2006: comparative analysis of innovation performance, PRO INNO EUROPE – INNO METRICS, Directorate General Enterprise and Industry

EC (2006c) "Country report: Greece" annex to the "Strategic evaluation on innovation and the knowledge based economy in relation to the structural and cohesion funds, for the programming period 2007–2013", Technopolis and Associates

Esposti R, Bussoletti S (2008) Impact of objective 1 funds on regional growth convergence in the European Union: a panel data approach. Reg Stud 42(2):159–173

ESYE (2008) Statistical database, available at: http://www.statistics.gr

Eurostat (2008a) High-tech statistics, available at: http://epp.eurostat.ec.europa.eu/statistics_explained/index.php/High-tech_statistics

Eurostat (2008b) Statistical database, available at: http://epp.eurostat.ec.europa.eu

Eurostat (2008c) Science, technology and innovation in Europe, News release, 34/2008

Fritsch M, Stephan A (2005) Regionalization of innovation policy-introduction to the special issue. Res Policy 34(8):1123–1127

Garcia-Aracil A, Fernandez De Lucio I (2008) Industry-university interactions in a peripheral European region: an empirical study of Valencian firms. Reg Stud 42(2):215–227

GSRT (2007) Strategic plan for the development of research, technology and innovation under the NSRF 2007-13, General Secretariat for Research and Technology, Hellenic Republic, Athens

GSRT (2008) Registry creation of Phd Holders in the Greek Labour Market – Final report (in Greek), Ministry of Development – General Secretariat for Research and Technology, 2008. Available at: http://www.gsrt.gr/default.asp?V_ITEM_ID=5462

Hatzikian J (2007) Research and technological development policy and innovative performance: the Greek case within EU. In: Korres G (ed) Regionalisation, growth, and economic integration. Heidelberg, Physica

Hollanders H (2007) 2006 European Regional Innovation Scoreboard (2006 RIS), MERIT, European Trend Chart on Innovation

Hospers G-J (2005) Best practices and the dilemma of regional cluster policy in Europe. Tijdschrift voor Economische en Sociale Geografie 96(4):452–457

Iammarino S (2005) An evolutionary integrated view of regional systems of innovation: concepts, measures and historical perspectives. Eur Plan Stud 13(4):497–519

Koschatzky K (2006) The regionalization of innovation policy: new options for regional change? In: Fuchs G, Shapira P (eds) Rethinking regional innovation and change-path dependency or regional breakthrough? Springer, The Netherlands, pp 291–312

Lagendijk A, Lorentzen A (2007) Proximity, knowledge and innovation in peripheral regions. on the intersection between geographical and organizational proximity. Eur Plan Stud 15(4): 457–466

Lundvall B-Å (ed) (1992) National systems of innovation: towards a theory of innovation and interactive learning. Pinter, London

Nelson R (1993) National innovation systems: a comparative analysis. Oxford University Press, Oxford

OECD (1997) National innovation systems. OECD, Paris

Oughton C, Landabaso M, Morgan K (2002) The regional innovation paradox: innovation policy and industrial policy. J Technol Transf 27:97–110

Overman H, Redding S, Venables AJ (2001) The economic geography of trade, production and income: a survey of empirics. CEPR Discussion Paper 2978

Petrakos G (2008) Regional inequalities in Europe: evidence, theory and policy. Town Plan Rev 79(5):i–vii

Petrakos G, Kallioras D (2006) Peripherality and integration: industrial growth and decline in the Greek regions. Paper presented in the 46th ERSA Congress in Volos, August 2006

Petrakos G, Pitelis Ch (2001) Peripherality and integration: the experience of Greece as a member of the EU and its implications for the Balkan economies in Transition. In: Petrakos G, Totev S (eds) The development of the Balkan region. Ashgate, Aldershot, pp 283–316

Petrakos G, Psycharis Y (2004) Regional development in Greece. Kritiki, Athens

Petrakos G, Rodriguez-Pose A, Anagnostou A (2004) Regional inequality in the EU. In: Bradley J, Petrakos G, Traistaru J (eds) Integration, growth and cohesion in an enlarged European Union. Springer, New York, pp 29–43

Petrakos G, Fotopoulos G, Kallioras D (2008) Economic integration and structural change in the Greek regions: a southern perspective. In: Krieger-Boden C, Morgenroth E, Petrakos G (eds) The impact of economic integration on regional structural change and cohesion. Routledge, London, pp 72–95

Prastacos G, Spanos Y, Kostopoulos K (2003) Innovation: determining factors and reflections on the future of the Greek economy, Report prepared for the General Secretariat of Research and Technology within the context of the project: Technology Foresight in Greece

RIP Thessaly (2008) A survey of the innovative characteristics of industrial firms in Thessaly, Regional Innovation Pole of Thessaly, University of Thessaly

Skayannis P (1990) The general conditions of production and infrastructure: the case of post-civil was Greece. D.Phil. Thesis, University of Sussex, UK

Skayannis P (2002) First innovation attempts in less developed European regions: contextual questions and policy issues. In: Bakouros Y, Skayannis P, Stamboulis Y (eds) Entrepreneurship and innovation policies in the European periphery: a research agenda. University of Thessaly Press, Volos

Skayannis P (2003) Spatial issues, infrastructures and new technologies in Greece. The Horizontal Sector of Greek Space and Regions, Report prepared for the General Secretariat of Research and Technology within the context of the project: Technology Foresight in Greece

Skayannis P (2005) Greece at a turning point: from the construction society to the knowledge society? AESOP 2005 Annual Congress, Technical University of Vienna, 12–17 July

Sofouli E, Vonortas N (2006) S&T Parks and business incubators in middle-sized countries: the case of Greece. J Technol Transf 32:525–544

Tödtling F, Trippl M (2005) One size fits all? Towards a differentiated regional policy approach. Res Policy 34(8):1203–1219

UNIDO (2007) Industrial statistics (cd), UNIDO

Uyarra E (2007) Key dilemmas of regional innovation policies, innovation. Eur J Soc Sci Res 20(3):243–261

World Bank (2008) Key development data and statistics, available at: http://web.worldbank.org

Zambarloukou S (2007) Is there a South European pattern of post-industrial employment? S Eur Soc Polit 12(4):425–442

Productivity Spillovers, Regional Spillovers and the Role of by Multinational Enterprises in the New EU Member States

Marcella Nicolini and Laura Resmini

1 Introduction

There is a widely held assumption that multinational enterprises (MNEs) generate benefits that spill over to the host economy, resulting in productivity growth. Several channels foster the diffusion of such spillovers. They include backward and forward linkages with local firms – through which multinational firms may encourage the entry and development of more efficient local suppliers and final goods producing firms (Markusen and Venables 1999), competition and demonstration effects (Wang and Blomstrom 1992; Glass and Saggi 2002), as well as movements of labour force from multinationals to local firms (Fosfuri et al. 2001). The transmission of spillovers from MNEs to domestic firms, however, is not automatic; rather, it is affected by several factors, most of which can be summarized in the concept of distance, broadly defined in order to encompass both the economic and the geographical dimension. Economic distance concerns relative backwardness and absorptive capacity and determines whether and to what extent local firms eventually benefit from Foreign Direct Investment (FDI)-induced spillovers (Findlay 1978; Glass and Saggi 1998).[1] Geographical distance, instead, affects the transmission mechanism, reducing the possibilities for indigenous firms located far from multinational enterprises to reap such benefits. Very recently, the literature has uncovered that other firm specific characteristics may affect the transmission of spillovers from foreign to indigenous firms, once controlling for absorptive capacity and distance. In particular, Nicolini and Resmini (2010) point out to the importance of indigenous firms' size and the activity of MNEs as potential factors able to affect

[1] In this paper, Multinational Enterprises (MNEs) and Foreign Direct Investment (FDI) are used as synonymous.

M. Nicolini
Fondazione ENI Enrico =Mattei, Corso Magenta 63, 20123 Milan, Italy

L. Resmini (✉)
Faculty of Political Science and International Relations, Università della Valle d'Aosta, Loc. Grand Chemin 73/75, 11020 Saint Christophe (AO), Italy
e-mail: l.resmini@univda.it

K. Kourtit et al. (eds.), *Drivers of Innovation, Entrepreneurship and Regional Dynamics*, Advances in Spatial Science, DOI 10.1007/978-3-642-17940-2_6,

the type of spillovers generated by foreign firms and indigenous firms' capacity to absorb them, not to mention the existence of possible country-specific patterns.[2] The lack of automaticness in the transmission of MNEs' spillovers at firm level, together with the consideration that spillovers may also be negative,[3] make it significant to examine whether and to what extent FDI spillovers are strong and systematic enough to have a net positive impact at aggregate levels.

In this paper we analyse empirically this issue. In particular, we ask whether and to what extent MNEs can enhance productivity in the domestic manufacturing sectors of three new EU member states, i.e. Bulgaria, Poland and Romania, and to what extent they foster growth and development at regional level.

The empirical analysis is based on an unbalanced panel with annual information on more than 40,000 domestic firms and about 10,000 foreign owned firms operating in Bulgaria, Poland and Romania during the period from 1998 to 2003.[4] Our analysis focuses on manufacturing firms only. According to the recent studies on productivity growth using longitudinal research data, a large proportion of aggregate productivity growth is attributable to resource allocation, which mainly occurs within manufacturing sectors (Bartelsman and Doms 2000). Moreover, the latter have been attracting a large number of MNEs since the early of the transition phase, while FDI flows in either primary or tertiary sectors have been quite scarce because of strict regulations, and other impediments removed very recently.

Also the choice of the countries to be included in the study was determined by the need to control for different degrees of foreign investment penetration and sectoral composition, in order to rule out the possibility that the observed effects may depend on FDI (sectoral) intensity only. At the beginning of the transition phase, Bulgaria, Poland and Romania had very similar technological levels and

---

[2]In particular, Nicolini and Resmini (2010) found that MNEs operating in traditional labour intensive sectors generate intra-sectoral spillovers that accrue to medium firms in Bulgaria, and small firms in Poland and Romania, while MNEs operating in high-tech manufacturing sectors generate inter-sectoral spillovers accruing to small firms in all the considered countries and large firms in Romania. These results suggest that at aggregate level, FDI induced spillovers depend, on the one hand, on the composition of each manufacturing sector in terms of the average size of indigenous firms, and, on the other hand, on the industry structure of each region, not to mention the regional distribution of high and low tech foreign firms within each host country.

[3]Negative effects arise when the entry of a (foreign) firms into the domestic market increases competition, and less competitive indigenous firms leave the market.

[4]These data come from Amadeus database published by Bureau Van Djik, which besides standard financial information gives also details on several other qualitative and quantitative variables, such as the structure of the ownership, industry classification, and geographical location within countries. Only firms whose ownership can be properly identified have been included in the sample. To this respect, firms with a share of foreign ownership greater than 10% have been classified as foreign affiliates, while all other firms with a percentage of foreign ownership below 10% have been classified as domestic. Although our sample does not consider the entire population of firms operating in the considered countries over period, its representativeness is fairly good, as it is shown in the Appendix.

managerial skills; despite that, they have followed very different paths towards open market economy, whereby Poland became a member of the European Union in 2004, while Bulgaria and Romania had to wait other three years before joining the EU. During the transition phase, FDI have flown into these countries according to the pace and the deepness of the structural reforms implemented in each country included in the sample. Consequently, Poland has rapidly become one of the most important FDI recipients in the area, while Bulgaria and Romania fail to attract a substantial stock of foreign capitals, at least till the end of the 1990s. Yet, the sectoral composition of foreign investments varies across countries according to their degree of development, with Poland attracting more medium high and high tech foreign firms than Bulgaria and Romania, which seem to be particularly attractive for MNEs operating in traditional labour intensive manufacturing sectors (Pusterla and Resmini 2007).

In order to achieve our research objective we apply key theoretical arguments concerning FDI induced spillovers and the measurement of externalities with spatial data. The existing empirical literature concerned with the determinants of such spillovers has largely ignored spatial correlation effects, despite the fact that these considerations are vital in modelling externalities, mainly when seeking to distinguish between local and interregional effects (Driffield 2006).[5] This paper is a first step in filling this gap.[6]

This integrated approach has several methodological advantages. First, the consideration of spatial dependence does not only provide additional insights on the geographical pattern of distribution of FDI spillovers, but it also changes the quantitative estimates of the marginal effect exerted by MNEs. Positive (negative) spatial spillovers induce dissemination and feedback effects that may magnify (reduce) the direct impact of MNEs on productivity growth rates. Therefore, disregarding spatial dependence may lead to an omitted variable bias and potentially misleading inference on the role played by MNEs in enhancing growth. Secondly, spatial autocorrelation allows to account for variations in the dependent variable arising from latent and unobservable variables. Therefore, it can act as a proxy for omitted variables and properly catch their potential effects. This is particularly useful in case of new EU member states, where explanatory variables at regional and sectoral level are quite scarce.

The rest of the paper is organized as follows. Section 2 describes the method we use to construct our measures of productivity at aggregate level and its corresponding growth rates. In Sect. 3 the estimated model and the econometric issues related to it are examined. Section 4 presents our empirical results and Sect. 5 concludes.

[5]To this respect, Keller (2002), Coe and Helpman (1995) are exceptions.

[6]Very recently, it has been recognized that spatial dependence may play a role as a determinant of MNEs location patterns (Bloningen et al. 2007; Coughlin and Segev 2000; Baltagi et al. 2008).

2 From Firm Level to Regional Level TFP

It is well acknowledged in the literature that MNEs possess both tangible and intangible assets, including intellectual property, technology, brand names and copyright as well as human capital embodied in these assets (Markusen 1995; Feldman 2003; Dunning 1977, 1981). Since knowledge, broadly defined, is embodied in these intangible assets, the ownership of these assets makes foreign firms a vehicle for the transfer of knowledge and a potential source of productivity spillovers in the host countries, provided that MNEs can not fully internalise the value of these benefits.

Productivity spillovers from MNEs may take place when local firms improve their efficiency by copying technologies of foreign affiliates operating in the local market, either based on observation or by hiring workers previously trained by the affiliates. Other kinds of spillovers occur if multinational entry leads to more severe competition in the host country market and forces indigenous firms to use their existing resources more efficiently or to search for new technologies. These are knowledge spillovers in nature and mainly occur when local firms benefit from the presence of foreign companies in their sector (horizontal spillovers). Finally, productivity spillovers may also take place, directly or indirectly, because of linkages between foreign firms and their local suppliers and clients (vertical spillovers).[7]

These brief theoretical considerations suggest that in order to learn whether FDI-induced spillovers are economically relevant, we should estimate the potential correlation between MNEs and Total Factor Productivity (TFP) at the sectoral and regional levels. Because official data does not provide this information directly, we need to estimate it starting from firm-level information.

Typically, plant level TFP is measured using the TFP residual ($ln\,\omega$) computed as the difference between the log of output ($ln\,Q$) and the contribution of inputs ($\beta' lnX$):

$$ln\omega = ln\,Q - \beta' lnX \quad (1)$$

where (1) represents either the gross output or the value-added production function.

The standard procedure for estimating (1) is to deflate the output or the value-added variable by replacing the unknown firm price index with the price index of the industry each firm belongs to (Aitken and Harrison 1999; Olley and Pakes 1996; Levinsohn and Petrin 2003). However, this solution has been considered not only imperfect, but mainly inappropriate, because the resulting measures are "*contaminated by variation in factor prices and demand shocks*" (Katayama et al. 2009,

[7]See Blomstrom and Kokko (1997) and more recently Barba Navaretti and Venables (2004) for an in depth discussion about FDI-induced spillovers.

p. 404).[8] The consideration that "*differentiating between productivity differences and differences in markups is difficult, if not impossible*" (Bartelsman and Doms 2000, p. 578), together with the observation that estimating a production function in "physical" terms may be meaningless, unless firms produce a unique homogenous good, yields us to assume (1) is a two factor Cobb-Douglas (not deflated) revenue function.

Following the approach most commonly used in the recent literature on the topic, we estimate it by applying the semi parametric estimation technique developed by Olley and Pakes (1996). This technique takes into account the simultaneity bias due to the endogeneity of the firm's input selection, which may arise if a firm responds to unobservable productivity shocks by adjusting its input choice. This would imply a correlation between the inputs and the error term which biases traditional OLS coefficient estimates. Olley and Pakes suggest as a solution to this problem the use of firm's investment decisions as a proxy for unobserved productivity shock.[9] By applying this two step procedure on a sectoral base, we obtained sector-specific labour and capital intensities.[10] We then fitted (1) and constructed the individual error terms, which were the logs of our estimated plant TFP.[11]

Discrete changes in (1) have been computed as log changes over the period (Petrin and Levinshon 2005), i.e.:

$$ln\omega_t - ln\omega_{t-1} \tag{2}$$

Starting from (2) an aggregate measure of TFP changes may be obtained. In so doing, a weight α_i is applied in order to take into account firm heterogeneity, thus yielding to the following approximation:

$$\sum\nolimits_i \alpha_i(ln\omega_{it} - ln\omega_{it-1}) \tag{3}$$

where i denotes the N plants in the sector/economy. For α_i, Tornqvist (1936) suggests averaging beginning and ending period shares in total output:

[8]Taking advantages of two unique panel-data samples which include information on firm price indexes, Mairesse and Jaumandreu (2005) demonstrate that the elasticities of factors of production included in a simple Cobb-Douglass production function vary more with the estimation procedures than with the particular specification of the production function equation, being the latter a real output function, a revenue function deflated either by individual prices or industry price, and a not deflated revenue function. Therefore, the omitted variable bias claimed by other scholars seems to be negligible.

[9]This implies that firms with zero or negative investment can not be considered when estimating input coefficients. Alternatively, Levinsohn and Petrin (2003) suggest that material inputs can be used as a proxy for the firm's reaction to productivity shocks.

[10]Two sectors, namely manufacturing of refined petroleum products (NACE 23) and recycling (NACE 37), were excluded because the small number of firms operating in these sectors made it impossible to apply the Olley and Pakes procedure.

[11]In so doing we recover the information on productivity of firms active in period t but with zero investments. In fact, omitting plants with zero investment would have meant omitting plants with low or declining productivity, thus introducing a sample bias in the next steps, i.e. the construction of an aggregate TFP measure and the analysis of the impact of FDI spillovers on it.

$\alpha_i = \dfrac{s_{it} + s_{it-1}}{2}$.[12] A principle feature of this approximation is that it allows to group plants in any subaggregates without affecting the measure of aggregate TFP growth. Given the structure of our databases, we consider three subgroups of firms, i.e., continuing firms, exiting firms and entrants. The former are all active within the period studied, therefore TFP changes and shares can be observed both in t and $t-1$. Exiting (entrant) firms, instead, contribute to aggregate TFP from $t-1$ to the time they exit (from the time they enter to t). Therefore, we can not observe either shares or TFP levels in t and $t-1$, respectively. As suggested by Petrin and Levinshon (2005), averaging shares is a good way to minimize potential errors, while missing information on TFP levels can be forecasted using values of $ln\omega$ observed immediately after (before) the entry (exit). Therefore, our measure for aggregate TFP changes assumes the following form:

$$\Omega_{t-1,t} = \sum\nolimits_{i\in C} \frac{s_{it} - s_{it-1}}{2}(ln\omega_{it} - ln\omega_{it-t}) + \sum\nolimits_{i\in E} \frac{s_{it}}{2}(ln\hat{\omega}_{it} - ln\omega_{it-t}) + \sum\nolimits_{i\in X} \frac{s_{it-1}}{2}(ln\omega_{it} - ln\hat{\omega}_{it-t}) \tag{4}$$

where C indicates continuing firms, E entrant firms and X exiting firms.

We use (4) to retrieve TFP changes in 21 manufacturing sectors in 30 NUTS II regions[13] belonging to Bulgaria, Poland and Romania over the period 1998–2003. Eventually we end up with 630 observations. In the econometric section we explore the role MNEs can play as a determinant of these changes.

3 Model Specification

This section presents and discusses the empirical specification used in this study. We specify a regression equation (log form) as follows:

$$\Delta TFP_{sr}^{03,98} = \alpha_0 + \beta_1 TFP_{sr}^{98} + \beta_2 LQ_{sr}^{98} + \beta_3 size_{sr}^{98} + \beta_4 MNE_{sr}^{98} + \beta_5 TRADE_{s}^{98} + \beta_6 (TRADE_{s}^{98} * LQ_{sr}^{98}) + \beta_7 \Delta PPI_{s}^{03,98} + \beta_8 HR_{r}^{98} + \varepsilon_{sr} \tag{5}$$

According to which the growth rate of TFP of sector s in region r over the 1998–2003 period ($\Delta TFP_{sr}^{03,98}$) is expressed as a function of a number of sectoral and regional characteristics measured at the beginning of the period in order to

[12]This approximation is preferred by a number of scholars. See Petrin and Levinshon (2005) for an in-depth discussion of the advantages of this approximation with respect to other possible aggregations available in the literature.

[13]Several studies have emphasized that the transition process yields to both regional and sectoral changes. These studies belong to an emerging body of literature focusing on regional performance following transition, i.e. detecting loosing and winning regions (Traistaru et al. 2003).

minimize possible endogeneity problems. Sectoral characteristics, always measured at regional level, include the initial level of TFP (TFP_{sr}^{98}), the relative concentration of the sector, measured by the location quotient (LQ_{sr}^{98}), the average size of firms ($size_{sr}^{98}$), a measure of MNE activity (MNE_{sr}^{98}), whose construction will be discussed below, a measure of trade openness ($TRADE_{s}^{98}$), obtained as the ratio of exports and imports over the output of the sector. This measure has been then interacted with the location quotient ($TRADE_{s}^{98} * LQ_{sr}^{98}$) in order to give it a regional dimension that would otherwise be impossible to obtain. Finally, we consider the variation of the production price index ($\Delta PPI_{s}^{03,98}$) in order to understand the role played by the price component in TFP changes, thus (partially) correcting for possible measurement errors in the construction of the dependent variable. Regional characteristics include the human capital endowment, proxied by the human resources devoted to science and technology activities (HR_{r}^{98}) while α_0 is the constant and ε_{sr} the error term.[14]

Existing literature is the basis for justifying (5). Technology diffusion models emphasize the importance of either absorptive capacity, that is, the ability of a country/region to adopt foreign technology for use in the domestic markets (Findlay 1978), or human capital in productivity growth (Nelson and Phelps 1966; Benhabib and Spiegel 1994). We use the initial level of TFP and the human resource variable to capture these effects. New technologies are diffused through a variety of channels, the most important of which are international trade and MNEs (Coe and Helpman 1995; Xu 2000; Xu and Wang 1999; Keller 2002). Finally, the findings that firms geographically clustered are more productive because of agglomeration economies (Henderson et al. 2001) justify the inclusion in the set of the explanatory variables of the location quotient, a traditional measure of industry concentration.

The measurement of inward investments that are used as potential sources of externalities are foreign firms' shares in sector s and region r's total employment ($E_{sr}^{f}/\sum_{s} E_{sr}$) weighted by the importance of the sector as a client/supplier of all other manufacturing sectors, itself included, as indicated by the input-output tables.[15] We thus end up with the following four measures for MNEs:

[14]The initial level of TFP, the relative concentration of each manufacturing sector and the average size of firms are drawn from the Amadeus database. Imports, GDPs at sector levels, Production price indexes as well as information on human resources devoted to science and technology at regional level come from Eurostat.

[15]We use the latest available national Input–Output tables at two digit level for each country. This strategy implies that supplier and client relationships occur within sectors as well. This is not unrealistic, given the level of aggregation we work with. This concept can be clarified if we consider the following two sub-sectors, i.e. cotton fibres and cotton fabrics. Although they reflect different stages of the same production chain, these two sub-sectors belong to the same two digit manufacturing sector, i.e. textiles (Nace 17). We are aware that this specification did not allow us to fully capture intra-sectoral spillovers, which also stem from foreign activity taking place at the same production stage as domestic firms. These spillovers derive from imitation and/or demonstration effects, as well as from personnel training and mobility. However, it is likely that multinational firms try to minimize them, because they involve the transmission of specific knowledge to their local competitors (Haskel et al. 2007).

$$MNE\_for^{98}_{sr} = \alpha_{ss} * \frac{E^f_{sr}}{\sum_s E_{sr}} \tag{6}$$

$$MNE\_for^{98}_{k \neq s,r} = \sum\nolimits_{k \neq s} \alpha_{sk} * \frac{E^f_{kr}}{\sum_k E_{kr}} \tag{7}$$

$$MNE\_back^{98}_{k \neq s,r} = \sum\nolimits_{k \neq s} \omega_{sk} * \frac{E^f_{kr}}{\sum_k E_{kr}} \tag{8}$$

$$MNE\_back^{98}_{sr} = \omega_{ss} * \frac{E^f_{sr}}{\sum_s E_{sr}} \tag{9}$$

Equations (7) and (8) measure foreign firm penetration in upstream and downstream industries, thus accounting for forward and backward linkages, respectively. $\alpha_{sk}(\omega_{sk})$ is the share of sector k output (input) that is supplied (sold) to sector s, as indicated by the input-output tables. Equations (6) and (9) have the same meaning as (7) and (8), respectively, but refer to foreign firms operating in the same sector (and region).

3.1 Econometric Issues

Within the approach outlined here three potential econometric issues may arise: multicollinearity among the explanatory variables, endogeneity between FDI spillovers variables and the dependent variable and spatial autocorrelation.

Two reasons suggest the presence of multicollinearity. Primarily, manufacturing sectors are strongly interrelated, as indicated by the input-output table.[16] This implies that (6)–(9) can not be included directly in (4). Therefore, we estimate the effects of downstream MNEs on TFP changes separately from those of upstream MNEs. Secondly, MNEs might be very sensitive to a number of variables included on the right hand side of (4), namely the initial level of TFP, the degree of relative concentration and openness of the manufacturing sector they belong to, and the human resource endowment of a region, as suggested by previous empirical studies on FDI location in transition countries (Campos and Kinoshita 2003; Resmini 2000; Pusterla and Resmini 2007). In order to avoid severe multicollinearity problems, we first regressed the MNE variables on the other explanatory variables and then used the residuals of these regressions as a proxy for MNEs in (4).[17]

[16]According to input–output tables, each sector has at the same time both client and supplier relationships with other manufacturing sectors; the degree of intensity of these relationships is often very similar.

[17]Residuals, by definition, are the portion of the variation of the dependent variable not explained by the explanatory variables. Thus, in our case, they pick up the effects of FDI not related to the other explanatory variables on changes in TFP proxy.

Concerning endogeneity, we can not exclude a priori that inward FDI is co-determined with productivity spillovers. Theoretically, foreign firms are attracted to industries and locations where there is potential for significant productivity improvements. We think that it is perhaps unlikely in transition countries, and the reverse regression supports this hypothesis.[18]

There are numerous approaches to spatial correlation. According to the relevant literature (see among many Anselin 2003; Le Sage 1998; Elhorst 2003) they can be classified into two broad groups, namely the spatial error models (SEM) and the spatial lag models (SAR). In the former, spatial dependence is restricted to the error term, which would have the following form:

$$\begin{aligned} \varepsilon_{sr} &= \lambda W \varepsilon_{sr} + \mu_{sr} \\ \varepsilon_{sr} &= (I - \lambda W)^{-1} \mu_{sr} \end{aligned} \tag{10}$$

where λ is the parameter indicating the extent of the spatial correlation between the errors, W is the squared matrix defining the distance among regions, and μ is an *i.i.d.* error component.

Spatial lag models, instead, assume that spatial dependence occurs through the dependent variable. This implies that TFP changes in one sector and region depend on a weighted average of TFP changes in neighbouring regions, in addition to the explanatory variables already identified. In order to account for this second form of spatial dependence, the set of the explanatory variables should also include the spatially lagged dependent variable, i.e. $\rho W \Delta TFP_{sr}^{03,98}$ where ρ is the parameter indicating the extent of the spatial dependence.

In both cases, the spatial weights, i.e. the elements of the matrix are proportional to the inverse of their bilateral great circle distance. By convention, the matrix has zero on the main diagonal and it is row standardized, so that it is relative and not absolute distance that matters. In formulas:

$$\begin{aligned} w_{ij}^* &= d_{ij}^{-1} \\ w_{ii}^* &= 0 \\ w_{ij} &= w_{ij}^* \Big/ \sum_j w_{ij}^* \end{aligned} \tag{11}$$

The estimation of (4) by OLS would result, in the first case, in unbiased but inefficient estimates, and in the second case in biased and inconsistent estimates. Instead, both models can be estimated using maximum likelihood (Anselin 2001). Specific tests (LM tests) help in choosing the right specification for spatial dependence.

[18] See the appendix for further details.

4 Empirical Findings

Table 1 shows the results for various specifications used in order to explore the existence and the nature of spatial dependence. Column (1) reports the results of the OLS estimation, while columns (2) and (3) the results of spatial lag and spatial error model, respectively. Columns (4) and (5) present the results of the spatial error model estimates with spatial fixed effects and sectoral fixed effects. The upper panel of the table shows estimates with a measure of MNEs in upstream manufacturing sectors, while the impact of MNEs in downstream sectors is shown in the lower panel of the table.

Spatial diagnostics are provided at the end of each of the columns. In the first column, we report two tests for spatial dependence: the Moran's *I* test, and the Lagrange multiplier (LM) tests for spatial dependence in OLS residuals. The Moran's *I* statistic for regression residuals is a general test for detecting spatial dependence, but it does not allow us to discriminate between spatial dependence and spatial error models. The statistics are significant at the 1% level in both specifications, indicating that we can reject the hypothesis of spatial independence. However, the LM lag and error tests are both significant, while the robust forms are not. For that reason, we estimate both the spatial lag and the spatial error models. Results are shown in columns (2) and (3) respectively. On the base of these results, the spatial error model seems to be the most appropriate, as confirmed by the R^2, the log likelihood, the BIC and AIC tests, which are all better in the SEM than in the SAR specification.

Our results indicate that changes in aggregate TFP are positively affected by MNEs operating in the same manufacturing sectors, but not by MNEs operating in complementary manufacturing sectors. This result holds for MNEs operating in both upstream and downstream sectors. It is also apparent that the presence of spatial dependence magnifies the impact of MNEs, as indicated by the magnitude of the coefficients, which are larger in the spatial error model than in the OLS model. This effect is more marked when MNEs operate in downstream sectors.

Although the other regressors are not directly concerned with the objective of this paper, they deserve some discussion. In particular, our results show that TFP changes respond positively to human resources devoted to science and technology activities, the average size of firms belonging to the sector, and, as expected, the variation in production price indexes.[19] This implies that regions' performance can be enhanced by improvements in the endowment of human capital, and by increasing the size of local firms. Changes in aggregate TFP, instead, seem to be independent from the relative specialization and the degree of openness of regions. The insignificance of the coefficients of these variables may indicate that local firms

[19]As discussed in Sect. 3, this variable has been included in order to correct possible error measurement in the dependent variable, which relies on nominal values.

Table 1 Estimation results: OLS vs. spatial models

	OLS		SAR-Pool		SEM-Pool	
	coeff.	t-stat	coeff.	t-stat	coeff.	t-stat
MNEs in upstream sectors						
Costant	−0.731**	−2.40	−0.763**	−2.54	−0.674**	−2.23
TFP level	−0.001	−0.05	−0.003	−0.32	−0.007	−0.66
Human resources	0.254***	2.91	0.248***	2.870	0.283***	2.66
Avg size	0.027**	0.32	0.025**	2.21	0.026**	2.18
Var PPI	0.144**	2.47	0.132**	2.29	0.142**	2.24
Openness* LQ	0.15	0.28	0.010	0.19	0.004	0.07
LQ	0.45	1.13	0.047	1.19	0.052	1.31
Openness	−0.066	−0.74	−0.057	−0.64	−0.065	−0.64
MNE same sect	0.027*	1.81	0.026*	1.83	0.030*	1.89
MNE other sect	0.015	1.15	0.017	1.31	0.017	0.26
Rho			0.23***	4.13	0.24***	4.37
Lambda						
n. obs	630		630		630	
R squared	0.042		0.041		0.055	
Adj. R suqared	0.028		0.027		0.041	
Log likelihood			−322.25		−322.09	
AIC			664.19		664.19	
BIC			708.65		708.65	
Moran I	3.08***					
LM (error)	6.54**					
Robust LM (error)	0.77					
LM (lag)	6.51**					
Robust LM (lag)	0.05					
MNEs in downstream sectors						
Costant	−0.734**	−2.42	−0.760**	−2.54	−0.64**	−2.22
TFP level	−0.022	−0.22	−0.005	−0.46	−0.008	−0.76
Human resources	−0.253**	2.89	0.245***	2.84	0.241***	2.80
Avg size	0.027**	2.34	0.025**	2.24	0.026**	2.20
Var PPI	0.144**	2.49	0.133**	2.33	0.144**	2.29
Openness* LQ	0.011	0.21	0.007	0.13	0.002	0.04
LQ	0.046	1.15	0.048	1.20	0.052	1.31
Openness	−0.066	−0.73	−0.055	−0.62	−0.062	−0.62
MNE sanie sect	0.024*	1.85	0.025*	1.94	0.030**	2.10
MNE other sect.	0.021	1.48	0.021	1.47	0.021	1.32
Rho			0.233***	4.12	0.234***	5.39
Lambda						
n. obs	630		630		630	
R squared	0.043		0.042		0.056	
Adj. R suqared	0.029		0.028		0.042	
Log likelihood			−321.92		−321.65	
AIC			663.84		663.29	
BIC			708.31		707.75	
Moran I	3.15***					
LM (error)	6.73***					
Robust LM (error)	0.27					
LM (lag)	6.45**					
Robust LM (lag)	0.00					

*, **, *** indicate significance level at 10, 5, and 1% respectively

Table 2 Estimation results: spatial panel fixed effects

	MNEs in upstream sectors				MNEs in downstream sectors			
	SEM spatial FE		SEM all FE		SEM spatial FE		SEM all FE	
	coeff.	t-stat	coeff.	t-stat	coeff.	t-stat	coeff.	t-stat
TFP level	0.003	0.28	0.004	0.35	0.000	0.02	0.002	0.15
HRST	–	–	–	–	–	–	–	–
Avg size	0.029**	2.54	0.032***	2.83	0.030***	2.64	0.32***	2.85
Var PPI	0.598**	3.85	0.600***	4.09	0.624***	3.99	0.649***	4.45
Openness* LQ	−0.039	−0.74	−0.036	−0.68	−0.041	−0.77	−0.033	−0.63
LQ	0.077**	1.96	0.074*	1.9	0.075*	1.92	0.072*	1.85
Openness	−0.002	−0.019	0.022	−0.19	0.006	0.05	−0.008	−0.07
MNE same sect	0.049**	2.67	0.058***	2.86	0.045***	2.77	0.072***	3.96
MNE other sect	0.046**	2.4	0.056***	2.67	0.049**	2.32	0.052**	2.04
Spat. Autocorr coeff.	0.26***	2.74	−0.30***	−2.80	0.25***	2.73	−0.034**	−2.54
Test F on fixed effects	F[38, 592] = 1.47***		F[58, 572] = 0.25		F[38, 592] = 2.05***		F[58, 572] = 0.13	
n. obs	630		630		630		630	
R squared	0.14		0.18		0.14		0.19	
Adj. R suqared	0.08		0.11		0.08		0.11	
Log likelihood	−511.78		−493.26		−511.57		−490.87	
AIC	1,040		1,025		1,039		998	
BIC	1,075		1,038		1,074		1,033	

*, **, *** indicates significance level at 10, 5, and 1% respectively

are not efficient enough to generate agglomeration forces able to positively affect regions' productivity, as well as to compete successfully on the international markets.[20]

In order to account for spatial heterogeneity, we include variable intercepts representing the effect of the omitted variables that are peculiar to each region. Results are shown in Table 2. To test the validity of the relaxation of these restrictions (i.e. if dummy variables should be introduced into the model) the F test has been performed. It allows us to reject the null hypothesis that the restricted and unrestricted specifications are the same. When regional fixed effects are considered, both intra and inter-sectoral spillovers become significant at the conventional levels, though the latter are weaker than the former. The position of the MNEs in the production value chain (i.e. upstream or downstream) does not seem to affect the transmission of spillovers. Finally, the restrictions are relaxed for sectoral dummies too, and the complete unrestricted model is estimated (Table 2). In this

[20]The insignificance of the coefficients of the openness variable may also be explained by a complementary relationship between trade and FDI. This implies that the degree of openness of a region is already captured by the presence of foreign enterprises. A spurious correlation may also affect the specialization variable and the average size of firms: the larger the size of the firms, the more concentrated is the sector.

case, however, the test F is not significant; therefore the constraint of not including sectoral dummies is valid.[21]

5 Summary and Conclusions

This paper has empirically analysed whether and to what extent multinational enterprises were able to enhance productivity in the domestic manufacturing sectors of three new EU member states, i.e. Bulgaria, Poland and Romania, and in so doing foster growth and development at the regional level. In order to achieve this research objective we applied theoretical arguments concerning FDI-induced spillovers with issues related to the measurement of externalities with spatial data. The previous empirical literature concerned with the determinants of such spillovers has largely ignored spatial correlation effects, despite the fact that these considerations are vital in modelling externalities. This paper has contributed to fill this gap.

Overall, our results, based on panel data estimation, support the role played by FDI as a vehicle for technology transfer, and its relationship with productivity growth. More specifically, we found that spillovers from MNEs occur both within and across complementary manufacturing sectors at regional level. The latter, however, are less significant than the former in almost all specifications, indicating that intra-sectoral spillovers are more robust than inter-sectoral spillovers.

Besides MNEs, there are other characteristics of sectors and regions that enable or affect changes in aggregate TFP. In particular, we found that relatively concentrated sectors with large firms in regions well endowed with human capital enjoy higher TFP growth rates. Spatial autocorrelation does exist in all countries, and seems to work through omitted variables with a spatial dimension, as indicated by diagnostics tests on spatial autocorrelation in the error term.

The use of spatial econometric techniques in the analysis of FDI-induced spillovers has enabled this study to provide useful insights into the linkages between FDI and regional economic development. The research could be further developed by considering also non manufacturing FDI for comparative purposes, together with the resultant implications for regional development and FDI promotion policies by national and regional governments. These refinements would help first to focus on potential trade-off between manufacturing and non-manufacturing FDI in enhancing regions' productivity and, secondly, to better assess the effectiveness of regional FDI promotion policies. The presence of positive spatial spillovers make irrelevant the location of foreign firms within a country, given that FDI-induced spillovers may cross regions' borders, and affect productivity of

[21]The poor explanatory power of specification (4) may be explained by the fact that sectoral heterogeneity is already captured by explanatory variables; therefore, adding sectoral fixed effects may generate multicollinearity problems that worsen the goodness of fit of the model.

neighbouring regions. This suggests, on the one hand, that national FDI promotion policies may also play a role in promoting increased linkages between FDI inflows and regional economic development; on the other hand, it casts some doubts on the efficiency of regional FDI promotion policies, given the incentive of regions to free ride on the efforts of other regions in attracting MNEs.

Appendix

Table A.1 Representativeness of the sample: distribution of manufacturing employment

	By region	By sector
1998	0.89 ($p > 0.000$)	0.89 ($p > 0.000$)
2003	0.97 ($p > 0.000$)	0.72 ($p > 0.000$)

Correlation with official data (EUROSTAT)

Table A.2 MNEs vs. TFP changes: the reverse regression

	Estimated coefficients for TFP changes			
	(1)		(2)	
	coeff.	t-prob.	coeff.	t-porb.
Dep. Var.				
MNE_forw same sect	0.184	0.1033	0.177	0.1083
MNE_forw other sect	0.076	0.5542	0.08	0.5104
MNE_back same sect	0.198	0.1223	0.191	0.1277
MNE_back other sect	0.107	0.3678	0.133	0.2613

OLS estimates (1) includes all the other regressors, while specification (2) includes as exogenous variable TFP changes only

References

Aitken B, Harrison A (1999) Do domestic firms benefit from direct foreign investment? Evidence from Venezuela. Am Econ Rev 89(3):605–618

Anselin L (2001) Spatial econometrics. In: Baltagi B (ed) A companion to theoretical econometrics. Basil Blackwell, Oxford, pp 310–330

Anselin L (2003) Spatial externalities, spatial multipliers and spatial econometrics. Int Reg Sci Rev 26(2):153–166

Baltagi B, Egger P, Pfaffermayr M (2008) Estimating regional trade agreement. Effects on FDI in an interdependent word. J Econometrics 145(1–2):194–208

Barba NG, Venables A (2004) Multinational firms in the world economy. Princeton University Press, Princeton, NJ

Bartelsman E, Doms M (2000) Understanding productivity: lessons from longitudinal microdata. J Econ Lit 38(3):569–594

Benhabib J, Spiegel M (1994) The roles of human capital in economic development: evidence from aggregate cross-country data. J Monetary Econ 34:143–173

Blomstrom M, Kokko A (1997) How foreign investment affects host countries, Policy Research Working Paper Series 1745, The World Bank

Bloningen B, Davis R, Waddell G, Naughton H (2007) FDI in space: spatial autoregressive relationships in foreign direct investment. Eur Econ Rev 51(5):1303–1325

Campos N, Kinoshita Y (2003) Why does FDI go where it goes? New evidence from the transition economies, IMF working paper no. 03/228

Coe DT, Helpman H (1995) International R&D spillovers. Eur Econ Rev 39:859–887

Coughlin C, Segev E (2000) Foreign direct investment in China: a spatial econometric study. World Econ 23(1):1–23

Driffield N (2006) On the search for spillovers from foreign direct investment (FDI) with spatial dependency. Reg Stud 40(1):107–119

Dunning J (1977) Trade, location of economic activity and MNE: a search for an eclectic approach. In: Ohlin B, Hesselborn P, Wijkman PM (eds) The international allocation of the economic activity. McMillan, London

Dunning J (1981) International production and multinational enterprise. Geroge Allen and Unwin, London

Elhorst JP (2003) Specification and estimation of spatial panel data models. Int Reg Sci Rev 26(3): 255–268

Feldman MP (2003) The locational dynamics of the US biotech industry: knowledge externalities and the anchor hypothesis. Ind Innovat 10(3):311–328

Findlay R (1978) Relative backwardness, direct foreign investment, and the transfer of technology: a simple dynamic model. Q J Econ 92(1):1–16

Fosfuri A, Motta M, Ronde T (2001) Foreign direct investment and spillovers through workers' mobility. J Int Econ 53:205–222

Glass A, Saggi K (1998) International technology transfer and the technology gap. J Dev Econ 55: 369–398

Glass A, Saggi K (2002) Multinational firms and technology transfer. Scand J Econ 104(4): 495–513

Haskel J, Pereira S, Slaughter M (2007) Does inward foreign direct investment boost the productivity of domestic firms? Rev Econ Stat 89(3):482–496

Henderson JV, Shalizi Z, Venables AJ (2001) Geography and Development, Journal of Economic Geography 1(1), pp 81–105

Katayama H, Lu S, Tybout J (2009) Firm-level productivity studies: illusions and a solution. Int J Ind Organiz 27(3):403–413

Keller W (2002) Geographic localization of international technology diffusion. Am Econ Rev 92(1): 120–142

Le Sage J (1998) The theory and practice of spatial econometrics, mimeo downloadable at: http://www.rri.wvu.edu/WebBook/LeSage/spatial/wbook.pdf

Levinsohn J, Petrin A (2003) Estimating production functions using inputs to control for unobservables. Rev Econ Stud 70:317–341

Mairesse J, Jaumandreu J (2005) Panel-data estimates of the production function and the revenue function: what difference does it make? Scand J Econ 107(4):651–672

Markusen J (1995) The boundaries of multinational enterprise and the theory of international trade. J Econ Perspect 9(2):169–189

Markusen J, Venables A (1999) Foreign direct investment as a catalyst for industrial development. Eur Econ Rev 43(2):335–356

Nelson R, Phelps E (1966) Investment in humans, technology diffusion, and economic growth. Am Econ Rev 56(1):69–75

Nicolini M, Resmini L (2010) FDI spillovers in new EU member states. Which firms create them and which firms really benefit? Econ Transit 18(3):487–511

Olley S, Pakes A (1996) The dynamics of productivity in the telecommunication equipment industry. Econometrica 64(6):1263–1297

Petrin A, Levinshon J (2005) Measuring aggregate productivity growth using plant-level data, NBER working paper no. 11887

Pusterla F, Resmini L (2007) Where do foreign firms locate in transition countries? An empirical investigation. Ann Reg Sci 41(4):835–856

Resmini L (2000) The determinants of foreign direct investment into the CEECs: new evidence from sectoral patterns. Econ Transit 8(3):665–689

Tornqvist L (1936) The bank of Finland's consumption price index. Bank Finland Mon Bull 10: 1–8

Traistaru I, Njikamp P, Resmini L (eds) (2003) The new emerging geography in transition countries. Ashgate, Aldershot

Wang J, Blomstrom M (1992) Foreign direct investment and technology transfer: a simple model. Eur Econ Rev 36:137–155

Xu B (2000) Multinational enterprises, technology diffusion, and host country productivity growth. J Dev Econ 62:477–493

Xu B, Wang J (1999) Capital goods trade and R&D spillovers in the OECD. Can J Econ 32: 1179–1192

Determinants of Entry and Exit: The Significance of Demand and Supply Conditions at the Regional Level

Jenny Grek, Charlie Karlsson, and Johan Klaesson

1 Introduction

Today there is an extensive research literature on entrepreneurship and firm demography. However, even if many studies have identified substantial and persistent variations in entrepreneurship rates across regions in a variety of countries (Georgellis and Wall 2000), most attempts trying to explain entrepreneurship variations have been restricted to industry determinants (Arauzo-Carod and Manjón-Antolín 2007). As a matter of fact, location factors are neglected in most studies trying to explain variations in entrepreneurship.[1] This is astonishing, since there are studies, which show that location factors matter.[2] There are numerous examples of location factors that might explain spatial variations in entrepreneurship in the literature on regional economics and economic geography including institutional framework, size of region, industrial and firm structure, in-migration, demand growth, employment specialization, unemployment rates, educational level, university R&D, the availability of financing, and population density. However, our understanding of the empirical structure of the regional variations in entrepreneurship as well as the underlying theoretical explanations is still incomplete. Thus, "the economics literature on regional dimensions of entrepreneurship looks ripe for further investigation and extension" (Parker 2004).[3] In a similar manner, Audretsch and Feldman (2004)

[1] "Entrepreneurial management, or the study of the creation and growth of new companies, has become a prominent field in the literature on management. This field has developed largely independently of location considerations". Porter (2000, 269).

[2] See, e.g., Reynolds et al. (1993), Audretsch and Fritsch (1994), Garofoli (1994), Guesnier (1994), Malecki (1993), Saxenian (1999), Fotopoulos and Spence (1999, 2001), Berglund and Brännäs (2001), Armington and Acs (2002), Arauzo and Teruel (2005), and Karlsson and Backman (2008).

[3] "It is surprising to observe that the geography of entrepreneurship has indeed received far less attention [than other aspects of entrepreneurship]". Nijkamp (2003).

J. Grek (✉), C. Karlsson, and J. Klaesson
Jönköping International Business School, Jönköping University,
P.O. Box 1026, SE-551 11 Jönköping, Sweden
e-mail: jenny.grek@jibs.hj.se

K. Kourtit et al. (eds.), *Drivers of Innovation, Entrepreneurship and Regional Dynamics*, Advances in Spatial Science, DOI 10.1007/978-3-642-17940-2_7,

remark that the life cycle of spatial units, such as agglomerations, clusters and regions, with regard to the role of entrepreneurship is a relatively unchartered area.

A better understanding of the drivers of entrepreneurship is important, since many authors assume a relationship between entrepreneurship and growth at both the national and the regional level (Plummer and Acs 2004; Fritsch and Mueller 2004; van Stel and Storey 2004; Audretsch et al. 2006).[4] It is imagined that the continual entry and exit of firms and plants is a necessary condition for regional growth. The entry of new firms and plants tend to be connected with productive innovation, i.e. the introduction of new or improved products and/or production methods (Baumol 2002). The entry of new actors in the market place increases competition and forces incumbents to become more effective, i.e. to increase their productivity, move to another region or to go out of business. Thus, when new innovative firms enter, some existing businesses are displaced in a process of creative destruction (Schumpeter 1934; Robinson et al. 2006). However, far from all, entry of new firms generates entrepreneur-driven Schumpeter-type creative destruction (Manjón-Antolin 2004). Instead, new firm formation is a heterogeneous phenomenon where innovative entrepreneurs are mixed with more normal entrepreneurs in the form of passive followers, over-optimist gamblers and people trying to find an alternative to unemployment (Vivarelli 2007). Cabral (2004) even claims that most entrepreneurial ventures are "entry mistakes". However, there seems to be a considerable consensus that entrepreneurship plays a critical role for the introduction of radical innovations.

However, disregarding the large diversity of entrepreneurial ventures, entrepreneurship has been identified by many authors as an important driver for economic growth, competitiveness and job creation (Thurik and Wennekers 2004; Friis et al. 2006), even if there are authors, such as Acs and Storey (2004), who claim that the evidence remains inconclusive. However, Karlsson and Nyström (2007) present a survey of empirical studies since 1996 of the relationship between entrepreneurship and productivity, employment and economic growth that shows that two thirds of the studies find a significant positive relationship.

The focus in this paper is how the conditions for entrepreneurship vary between regions. Our starting point is the functional region approximated by the commuting region. Our prime motivation for this is that the functional region for almost all businesses makes up their home market but also offers most of the supply conditions in terms of labour, business services, infrastructure, that are critical for both the start-up of firms and for the efficient operation of firms. Since both the market and the supply conditions vary with the size of functional regions, we have strong reasons to believe significant variations in the entry and exit of firms between regions of different size but also differences in terms of the sectoral composition, the size distribution, of entering and exiting firms. A better understanding of how and why the entry and exit of firms vary between different functional regions is important as a knowledge base to develop regional policies that support the entry and exit of firms in regions of various

[4]Already Chinitz (1961) argued that the existence of many small firms and a culture of entrepreneurship could explain why New York was much more successful than Pittsburgh.

sizes. Since the background conditions vary with the size of functional regions so must the regional policies that directly and indirectly are targeted at supporting the entry and exit of firms to stimulate regional economic growth.

The theoretical and empirical literature gives a strong support to the idea that the agglomeration of population and jobs in large urban regions offers favourable incubator conditions for creative entrepreneurship and innovation (Hoover and Vernon 1959; Thompson 1968; Leone and Struyck 1976; Pred 1977; Davelaar 1991) and has a positive effect on the location of firms and entrepreneurship (Hansen 1987; Guimarães et al. 2000; Figueiredo et al. 2002). The clustering of economic activities in a geographical space (Audretsch and Feldman 2004; Porter 1998) suggests that there are positive agglomeration economies influencing the location of new firms and that these economies may compensate for the negative effects in terms of higher rent, and wage costs in agglomerations and of potential competition from incumbent firms located in the agglomeration. However, earlier studies use administrative regions rather than functional regions as their spatial units. We claim that a better understanding of the role of size of regions can be achieved if the empirical analysis is based on functional regions. We also claim that incomes are a better measure of size than either population or jobs, since they represent a measure of the potential market demand in different regions. Furthermore, this potential market demand can for each locality within a functional region be divided into three components: the intra-locality, the intra-regional, and the inter-regional demand potential, respectively. This offers an opportunity to highlight the relative importance of these three demand (=size) measures in general for entrepreneurship as well as for entrepreneurship in different sectors.

The purpose of the current paper is to analyse the theoretical arguments as to why large regions generally should generate more entrepreneurship and empirically analyze the role of regional size in explaining variations in total entrepreneurship and entrepreneurship in different sectors across functional regions using data from Sweden.

The paper is organized as follows: In Sect. 2 we present our theoretical framework and our hypotheses. Our empirical approach and our data are presented in Sect. 3. The empirical analyses are performed in Sects. 4 and 5 provides the conclusions.

2 Regional Size as a Stimuli for Formation of New Firms: The Theoretical Arguments

The variations in entrepreneurship between large and small functional regions are related to variations in on the one hand, demand conditions and, on the other hand, supply conditions. Regional variations in demand conditions, in terms of regional market potential and regional demand for new products, generate spatial variations in entrepreneurial opportunities. On the supply side, there are similar differences between regions of different sizes in the number of economic agents with a capacity to discover, create and exploit entrepreneurial opportunities due to differences in educational achievements, work experiences, networks and so on. On the supply side there are also variations between regions of various sizes as regards conditions such as the knowledge base, information supply, industrial structure, company structure, supply of inputs and producer services, supply of financial services and capital, infrastructure supply. At an aggregated level the differences in demand and supply conditions between regions of different size create differences in the agglomeration economies offered by different regions.

The interest in the effects of agglomeration economies goes back at least to Marshall (1920), who identified three types of external economies: a common labour market, information spillovers, and the development of specialised input suppliers. However, it was Ohlin (1933), who by his classification of agglomeration economies illustrated how individual firms and thus potential entrepreneurs are affected by being localised to regions of varying size[5]:

- Internal economies of scale associated with the production function of the firm. Since many products are distance-sensitive or have high geographical transaction costs, the size of the regional market is critical for many firms to be able to reach a break-even and make a profit.
- Location economies, which are external to the individual firm but are derived from the size of its industry in the region where it is located.
- Urbanisation economies, which are external to both the firm and its industry, and which are dependent upon the size of the total regional economy.
- Inter-industry linkages of input-output type where proximity to input suppliers reduces the price of inputs due to scale advantages and reduced transport costs.

These factors explain why entrepreneurs may choose to locate in large urban regions, and accept increased land and labour prices, in situations where simple location analysis would suggest a decentralised location. However, the localisation and urbanisation economies are in principle static concepts. If we ask about the spillovers entrepreneurs look for we need to turn to the dynamic version of the localisation economies, i.e. the Marshall-Arrow-Romer's (MAR) and Porter's theories of specialisation, and the dynamic version of urbanisation economies, i.e. Jacobs' theory of diversity (Glaeser et al. 1992). The MAR's and Porter's theories stress that the industrial specialisation of urban regions is the most important source of spillovers, an attractor for entrepreneurs in the same industry. Jacobs, on the other hand, claims that the most important spillovers come from outside the industry with industrial diversity being an attractor for entrepreneurs. Henderson et al. (1995) show that the importance of the two types of spillovers varies between industries. Regarding manufacturing industry, they find that diversity doesn't really matter. However, the diversity of an urban region attracts new industries and more particularly high technology industries. In the sequel, we discuss in more detail how agglomeration economies in general and urbanisation economies in particular influence entrepreneurial behaviour.

[5]A similar classification of agglomeration economies has been provided by Hoover (1948).

2.1 *Entrepreneurship and Regional Variations in Demand Conditions*

The best opportunities for entrepreneurial initiatives are offered by functional regions, which combine a large home market with a high accessibility to markets in other regions, i.e. large functional regions. Generally speaking, the larger the functional region, the higher the per capita income. Entrepreneurs that make start-ups in large functional regions may take advantage of close proximity to a concentration of (potential) customers, i.e. of purchasing power, which of course can be other firms.

Under certain conditions, entrepreneurs may grab incumbents' market share if they locate near them (Hotelling 1929). Admittedly, this gain may be short-lived if more entrepreneurs enter, or if incumbents react to this unwanted competition. When the competition in the product market is imperfect, geographical proximity increases competition in the product market (Fujita et al. 1999), which implies that entrepreneurs may suffer from proximity to firms offering similar products, in particular if the demand is thin. However, when there is a demand for variety among customers, large regions always offer entrepreneurs a greater possibility to successfully launch a new product.

A third motive for entrepreneurs to locate start-up firms in large functional regions may be more long-term. Entrepreneurs may choose to locate in a large functional region because they are more likely to be better exposed to customers. The underlying reason is that searching is costly for customers who, *ceteris paribus*, prefer to minimize search cost by purchasing in areas of concentrated (and varied) supply. This is particularly relevant in product markets with discerning potential customers with specific demands and requirements, who wish to search and inspect before purchasing. Such product markets are to a high extent concentrated in large functional regions. It is these regions, which host the most demanding customers in the form of company headquarters, R&D and product development units of large companies as well as other advanced establishments, such as research universities, university hospitals, R&D institutes, and specialised R&D firms.

A fourth advantage for entrepreneurs of locating in large functional regions is the positive information externality in such regions, through which individual (potential) entrepreneurs receive signals about the strength of the regional demand by observing the successful trades of incumbent firms. Such observations also provide information about varieties of existing products, which also might trigger the development of new varieties that might be complements or substitutes to existing product varieties. One further advantage for an entrepreneur of a location in a large functional region might be the signalling effect to potential customers in the form of an indication or image of quality.

A sixth advantage for entrepreneurs to locate in large functional regions is risk reduction (Mills and Hamilton 1984). To the extent that fluctuations in demand are imperfectly correlated across customers, demand can be stabilized since some customers are buying while others are not. Finally, when an entrepreneur chooses a location in a large functional region, he can expect a local economic milieu of qualified and demanding customers, which is important for entrepreneurs engaged in innovation processes and product development.

2.2 *Entrepreneurship and Regional Variations in Supply Conditions*

On the supply side, the regional economic milieu (including its culture, knowledge-base, and attitudes to businesses) often appears to act as a critical success factor for new forms of entrepreneurship (Camagni 1991). Large functional regions offer advantages to entrepreneurs in terms of knowledge flows that are particularly important when the product and/or process knowledge is complex and perhaps tacit in nature (Jaffe et al. 1993; Karlsson and Manduchi 2001). Tacit knowledge, which is vital to found new firms, is best communicated informally through frequent face-to-face interactions, hence the importance of the proximity offered by large functional regions. Major research universities and research laboratories are normally located to large functional regions. This implies that these regions offer a larger and more varied knowledge base for potential entrepreneurs to draw upon (Audretsch et al. 2006).

A special type of information externality accrues to potential entrepreneurs from observing a large number of successful incumbents in large functional regions, i.e. there is large potential for product and production knowledge to spill-over in large regions.[6] This implies that the start-up rate for each industrial sector should increase with the existing density of firms in each sector. Large and dense regions offer physical proximity, which facilitates the integration of multi-disciplinary knowledge that are tacit and therefore person-embodied rather than information-embodied as well as allowing the rapid decision making needed to cope with uncertainty (cf., Patel and Pavitt 1991). Due to urbanisation economies, these regions also offer diversity, that is, economies of scope, in information, skills, knowledge, competence, producer services, and other inputs, which are crucial for innovative entrepreneurial processes. Furthermore, they offer advantages in terms of access to a large pool of well-educated and specialized labour (Marshall 1920), particularly specialized workers in accounting, law, advertising and different technological fields.[7] This is partly a result of the fact that most leading higher education institutions are located in large functional regions. This reduces the costs of starting-up and expanding new businesses (Krugman 1993). It probably also leads to a higher

---

[6]We thus have a spatial version of the so-called "knowledge spillover theory of entrepreneurship" (Audretsch and Lehman 2005).

[7]Florida (2002) has suggested that creative capital rather than human capital is the source of entrepreneurship and economic growth in regions.

proportion of high-quality start-ups but may also be an important success factor for new firms following the start of operations.

Large, densely populated functional regions are also conducive to a greater provision of non-traded inputs, i.e. their service infrastructure is more developed. These regions provide these inputs both in greater variety, with a higher degree of specialization at lower costs (Krugman 1991a, b). This implies that entrepreneurs that start firms in large regions can take full advantage of all potential substitution possibilities inherent in available production functions, i.e. taking full advantage from outsourcing of functions and activities. One special aspect of the greater provision of non-traded inputs is the larger and more varied supply of financial services in large regions. Risk capital investors prefer to locate in large urban regions since they try to lower their risks by investing in several new projects and by keeping hands on relationship with the entrepreneurs and their new companies, which demands geographical proximity and frequent face-to-face interactions (Thornton and Flynn 2003). A larger and more varied supply of financial services in large regions tends to stimulate the start-up of new firms in such regions (Backman 2008). The reason is that face-to-face contacts normally are required to obtain funding for starting new firms, since there are normally no physical assets that can serve as collateral.

There also exist physical infrastructure benefits for entrepreneurs to locate in large functional regions in terms of, for example, access to major highways, international airports and broadband capacity. A final reason for large functional regions providing advantages for entrepreneurs arises from the reductions in spatial transaction costs that is made possible by locating in a large and dense region (Quigley 1998).[8] In particular, search costs for customers, suppliers, services, and knowledge are lower in large and dense functional regions.

The information flow economies (Acs et al. 1992) on both the supply and the demand side are greater in large functional regions than small ones. Information and knowledge is a *sine qua non* for entrepreneurial success. Learning-by-doing supported by inter-firm network collaboration enhances, for example, the competitive potential of new firm initiatives (Malecki and Poehling 1999). Since large urban regions are the leading communication and transport network nodes, they are the primary locations of emission and reception of information and knowledge. Because these regions contain concentrations of complex communication equipment and transport terminals, much information and knowledge diffuse more easily from urban region to urban region, than to the hinterland around each urban region (Florax and Folmer 1992; Guillain and Huriot 2001). Thus, new firms are likely to be started where such economies are greatest. Large functional regions, in particular, offer favourable conditions for innovative entrepreneurship as a result of a larger and more varied access to knowledge and to skilled and educated labour,

[8]It may be observed that while in some respects spatial transaction costs have fallen over time, there are other aspects in which spatial transaction costs appear to have actually increased over time (McCann and Sheppard 2003).

economics of density and entrepreneurial opportunities generated by the large functional regions being a nucleus of innumerable networks with a scale ranging from local to global (Nijkamp 2003). Regional economic milieus which offer a rich supply of various types of networks (i.e. mainly large functional regions) tend to encourage entrepreneurship, since participation or involvement in regional or broader economic networks makes it possible to externalize some of the risks involved (Shapero 1984).[9] This implies that conditions that can generate potentially synergetic situations and support for learning are mostly available in large functional regions.

Besides general demand and supply conditions, one may assume that the larger and richer a functional region, the larger the number of potential entrepreneurs, since individuals in such regions in general are better educated, have a more varied work experiences, and so on. We may even assume that large (and dense) functional regions offer increasing returns in the acquisition of entrepreneurial skills and entrepreneurial competence due to more numerous, more varied and more effective interactions between individuals in such regions (Glaeser 1999; Desmet 2000). Since larger functional regions offer larger opportunities and higher capacity for entrepreneurship, they bound to experience a build-up of entrepreneurial knowledge, which will stimulate further entrepreneurial action.

Furthermore, entrepreneurs are change agents who will not only make decisions to start firms but they will also actively try to shape the regional economic milieu and regional institutions by trying to influence political decision makers in the region or try to take their own initiatives (Stimson et al. 2006). They will spend resources and develop relationships that further their own interests as well as the interests of potential entrepreneurs, through the creation of a positive regional entrepreneurial environment (Feldman 2001). Good conditions for entrepreneurial actions and good conditions in general in large functional regions will stimulate potential entrepreneurs, often well-educated people, in smaller regions to move to larger regions.[10] When more potential entrepreneurs gather in larger functional regions, the conditions for entrepreneurial actions will improve due to an increased availability of entrepreneurial knowledge. This will further induce entrepreneurial initiatives and encourage in-migration of potential entrepreneurs from other (smaller) regions. In this sense, entrepreneurial spatial behaviour generates a dynamic cumulative concentration process of entrepreneurial knowledge, skills and competence.

---

[9] This does not exclude the possibility that some smaller functional regions may offer favourable seed-bed conditions for entrepreneurship within, for example, specialized industrial clusters.

[10] The concept potential entrepreneurs is used here to stress that when well-educated people move into larger regions from smaller ones the major attractor is probably the dynamic labour market in larger functional regions. However, as soon as the in-migrants are established in the larger region, they become potential entrepreneurs that sometimes are better at discovering business opportunities than people who have lived in the larger region for a long time. It seems, on the other hand, to be well-established in the literature that entrepreneurs rarely move when they establish new (Stam 2007) and, in particular new high-tech firms (Cooper and Folta 2000). However, they may have migrated to the region well before they become entrepreneurs.

It is clear that new enterprises in a functional region contribute significantly to its economy and employment but, in particular, to its renewal. It is important in this connection to understand that all entrepreneurial start-ups can be seen as experiments. All potential and active entrepreneurs benefit from learning the outcome of such experiments and the knowledge spillovers from such experiments are proportional to the number of experiments. Since larger functional regions normally host many more such experiments than smaller ones, they also benefit proportionally more from such entrepreneurial learning. Thus, they accumulate a larger stock of entrepreneurial knowledge over time, which implies that externalities from entrepreneurial knowledge are dynamic in nature. In a multi-regional context, each functional region may build up a stock of regional entrepreneurial knowledge based upon past and current entrepreneurial activity (Glaeser et al. 1992), involving sets of cumulative experiments. Such regional accumulation of entrepreneurial knowledge affects the probability that entrepreneurial actions will take place and be successful and since large functional regions accumulate more entrepreneurial knowledge, they will also generate more entrepreneurial activities.

Even if there are many reasons why larger regions should be expected to offer better supply side opportunities for entrepreneurship, it is not clear in the available literature whether more diversified or more specialised regional economic milieus offer the best conditions for entrepreneurship and firm growth (Glaeser et al. 1992; Henderson et al. 1995). It has in this connection been suggested that firms and thus entrepreneurs prefer different regional economic milieus depending upon the stage of the life cycle of their products (Duranton and Puga 2001). However, the distinction between diversified and specialised regions is no easy issue, since large diversified regions also exhibit many different specialisations.

2.3 *Entrepreneurship and Agglomeration: The Empirical Picture*

The general relationship between agglomeration economies and location has been analysed extensively from both a theoretical and an empirical point of view.[11] However, the more specific relationship between agglomeration economies and entrepreneurship has been less extensively analysed.

In van Ort and Stam (2006) the relationship between agglomeration economies and entrepreneurship is analysed. The authors find evidence that agglomeration effects have a stronger effect on new firm formation than on growth of incumbent in the ICT industry. According to van Ort and Stam (2006), there are two reasons that can explain this relationship. The first reason, incumbent firms usually have a wider spatial orientation, and the second reason is that the incumbent firms usually tend to

[11] See, e.g., McCann (1995), Guimarães et al. (2000), Fujita and Thisse (2002), Rosenthal and Strange (2003), McCann and Sheppard (2003), Holl (2004), and Viladecans (2004).

keep their knowledge inside the company as much as possible in comparison to new firms. According to Acs (2000), the geography of innovation and the new economic geography suggest that the extent to which a country is entrepreneurial and its economic system is agglomerated can be a factor that explains technological change. He concludes that, entrepreneurial activity and agglomeration have a positive effect on technological change in the EU. McCann (1995) analyses the characterization of various types of agglomeration economies, i.e. internal returns to scale, localization economies and urbanization economies. Intermediate locations are indeed the norm regarding agglomeration economies. Agglomeration factors are available in a larger metropolitan area, like sheer size of medium-sized agglomerations and access to high-level knowledge in local universities. Agglomeration economies, foot looseness, are probably to increase in the future, based on growing globalization. This would mean that medium-sized agglomerations at some distance from the large cities are facing additional opportunities for economic growth.

2.4 Hypotheses

Based upon the above discussion we now launch our first hypothesis:

H1: The higher the regional income level the larger the gross rate of firm formation *However, it must be stressed that start-ups and newly established firms face substantial risks for numerous reasons. As a consequence, the death rate among start-ups is relatively high but tends to decrease rather rapidly over time.*[12] *A high death rate among newly formed firms is natural since they are involved in the introduction of new products and/or new production processes in the market place. Accordingly, they provide a major challenge to established firms and encourage or force them to improve product quality and services, or to reduce prices or to leave the market. This means that entrepreneurs play a fundamental role in the renewal of regional economies by strengthening competition and initiating competitive processes at the ultimate end in a creative destruction of existing modes of production. Thus, we shall expect regions with extensive entrepreneurial activities to also be characterized by extensive firm exits, i.e. we expect entry and exit rates across regions of various sizes to be correlated (Cf., Keeble and Walker 1994; Reynolds et al. 1994). Earlier studies show that high population density leads to relatively low survival rates of new businesses, but to higher average employment in those start-ups that manage to survive (Fritsch et al. 2006; Weyh 2006). This implies that higher intensity of competition in larger agglomerations results in a more rigorous market selection in these regions with the surviving businesses performing relatively well there.*

[12]Naturally, the survival or success rates of new entrepreneurs show large variations between sectors and regions (Acs 2000).

This discussion allows us to formulate our second hypothesis:

H2: The larger the regional income the larger the rate of firm deaths *The idea here is that entries, among other things, can be expected to generate a displacement effect that causes exits to increase. However, exits might free niches in the market place and economic resources that might increase the ability and willingness of potential entrepreneurs to enter the market (Acs and Audretsch 1990; Nyström 2006; Arauzo et al. 2007). If we consider hypotheses H1 and H2 together, we run into uncertainty concerning the effects of regional size on the net entry of firms. However, based upon our basic arguments about the importance of regional size for new firm formation we launch the following hypothesis concerning the net entry of firms.*

H3: The larger the regional income the larger the net entry of firms *In the introduction to this paper, we made the remark that it is possible to work with a more sophisticated representation of the regional income and its influence on entrepreneurship. Our starting point is that the geographical space of any nation can be divided into a number of functional regions, where each functional region contains a number of localities. For practical reasons these functional regions can be approximated by labour market, i.e., commuting, regions. Entrepreneurs start their firms in a locality and, depending upon their type of product, they market their product within the locality, within the region and/or within the whole national economy.*[13] *Thus, the different markets are not of equal importance for all products. In this paper, excluding the primary sector and the public sector, we make a distinction between four types of products or industries: primary products, manufacturing products, ordinary services and advanced business services, where ordinary services have a high content of household services. For these four product groups, we now launch the hypotheses H4–H6.*

H4: We expect entrepreneurship in advanced business services to have the highest dependency on the size of the intra-regional market potential *The motivation for this hypothesis is the general tendency among advanced business services to locate in large functional (urban) regions, where their major customers are located and where they can interact face-to-face with their customers.*

H5: We expect entrepreneurship in manufacturing to have the highest dependency on the size of the inter-regional market potential *Producers of manufacturing products on average have a lower need for face-to-face contacts with their customers than producers of advanced business services. Manufacturing production is also more space demanding. This induces entrepreneurs in manufacturing to start their enterprises in regions with plenty of space but with a good location in the logistical networks to be able to deliver efficiently to the customers in the most important markets.*

---

[13]The international market is of course also an alternative but an alternative that we disregard here.

H6: We expect entrepreneurship in ordinary services to have the highest dependency on the size of the intra-locality market potential *Most enterprises in the household service sector operate with a very small market area. Thus, entrepreneurs starting firms delivering household services will prefer to locate in localities with a high market potential.*

3 Empirical Approach and Data

This paper uses FEVD regressions, i.e. fixed effects vector decomposition regressions, to investigate the market potential and new firm formation in industrial sectors at the local level.[14]

The regional concept in focus in this study is the functional region, which for practical purposes is approximated with the commuting region. A functional region normally consists of a central municipality surrounded by a number of hinterland municipalities.

To model size we measure for each municipality its total accessibility to Gross Regional Product (GRP), which we interpret as accessibility to market potential and thus to potential demand. For each municipality we break down the total accessibility to GRP into two components:

- Local accessibility to GRP
- External accessibility to GRP

This makes it possible for us to make statements about which accessibility that matters most for gross entries, exits and net entries of firms in different sectors in the economy. In this study we make a distinction between four sectors: (a) the primary sector, (b) the manufacturing sector, (c) the ordinary service sector and (d) the advanced service sector. Gross entry of firms in a municipality is defined as (number of entering firms +1)/(population in working age in 1,000s). Exit of firms in a municipality is defined as (number of exiting firms +1)/(existing firms). Net entry of firms in a municipality is defined as (number of entering firms – number of exiting firms)/(existing firms +1).[15]

To control for the influence on gross entries, exits and net entries from other factors, we also include a number of control variables in our analysis. These control variables are (a) the employment rate, (b) the share of the labour force with more than 3 years of university schooling, (c) the share of small firms with one to four employees in all sectors, (d) local accessibility to GRP, (e) external accessibility to GRP.

[14]See Plümper and Troeger (2007) for further details.

[15]We add one to all observations to avoid any zeros in entries and exits in any municipality. This "transformation" will only have a very slight influence on our econometric results.

The data used in this paper comes from Statistics Sweden and spans over the years between 1993 and 2004.[16] Tables 1 and 2 below give a short presentation of the data.

Table 1 Variable descriptions

Variable	Description/explanation
Population in working age	Population in the ages between 20 and 64 years of age
Entry	(Number of entering firms + 1)/(population in working age in 1,000s)
Exit	(Number of exiting firms + 1)/(existing firms + 1)
Net entry	(Number of entering firms – number of exiting firms)/ (existing firms + 1)
Local accessibility to GRP	Accessibility to Gross Regional Product in municipality *m* coming from economic activity in municipality *m*
External accessibility to GRP	Accessibility to Gross Regional Product in municipality *m* coming from economic activity in all other Swedish municipalities
Education	Population with more than 3 years of university schooling/population in working age
Employment rate	Population holding a job/population in working age
Share of small firms	Number of firms with 1–4 employees/all firms

Table 2 Descriptive statistics

Variable	Mean	Median	Std. dev.	Min	Max
Entry in primary sector	0.346	0.256	0.310	0.012	3.256
Entry in the manufacturing sector	0.980	0.885	0.489	0.092	4.889
Entry in the ordinary service sector	1.857	1.741	0.780	0.202	8.160
Entry in the advanced service sector	1.533	1.395	0.808	0.170	8.537
Exit in the primary sector	0.228	0.179	0.199	0.014	3.000
Exit in the manufacturing sector	0.108	0.098	0.051	0.012	0.571
Exit in the ordinary service sector	0.137	0.132	0.045	0.017	0.533
Exit in the advanced service sector	0.193	0.175	0.086	0.016	1.000
Net entry in the primary sector	−0.006	0.000	0.365	−0.857	0.889
Net entry in manufacturing sector	−0.059	−0.060	0.255	−0.900	0.875
Net entry in the ordinary service sector	−0.051	−0.046	0.189	−0.857	0.857
Net entry in advanced service sector	0.037	0.037	0.237	−0.889	0.900
Local accessibility to GRP	0.006	0.003	0.017	0.001	0.289
External accessibility to GRP	0.026	0.011	0.044	0.001	0.319
Education	0.098	0.087	0.043	0.040	0.366
Employment rate	0.749	0.750	0.044	0.514	0.882
Share of small firms	0.606	0.602	0.056	0.417	0.809

Note: N = 3,420

[16] See tables of correlation matrix in the appendix for further details/characteristics of the variables.

4 Empirical Results

Here we present our empirical results. We start by analyzing the determinants of firm entry in our four different sectors during the time period 1993–2004.[17] Then we analyze the determinants of firm exit in our four different sectors and thereafter we investigate the determinants of net entry of firms in the different sectors.

4.1 *Firm Entry*

The dependent variable is defined as entry of firms in each one of our four sectors, for the time period 1993–2004 (Table 3).

Local accessibility to GRP has a strong significant impact on entry of new firms. For the primary sector and the manufacturing sector this impact is negative and for the ordinary service sector and the advanced service sector it is positive. External accessibility to GRP also has a strong significant impact on entry of new firms. However, the negative impact is only valid for the primary sector and positive for the three other sectors. Further, the strong significant impact on new firm formation can be interpreted as follows. In municipalities with a high external accessibility to GRP, potential entrepreneurs tend to form new firms. A high employment rate

Table 3 Results from the FEVD regressions for firm entry

Dependent variable	FEVD (fixed effects vector decomposition)			
	Entry (1) (primary sector)	Entry (2) (manufacturing sector)	Entry (3) (ordinary service sector)	Entry (4) (advanced service sector)
Local accessibility to GRP	−2.547 (11.08)***	−0.979 (2.35)**	9.274 (16.18)***	12.991 (24.38)***
External accessibility to GRP	−1.590 (15.50)***	0.577 (3.08)***	3.292 (12.89)***	5.556 (23.33)***
Education	−0.666 (5.92)***	−3.266 (16.03)***	−6.380 (22.26)***	1.643 (5.86)***
Employment rate	−1.073 (12.17)***	−3.192 (19.19)***	−3.778 (17.30)***	−2.385 (11.94)***
Share of small firms	0.997 (15.31)***	1.687 (14.22)***	2.227 (13.82)***	2.484 (16.79)***
Constant	0.669 (8.54)***	2.660 (18.11)***	3.817 (19.56)***	1.421 (7.97)***
Observations	3,420	3,420	3,420	3,420
R-squared	0.61	0.48	0.62	0.70

**Significant at 5%, ***significant at 1%

[17]The four sectors used in this paper are: the primary sector, the manufacturing sector, the ordinary service sector and the advanced service sector.

implies that there is a strong negative impact of firm entry in all sectors. This is in line with what one could expect, since there is less need for individuals to start their own business when compared to situations where the employment rate is low. If a municipality or a region has a high level of education among its citizens, there is a strong negative impact on the entry of new firms in all sectors except in the advanced service sector where the impact is positive. Further, the presence of many small firms in different sectors also has a strong positive significant impact on new firm formation in all sectors.

4.2 Firm Exit

The dependent variable is defined as exit of firms in each one of our four sectors, for time period 1993–2004 (Table 4).

Local accessibility to GRP has a strong negative significant impact on firm exit in all sectors except in the ordinary service sector. External accessibility to GRP also has a strong significant impact on entry of new firms for all sectors. In this case the impact is positive, which means that given that a municipality has a high degree of external accessibility to GRP, firm exit is assumed to increase. As in the case of firm entry, a high employment rate and a high level of education implies that there is a strong negative impact of firm exit in all sectors. Further, the presence of many small firms in the different sectors also has a strong positive significant impact on firm exit in all sectors, i.e. the higher share of small firms the higher the exit of firms, for all sectors.

Table 4 Results from the FEVD regressions for firm exit

Dependent variable	FEVD (fixed effects vector decomposition)			
	Entry (1) (primary sector)	Entry (2) (manufacturing sector)	Entry (3) (ordinary service sector)	Entry (4) (advanced service sector)
Local accessibility to GRP	−0.877	−0.134	0.071	−0.355
	(4.80)***	(2.84)***	(1.63)	(4.19)***
External accessibility to GRP	0.945	0.197	0.293	0.116
	(11.50)***	(9.22)***	(14.60)***	(3.04)***
Education	−0.492	−0.112	−0.267	−0.253
	(5.50)***	(4.84)***	(12.21)***	(6.13)***
Employment rate	−0.997	−0.718	−0.463	−0.673
	(14.10)***	(33.92)***	(25.33)***	(20.00)***
Share of small firms	0.060	0.237	0.142	0.191
	(1.16)	(17.58)***	(11.44)***	(7.95)***
Constant	0.968	0.508	0.416	0.604
	(15.41)***	(27.85)***	(26.04)***	(20.49)***
Observations	3,420	3,420	3,420	3,420
R-squared	0.40	0.38	0.32	0.31

***Significant at 1%

Table 5 Results from the FEVD regressions for net entry of firms

Dependent variable	FEVD (fixed effects vector decomposition)			
	Entry (1) (primary sector)	Entry (2) (manufacturing sector)	Entry (3) (ordinary service sector)	Entry (4) (advanced service sector)
Local accessibility to GRP	0.851 (2.02)	1.927 (6.03)***	1.278 (5.78)***	0.694 (2.52)
External accessibility to GRP	0.199 (1.05)	0.618 (4.54)***	0.649 (6.64)***	0.528 (4.23)***
Education	−0.568 (2.74)***	−1.683 (8.97)***	−0.748 (6.25)***	−0.656 (4.64)***
Employment rate	0.273 (1.68)	1.237 (8.77)***	0.438 (5.10)***	−0.095 (0.91)
Share of small firms	0.129 (1.07)	0.283 (3.38)***	0.159 (2.61)***	0.137 (1.76)
Constant	−0.244 (1.69)	−1.021 (8.78)***	−0.426 (5.70)***	0.071 (0.76)
Observations	3,420	3,420	3,420	3,420
R-squared	0.05	0.06	0.08	0.05

Significant at 10%, significant at 5%, ***significant at 1%

4.3 *Net Entry of Firms*

The dependent variable is defined as net entry of firms in our four sectors, i.e. firm entry – firm exit in each sector, for the time period 1993–2004 (Table 5).

Both local and external accessibility to GRP have a strong positive significant impact on net entry of new firms in all sectors. There is no such significant impact for the primary sector in the case of external accessibility to GRP. A high employment rate implies that there is a strong positive impact of net firm entry in the primary sector, the manufacturing sector, and in the ordinary service sector. A high level of education implies a strong negative significant impact on net entry of new firms in all sectors. As in the case of new firm formation and firm exit, the presence of many small firms in the different sectors also has a strong positive significant impact on new firm formation in all sectors except for the primary sector.

5 Conclusions

The purpose of the paper is to show how entrepreneurship conditions vary between regions of various sizes, and to analyse the theoretical arguments as to why large regions generally should generate more entrepreneurship. The purpose also covers analyzing empirically the role of regional size in explaining variations in total entrepreneurship and entrepreneurship in different sectors across functional regions, using data from Sweden for the period 1993–2004.

Using FEVD regressions, we estimate how the conditions for entrepreneurship vary between regions of various sizes. The empirical results from the estimated FEVD regressions show that local and external accessibility to GRP have a strong significant impact on both entry of new firms and on firm exit. For the primary sector and the manufacturing sector this impact is negative and for the ordinary service sector and the advanced service sector it is positive. Hence, both local and external accessibility to GRP are of great importance for the two different service sectors. However, accessibility to GRP is not that important for new firm formation in the primary sector and the manufacturing sector. Further, a high employment rate implies that there is a strong negative impact of firm entry in all sectors. This is in line with what one could expect, since there is less need for individuals to start their own businesses compared to times when employment rate is low. The presence of many small firms in the different sectors also has a strong positive significant impact on new firm formation in all sectors. Hence, the more small firms there are the higher the potential for new firm formation.

Appendix

Table A.1 Correlation matrix for the primary sector

	Entry (1)	Exit (1)	Net entry (1)	Local accessibility to GRP	External accessibility to GRP	Share of small firms	Employment rate	Education
Entry (1)	1.0000							
Exit (1)	0.0798	1.0000						
Net entry (1)	0.3344	−0.3124	1.0000					
Local accessibility to GRP	−0.2259	−0.0614	0.0119	1.0000				
External accessibility to GRP	−0.3087	−0.0843	0.0043	0.1571	1.0000			
Share of small firms	0.1571	0.0396	0.0119	−0.1655	0.1445	1.0000		
Employment rate	−0.1571	−0.0887	0.0054	−0.1175	0.2449	0.0462	1.0000	
Education	−0.3581	−0.0436	−0.0088	0.3806	0.5519	0.0926	0.2695	1.0000

Note: N = 3,420

Table A.2 Correlation matrix for the manufacturing sector

	Entry (1)	Exit (1)	Net entry (1)	Local accessibility to GRP	External accessibility to GRP	Share of small firms	Employment rate	Education
Entry (1)	1.0000							
Exit (1)	0.3088	1.0000						
Net entry (1)	0.4577	−0.4388	1.0000					
Local accessibility to GRP	−0.1341	−0.0255	0.0053	1.0000				
External accessibility to GRP	−0.1550	−0.0057	0.0319	0.1571	1.0000			
Share of small firms	0.1677	0.2565	0.0396	−0.1655	0.1445	1.0000		
Employment rate	−0.1841	−0.3346	0.0223	−0.1175	0.2449	0.0462	1.0000	
Education	−0.2994	−0.0959	0.0163	0.3806	0.5519	0.0926	0.2695	1.0000

Note: N = 3,420

Table A.3 Correlation matrix for the ordinary service sector

	Entry (1)	Exit (1)	Net entry (1)	Local accessibility to GRP	External accessibility to GRP	Share of small firms	Employment rate	Education
Entry (3)	1.0000							
Exit (3)	0.2709	1.0000						
Net entry (3)	0.4365	−0.4410	1.0000					
Local accessibility to GRP	0.1019	−0.0004	0.0573	1.0000				
External accessibility to GRP	−0.0062	0.0625	0.1068	0.1571	1.0000			
Share of small firms	0.1108	0.1707	0.0385	−0.1655	0.1445	1.0000		
Employment rate	−0.2040	−0.2006	0.0329	−0.1175	0.2449	0.0462	1.0000	
Education	−0.0097	−0.0594	0.0983	0.3806	0.5519	0.0926	0.2695	1.0000

Note: N = 3,420

Table A.4 Correlation matrix for the advanced service sector

	Entry (1)	Exit (1)	Net entry (1)	Local accessibility to GRP	External accessibility to GRP	Share of small firms	Employment rate	Education
Entry (1)	1.0000							
Exit (1)	0.0415	1.0000						
Net entry (1)	0.3619	−0.4745	1.0000					
Local accessibility to GRP	0.3484	−0.0909	0.0180	1.0000				
External accessibility to GRP	0.3857	−0.0889	0.0401	0.1571	1.0000			
Share of small firms	0.1723	0.1179	0.0263	−0.1655	0.1445	1.0000		
Employment rate	−0.0069	−0.0069	−0.1916	0.0050	−0.1175	0.0462	1.0000	
Education	0.5701	−0.1809	0.0202	0.3806	0.5519	0.0926	0.2695	1.0000

Note: N = 3,420

References

Acs ZJ (ed) (2000) Regional innovation, knowledge and global change. Frances Pinter, London

Acs ZJ, Audretsch DB (1990) Innovation and small firms. MIT, Cambridge, MA

Acs ZJ, Storey DJ (2004) Introduction: entrepreneurship and economic development. Reg Stud 38:871–877

Acs ZJ, Audretsch DB, Feldman MP (1992) Real effects of academic research: comment. Am Econ Rev 82:363–367

Arauzo JM, Teruel M (2005) An urban approach to firm entry: the effect of urban size. Growth Change 36:508–528

Arauzo JM et al (2007) Regional and sector-specific determinants of industry dynamics and the displacement-replacement effects. Empirica 34:89–115

Arauzo-Carod JM, Manjón-Antolín MC (2007) Entrepreneurship, industrial location and economic growth: an appraisal. Edward Elgar, Cheltenham, pp 3–17

Armington C, Acs ZJ (2002) The determinants of regional variation in new firm formation. Reg Stud 36:33–45

Audretsch DB, Feldman MP (2004) Knowledge spillovers and the geography of innovation. In: Henderson VJ, Thisse JF (eds) Handbook of urban and regional economics. Elsevier, Amsterdam, pp 2713–2739

Audretsch DB, Fritsch M (1994) The geography of firm births in Germany. Reg Stud 28:359–365

Audretsch DB, Lehman EE (2005) Does the knowledge spillover theory hold for regions? Res Policy 34:1191–1202

Audretsch DB, Keilbach MC, Lehmann EE (2006) Entrepreneurship and economic growth. Oxford University Press, Oxford

Backman M (2008) Financial accessibility and new firm formation. Department of Economics, Jönköping International Business School (mimeo)

Baumol W (2002) The free-market innovation machine: analysing the growth miracle of capitalism. Princeton University Press, Princeton

Berglund E, Brännäs K (2001) Plants' entry and exit in Swedish municipalities. Ann Reg Sci 35:431–448

Cabral L (2004) Simultaneous entry and welfare. Eur Econ Rev 48:161–172

Camagni R (1991) Innovation networks: spatial perspectives. Belhaven, London

Chinitz B (1961) Contrasts in agglomeration: New York and Pittsburgh. Am Econ Rev 71:279–289

Cooper A, Folta T (2009) Entrepreneurship and high-technology clutsers. In: Sexton DL, Landström H (eds) The Blackwell handbook of entrepreneurship. Blackwell, Malden, MA, pp 348–367

Davelaar EJ (1991) Incubation and innovation. A spatial perspective. Ashgate, Aldershot

Desmet K (2000) A perfect foresight model of regional development and skill specialization. Reg Sci Urban Econ 30:221–242

Duranton G, Puga D (2001) Nursery cities: urban diversity. Process innovation and the life cycle of products. Am Econ Rev 91:1454–1477

Feldman MA (2001) The entrepreneurial event revisited: firm formation in a regional context. Ind Corp Change 10:861–891

Figueiredo O, Guimarães P, Woodward D (2002) Home-field advantage: location decisions of Portuguese entrepreneurs. J Urban Econ 52:341–361

Florax R, Folmer H (1992) Knowledge impacts of universities on industries: an aggregate simultaneous investment model. J Reg Sci 32:437–466

Florida R (2002) The rise of the creative class. Basic Books, New York

Fotopoulos G, Spence N (1999) Spatial variations in new manufacturing plant openings: some empirical evidence from Greece. Reg Stud 33:219–229

Fotopoulos G, Spence N (2001) Regional variations of firm births, deaths and growth patterns in the UK, 1980–1991. Growth Change 32:151–173

Friis C, Karlsson C, Paulsson T (2006) Relating entrepreneurship to growth. In: Johansson B, Karlsson C, Stough RR (eds) The emerging digital economy, entrepreneurship, clusters, and policy. Springer, Berlin, pp 83–111

Fritsch M, Mueller P (2004) Effects of new business formation on regional development over time. Reg Stud 38:961–976

Fritsch M, Brixy U, Falck O (2006) The effect of industry, region and time on new business survival: a multi-dimensional analysis. Rev Ind Organ 28:285–306

Fujita M, Thisse JF (2002) Economics of agglomeration: cities, industrial location, and regional growth. Cambridge University Press, Cambridge

Fujita M, Krugman P, Venables A (1999) The spatial economy. Cities, regions and international trade. MIT, Cambridge, MA

Garofoli G (1994) New firm formation and regional development: the Italian case. Reg Stud 28:381–393

Georgellis Y, Wall HJ (2000) What makes a region entrepreneurial? Evidence from Britain. Ann Reg Sci 34:385–403

Glaeser E (1999) Learning in cities. J Urban Econ 46:254–277

Glaeser E et al (1992) Growth in cities. J Polit Econ 100:1126–1152
Guesnier B (1994) Regional variations in new firm formation in France. Reg Stud 28:347–358
Guillain R, Huriot J-M (2003) The local dimension of information spillovers: a critical review of empirical evidence in the case of innovation. Can J Reg Sci 24:313–338
Guimarães P, Figueiredo O, Woodward D (2000) Agglomeration and the location of foreign direct investment in Portugal. J Urban Econ 47:115–135
Hansen ER (1987) Industrial location choice in São Paulo, Brazil: a nested logit model. Reg Sci Urban Econ 17:89–108
Henderson JV, Kuncoro A, Turner M (1995) Industrial development in cities. J Polit Econ 103:1067–1085
Holl A (2004) Transport infrastructure, agglomeration economies and firm birth: empirical evidence from Portugal. J Reg Sci 44:683–712
Hoover EM (1948) The location of economic activity. McGraw Hill, New York, NY
Hoover EM, Vernon R (1959) Anatomy of a metropolis. Harvard University Press, Cambridge, MA
Hotelling H (1929) Stability in competition. Econ J 39:41–57
Jaffe A, Trajtenberg M, Henderson R (1993) Geographical localization of knowledge spillovers as evidenced by patent citations. Q J Econ 108:577–598
Karlsson C, Backman M (2011) Accessibility to Human Capital and New Firm Formation, International Journal of Foresight and Innovation Policy (Forthcoming)
Karlsson C, Manduchi A (2001) Knowledge spillovers in a spatial context: a critical review and assessment. In: Fischer MM, Fröhlich J (eds) Knowledge, complexity and innovation systems. Springer, Berlin, pp 101–123
Karlsson C, Nyström K (2007) Nyföretagande, näringslivsdynamik och tillväxt i den nya världsekonomin, Underlagsrapport nr 5 till Globaliseringsrådet, Regeringskansliet, Västerås
Keeble D, Walker S (1994) New firms, small firms and dead firms: spatial patterns and determinants in the United Kingdom. Reg Stud 28:411–427
Krugman P (1991a) Increasing returns and economic geography. J Polit Econ 99:483–499
Krugman P (1991b) History and industry location: the case of the manufacturing belt. Am Econ Rev 81:80–83
Krugman P (1993) First nature, second nature and metropolitan location. J Reg Sci 33:129–144
Leone RA, Struyck R (1976) The incubator hypothesis: evidence from five SMSAs. Urban Stud 13:325–333
Malecki EJ (1993) Entrepreneurs, networks, and economic development: a review of recent research, vol 3, Advances in entrepreneurship, firm emergence and growth. JAI Press, Greenwich, CT, pp 57–118
Malecki EJ, Poehling RM (1999) Extroverts and introverts: small manufacturers and their information sources. Entrepren Reg Dev 11:247–268
Manjón-Antolin MC (2004) Firm size and short-term dynamics in aggregate entry and exit. Center Discussion Paper, Tilburg University, Tilburg
Marshall A (1920) Principles of economics, 8th edn. Macmillan, London
McCann P (1995) Rethinking the economics of location and agglomeration. Urban Stud 32:563–578
McCann P, Sheppard S (2003) The rise, fall and rise again of industrial location theory. Reg Stud 37:649–663
Mills ES, Hamilton BW (1984) Urban economics, 3rd edn. Scott Foresman, Glenview, IL
Nijkamp P (2003) Entrepreneurship in a modern network economy. Reg Stud 37:395–405
Nyström K (2006) Entry and exit in Swedish industrial sectors. JIBS Dissertation Series No. 32, Jönköping International Business School, Jönköping
Ohlin B (1933) Inter-regional and international trade. Harvard University Press, Cambridge, MA
Parker SC (2004) The economics of self-employment and entrepreneurship. Cambridge University Press, Cambridge
Patel P, Pavitt K (1991) Large firms in the production of the world's technology: an important case of 'non-globalisation'. J Int Bus Stud 22:1–21

Plummer LA, Acs ZJ (2004) Penetrating the 'knowledge filter' in regional economics. Discussion papers in entrepreneurship, growth and public policy, Max Planck Institute for Research into Economic Systems, Jena

Plümper T, Troeger VE (2007) Efficient estimation of time-invariant and rarely changing variables in finite sample panel analyses with unit fixed effects. Polit Anal 15:124–139

Porter M (1998) Clusters and the new economics of competition. Harv Bus Rev 76(6):77–90

Porter ME (2000) Location, clusters and company strategy. In: Clark GL, Feldman MP, Gertler MS (eds) The Oxford handbook of economic geography. Oxford University Press, Oxford, pp 253–274

Pred A (1977) City-systems in advanced economies. Hutchinson, London

Quigley JM (1998) Urban diversity and economic growth. J Econ Perspect 12:127–138

Reynolds P, Miller B, Maki W (1993) Regional characteristics affecting business volatility in the United States, 1980–1984. In: Karlsson C, Johannisson B, Storey DJ (eds) Small business dynamics. Routledge, New York, pp 78–115

Reynolds P, Storey DJ, Westhead P (1994) Cross-national comparisons of the variation in new firm formation rates. Reg Stud 28:443–456

Robinson C, O'Leary B, Rincon A (2006) Business start-ups, closures and economic churn: a review of the literature. Enterprise Directorate, BERR, UK

Rosenthal SS, Strange WC (2003) Geography, industrial organisation and agglomeration. Rev Econ Stat 85:377–393

Saxenian A (1999) Silicon valley's new immigrant entrepreneurs. Public Policy Institute of California, San Fransisco

Schumpeter JA (1934) The theory of economic development. Harvard University Press, Cambridge, MA

Shapero A (1984) The entrepreneurial event. In: Kent CA (ed) The environment for entrepreneurship. Lexington Books, Lexington, MA, pp 21–40

Stam E (2007) Why butterflies don't leave: locational behaviour of entrepreneurial firms. Econ Geogr 83:27–50

Stimson RJ, Stough RR, Roberts BH (2006) Regional economic development. Analysis and planning strategy, 2nd edn. Springer, Berlin

Thompson WR (1968) Internal and external factors in the development of urban economies. In: Perloff HS, Wingo L (eds) Issues in urban economics. John Hopkins, Baltimore, MD, pp 43–62

Thornton PH, Flynn KH (2003) Entrepreneurship, networks and geographies. In: Acs ZJ, Audretsch DB (eds) Handbook of entrepreneurship research. Kluwer Academic, Dordrecht, pp 401–433

Thurik R, Wennekers S (2004) Entrepreneurship, small business and economic growth. J Small Bus Enterprise Dev 11:140–149

Van Ort FG, Stam E (2006) Agglomeration economies and entrepreneurship in the ICT industry, ERIM report series, 016-ORG. Erasmus University

Van Stel A, Storey DJ (2004) The link between firm births and employment: is there an Upas tree effect? Reg Stud 38:893–909

Viladecans E (2004) Agglomeration economies and industrial location: city-level evidence. J Econ Geogr 4:565–582

Vivarelli M (2007) Entry and post-entry performance of newborn firms. Routledge, London

Weyh A (2006) What characterizes successful start-up cohorts? In: Fritsch M, Schmuck J (eds) Entrepreneurship in the region. Springer, Berlin, pp 61–74

Creativity and Diversity: Strategic Performance Management of High-Tech SMEs in Dutch Urban Areas

Karima Kourtit and Peter Nijkamp

1 The Role of High-Tech SMEs: Introduction

The world of business environments in modern economies and cities has changed dramatically the way of pursuing business (Spence 2004), and depends nowadays heavily on the performance in generating and utilizing new knowledge, imagination, creativity, innovations and technologies. This holds even more so in the high technology industry which is usually characterized by an extremely volatile, dynamic and uncertain business climate.

The trend in this high technology sector is that – next to the presence of large multinationals corporations – small and medium-sized enterprises (SMEs) emerge and grow continuously, thus increasing their employment, while large firms tend to decline in number (down-size and focus on their core business activities) and to cut their employment (Tether and Storey 1998). The decline in the manufacturing employment in large firms in the West corresponds with the industrialization of China and India and the relocation of many large firms to these and other nations which act as a magnet for manufacturing.

High-tech firms are the most active forces of the dynamics in SMEs. High-tech SMEs are creating and implementing technological innovations (a major source of developing the high-tech industry) and represent a powerful medium for the creation of new jobs and wealth for society as suggested often in the literature (Lee et al. 2004; McGranahan and Timothy 2007). Policy makers increasingly view high-tech SMEs as key contributors to industrial creativity and innovation performance, technological change, social development and (building and sustaining) economic growth and progress (Jones-Evans and Klofsten 2005; Bommer and Jalajas 2002).

The increasingly vital role of high-tech SMEs in creativity and innovation – in both regional and national economic growth and social development and

K. Kourtit (✉) and P. Nijkamp
Department of Spatial Economics, VU University Amsterdam, De Boelelaan 1105, 1081 HV Amsterdam, The Netherlands
e-mail: k.kourtit@vu.nl; p.nijkamp@vu.nl

K. Kourtit et al. (eds.), *Drivers of Innovation, Entrepreneurship and Regional Dynamics*, Advances in Spatial Science, DOI 10.1007/978-3-642-17940-2_8,

global competitiveness – has been widely acknowledged in the economic and entrepreneurship literature (Berry and Taggart 1998; 2007; Cooper and Park 2008). High-tech SMEs are designed to be flexible and innovative and are often built around a successful innovative product (Trumbach et al. 2006). However, the death rate of high-tech SMEs is higher than that of large enterprises due to capital scarcity and their small scale. They operate and compete in continually changing business environment where innovation is continuous (Torraco and Swanson 1995) and the competition is often so intense that there is no breathing space for relaxation and strategy development (Sureshchandar and Leisten 2005).

The above sketched uncertainty, whether in terms of competition, technology advancements or business culture (Sureshchandar and Leisten 2005) warrants the design of a valid and tailor-made model for these firms that shows how they are performing and that offers the means to improve their creative and innovative performance and to support (control and manage) them better in the challenging business environment. This, in order to enhance their performance by remaining viable and to realize sustainable competitive advantages associated with their human capital.

The growing importance of external and environmental changes puts much emphasis on entrepreneurship (information and knowledge-based activities) and has further intensified and supported the need for efficient and effective management techniques which encourage businesses to stay competitive and profitable (Zeng and Zhao 2005).

Currently the most popular management technique in business practice is Strategic Performance Management (SPM) (Davis and Albright 2004). There is no universally valid definition of SPM, however. Cummings and Worley (2005) define SPM as an "*integrated process of defining, assessing, and reinforcing employee work, behaviours and outcomes*". In this respect, SPM can be understood as *a "business supporting process where steering of the organization takes place through the systematic definition of mission, strategy and objectives of the organization, making these measurable through critical success factors and key performance indicators, in order to be able to take corrective actions to keep the organization on track*" (de Waal 2007).

Following this definition, the SPM concept offers opportunities not only for managing human capital, but also for acquiring a sustainable competitive advantage through providing an environment that fosters entrepreneurship. SPM is often seen as a sine qua non for executing an effective business strategy (Mohrman and Mohrman 1995), and offers aids and knowledge to evaluate and monitor how a business performs, provides reliable and robust steering measures at both the management level and at the workforce level that are in tune with the manifold circumstances of firms.

Against this background, the first aim of the present paper is to review the current state-of-art knowledge on SPM, with a particular view to the high-tech SME sector. To that end, a systematic framework will be offered to position various contributions from the recent literature and to create a frame of reference for the comparative studies that follow later in this paper. Next, the empirical part consists of amuse of findings from a previously undertaken study on corporate firms, followed by the successes and failures of SPM strategies of high-tech SMEs.

The paper will be concluded with an outline of the future research question: "*do the SPM use and outcomes depend on business geographic clustering*?" This research examines the relation between business performance and location choice factors that high-tech enterprises concern in the choice of location.

2 Strategic Performance Management

SPM has become a critical and an increasingly important approach for business management which not only reflects the growing complexity in today's unpredictable, open, diverse and dynamic business world, but also monitors the firm's strategic response to this complexity. Today's turbulent business environment demands regular adaptation of organizational strategies based on market and economic conditions. And therefore, it is important to understand and recognize that firm's strategic objectives can change constantly and to anticipate changing circumstances throughout the organization, from top-level further to operational level. This also demands more clarity of people about their role and contribution (transparency in the added value is of the employees) towards the achievement of the strategy and organizational goals and a better strategic alignment of individual objectives to organizational objectives (better operationalization of the strategy through further cascading of SPM) to improve the performance of the firm and to ensure a sustainable competitive advantage in regards to its chosen organizational strategies, in a dynamic environment.

The development of today's business and managerial strategies have been influenced by Sun Tzu's "*Art of War*" (1910) to understand the importance of competition, competitive advantages and positioning in strategy to make the correct decisions and to create innovations in the competitive business environment to ensure financial viability. The findings in the book, which have influenced government and military policy, illustrated and recommended a (military) strategic method "*How to achieve a mission*" without conflict, an approach which had already become favourite in the Napoleonic time. To link the company's long-term strategy to the day-to-day operations in modern business, an operational Performance Measurement System (PMS) has to be designed (Kald and Nilsson 2000). PMS may be defined as: "*the formal, information-based routines, procedures and process of collecting and tracking data used in Performance Management by managers to maintain or alter pattern in organizational activities*" (adapted from Simons 2000; de Waal 2002).

There are numerous, operational methods for PMS, which can be used by management to regularly assess the performance of the firm e.g., shareholder value, human resource accounting, activity-based costing, knowledge management scorecards. The currently most popular PMS method in business practice is the Balanced Scorecard (BSC), created by Kaplan and Norton (1992, 1996a, 2001a, 2001b). It is a strategic management system that uses Critical Success Factors (CSFs) and Key Performance Indicators (KPIs) for translating an organization's mission and strategy into a balanced set of integrated performance measures (Ho and

Chan 2002; Brignall 2002). The performance measures provide a complete picture of an organization's progress towards its mission and goals (Ho and McKay 2002).

The BSC contains a varied set of performance measures, a combination of non-financial measures (leading indicators) and financial measures (lagging indicators), organized according to four distinct perspectives, namely financial performance, customer relations, internal business processes, and the organization's learning, innovation and growth activities (Kaplan and Norton 1992; Lipe and Salterio 2000). The BSC essentially follows a linear one-way approach (Kaplan and Norton 1992); it assumes the following causal relationship: it starts with the learning, innovation and growth perspective and culminates in financial results or outcomes (Brignall 2002). But this research does not only concentrate on the measurement tool BSC, but on all types of measurements tools, and therefore, the more general term SPM is used in this type of research.

The effectiveness of the SPM process is defined as the achievement of financial as well as non-financial targets, the development of skills and competencies, and the improvement of customer care and process quality (de Waal 2007). In order to get the right approach, firms need to start thinking in terms of strategy and business activities instead of IT systems. They need to get straight into the necessary detail to describe what they are doing (their market position), what they want to do and how they want to achieve it (a top-down framework). The starting point of the process of SPM begins with:

1. Reviewing the mission and strategy

Firms have to establish a mission and determine a strategy, they need to ask: "*what do we want to do and accomplish*" and "*how can we achieve that mission*?"

2. Formulating strategic objectives

In order to make a firm's strategy concrete and tangible, strategic objectives need to be formulated, so that it becomes clear which activities have to be undertaken in order to implement the organization's strategy.

3. Monitoring with critical success factors (CSFs) and key performance indicators (KPIs)

Finally, whether or not objectives and value creation are being achieved (guide and improve all the business functions within the firm) can be monitored with the help of indicators that are expressed in the form of critical success factors (CSFs) (which factors define their success), and measured by key performance indicators (KPIs). Thus, firms need to ask: "*What is the measurement for the objective?*" and "*how can this objective be measured?*" The use of CSFs and KPIs enables the measurement, and thus the control of strategic objectives and value creation of a firm. Figure 1 gives an illustration of the development of these measures.

SPM involves performance measurement at the following levels: mission, strategies, objectives, critical success factors and finally key performance indicators and to establish a clear link between performance and strategy. Thus, firms need a starting point of access to view, monitor and measure their performance, which in turn can improve performance at all levels of a firm; firms and their agents need to

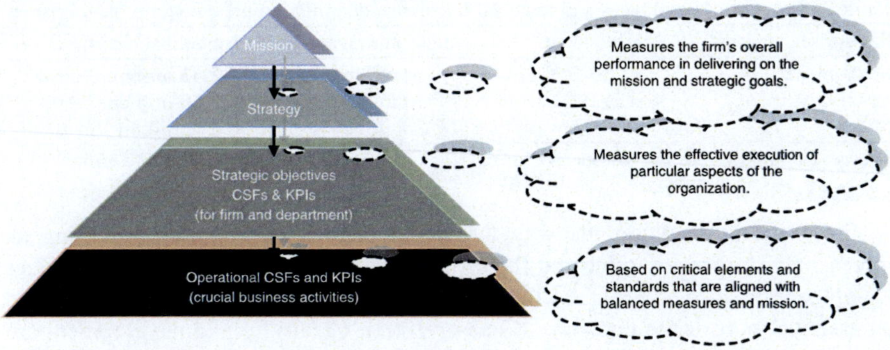

Fig. 1 The development of CSFs and KPIs illustration

know and understand the strategy and their role and contribution in it. Thus, it is more a process that starts with understanding where the organization is today (positioning), which direction it wants to take, what targets should be set, and how resources should be allocated to achieve those targets. Furthermore, it helps firms to continuously anticipate changing circumstances within their industry and build a flexible capacity for continuous adaptation of their firm. As Sun Tzu said: "*All men can see the tactics whereby I conquer, but what none can see is the strategy out of which victory is evolved*". We will now address the link between SPM and the development and performance of SMEs.

3 High-Tech SMEs and SPM

There is no commonly accepted definition of high-tech SMEs, regardless of the overall recognition of their important contribution to (regional) economic growth. For the selection of these firms or sectors within the industry, our study draws on the broad conceptual approach of the Commission of the European Communities (2003). Table 1 below outlines the key aspects of the definition as recommended by the European Commission; which are those enterprises with fewer than 250–300 employed persons.

These high technology enterprises operate in an extremely volatile, dynamic and uncertain business climate with continually changing technologies, markets, and business strategies, and shifting consumer needs for products and services (fundamental transformations). They are engaging in development, manufacturing, and distribution of high-tech products, technology transfer and consultation. The Dutch SMEs include 14,500 technology firms, concentrated in high-tech sectors, as well as in the machine and devices industry, the chemical industry, the food and nutrition industry, the engineering and architecture branch and the ICT-branch (CBS 2011). They are frequent users of the creative and innovation policy with a particularly high degree of professional specificity to generate significant new value for the stakeholders and their firm, both economically and morally.

Table 1 Number of employees and financial thresholds determining the categories of enterprise

Enterprise category	Headcount	Annual turnover	Annual balance sheet total
Medium-sized	<250–300	≤ € 50 million	≤ € 43 million
Small	<50	≤ € 10 million	≤ € 10 million
Micro	<10	≤ € 2 million	≤ € 2 million

Source: Commission of the European Communities (2003)

Creativity is changing the way the world does business; it is becoming an increasingly important input into the production process of all goods and services – and therefore also critical for a business's long-term viability over the years – and an essential part of the economy (Glaeser 2005). Creativity and development are the key elements of innovation and are improving competitiveness, particularly high-tech SMEs, in a rapidly changing global economy. They are challenged to continually innovate and improve – continual process of improvement – the quality of their already existing products and services, and methods and techniques to stay ahead of the heavy – often global – competition in which the combined pressures of economic liberalization, technological change and shifts in regulatory systems lead to a complex action space and create a changing and challenging business environment.

High-tech SMEs have to embrace innovation in their business strategies; their strategic goals have to be growth oriented and to search for new (long- and short-term) opportunities or even create new products and services or business models that change their strategies and generate significant new value for the stakeholders and the firm (strategic innovation). They need to do things differently by using performance management strategies for executing their business strategies (Mohrman and Mohrman 1995), which not only affects the current organizational performance, but also their future performance and (strategic) directions (Millett 1998; de Haas and Kleingeld 1999; Norreklit 2000); and bring people together, aiming to achieve more teamwork. Innovation is a "magic word", in particular when it becomes a part of management and when it is managed further to an operational level, because fast (integration of) innovation (operationalizing) in the entire organization increases through a structural approach e.g. with high interaction with different actors, collective goals and creativity which increases also the employees' involvement by triggering discussions and calling attention of organizational members.

The SPM concept is not new and has only been acknowledged more recently, because of the changing business environments. Firms are now also recognizing the importance of non-financial information which can be related to the strategic indicators, such as customer and employees satisfaction, research and development, changes in external environment of the corporate organization, manufacturing and production, sales and marketing. The traditional approach focused mainly on financial results (e.g., sales, turnover, profit, costs) and lacked match between the company's competences and its dynamic business environment; there was no strategic alignment, and management had a short-term vision overwhelmed with data (Kald and Nilsson 2000; Bourne et al. 2003; Kanji 2005). These shortcomings enticed organizations to search for performance measurement systems that supported them better in the challenging business environment (Waal and Counet 2006). As a result, the traditional financial indicators (past performance) were complemented with "non-traditional soft indicators" – the drivers of future performance – about e.g. customer and employees satisfaction, innovation and growth for translating a firm's mission and strategy into a balanced and comprehensive set of integrated performance measures (Brignall 2002; Ho and Chan 2002) "*blending traditional and non-traditional approaches to business strategy*" (Fernandes et al. 2006), e.g. a balance between short- and long-term objectives, between financial and non-financial measures, between leading and lagging indicators, and between internal and external performance perspectives (Fernandes et al. 2006). The BSC has become one of the first SPM methods that really succeeds in translating mission and strategy into financial (lagging indicators) and non-financial indicators (leading indicators) that can lead to action (Waal and Counet 2006).

Nowadays SPM has become "paramount" to the overall success of organizations in the today's business environment (innovative strategic business practice). SPM offers opportunities to encourage entrepreneurial spirit and competition in the longer term, creativity and development; to contribute to the growth and progression of knowledge needs; to attract and maintain competent human capital (qualified and technological workers), and to achieve superior financial performance (survival and growth strategies). Both the popular and scientific literature (Berry et al. 1995; Kaplan and Norton 1996b; Chow et al. 1998; Zairi and Jarrar 2000; Niven 2002; Andersen et al. 2006) indicates a broad coverage of SPM e.g. to perform health checks throughout the organization; to clarify and translate vision into operational strategy; to clarify the objectives of the organization; to facilitate the setting of targets for the organisation and its managers; to understand the processes within the organisation; to communicate and link strategic objectives and business measures; to enhance strategic feedback and learning; to use performance levels to conduct detailed operational planning of activities and processes; to establish an early warning system through monitoring of key indicators; to influence and alter employee behavior to promote desired changes, and to promote the accountability of the organization to its stakeholders.

It is estimated that around 60–70% of medium-to-large sized for-profit firms in the US and Europe – as well as many governmental departments – have adopted (with varying degree of adaptation rates) – or are familiar with – the SPM concept (Silk 1998; Rigby 2001; Neely et al. 2004; Marr et al. 2004). SPM is mainly adopted by large organizations (Fernandes et al. 2006). Translating SPM into SMEs requires light to be shed on the following questions:

1. To what extent have firms practiced SPM?
2. What are the implications of SPM use:
   (a) Advantages
   (b) Disadvantages
   (c) Reasons for implementation?
3. How do the reasons for SPM influence advantages and disadvantages?

Various literature, case studies and practical experience (Hronec 1993; Lynch and Cross 1995; McDonald and Smith 1995; Lingle and Schiemann 1996, 1999; Kaplan and Norton 1996b; Atkinson et al. 1997; Armstrong and Baron 1998; Waal 2001; Lawson et al. 2003) show that companies that have implemented SPM perform better, financially as well as non-financially, than companies that are less SPM-driven over a longer period of time. Nonetheless, many of these studies of SPM's positive impact on performance is anecdotal (Neely et al. 2004) and stems from case studies that are less based upon a rigorous, systematic, scientific analysis of empirical facts or solid business management theory, and focus on research in large firms in general, and are very limited within high-tech SMEs. This is mainly due to:

- The gap in the literature of the perception and knowledge of high-tech SMEs regarding SPM
- The understanding and application of SPM concepts by high-tech SMEs is often unknown e.g. no linking to business strategy, unstructured assessments, inconsistent measures throughout the firm
- These firms tend to focus only on the core business (purely technological side of their business) and neglect other key strategic issues.

Furthermore, despite recent empirical research, little is actually known about the specific reasons why high-tech SMEs (or normal organizations) have implemented SPM (Robinson 2004), because specific reasons for using SPM can yield automatically particular advantages (positive relationship) and disadvantages (negative relationship).

Because SPM has attracted much research interest in recent years from both the academic and business communities, it is essential to know whether the implementation of SPM in high-tech SMEs will yield particular advantages and disadvantages, as predicted by the literature (Dumond 1994; Groves and Valsamakis 1998; Haas and Kleingeld 1999; Lovell et al. 2002; Hoque and James 2000; Kald and Nilsson 2000; Norreklit 2000; Malina and Selto 2001; Shulver and Antarkar 2001; Sim and Koh 2001; Hoque 2003; Braam and Nijssen 2004; Davis and Albright 2004; Papalexandris et al. 2004; Robinson 2004; Scheipers et al. 2004; Lawson et al. 2005; Tapinos et al. 2005). To provide answers to these three questions it is necessary to undertake empirical research. The research methods adopted in this paper include:

(a) Analysis of the literature
(b) Structured interviews
(c) Common Factor Analysis (1st-stage: identify the main components: advantages, disadvantages and reasons for use)
(d) Multivariate Analysis (2nd-stage: identify the various relations between the factors).

This paper has nine sections. Section 3 examines the advantages, disadvantages and reasons for SPM use, as identified in the literature. These are reviewed, summarized and then tested on 43 high-tech SMEs in Netherlands. Additional testing using common factor analysis and multiple regression analysis is presented in Sects. 6–8. Finally, the last section provides a summary and discusses the limitations of our research.

4 SPM Advantages, Disadvantages and Reasons for Use: Analysis of Key Literature

An analysis of key literature focusing on advantages, disadvantages and reasons for use SPM in practice was undertaken. This involved a review of a wide range of measurements tools relevant to the application of SPM. The analysis identified 3 financial advantages, 22 non-financial advantages, 8 non-financial disadvantages associated with SMP and 41 reasons for using SPM. The financial and non-financial advantages indicators refer to *People (P)*,*Results (R)*,*Information (I)* and *Organization (O)*, the disadvantages indicators refer to *Systems* (S) and *Information (I)*, and the reasons for use SPM indicators refer to *Control (C)* and *Strategy (S)*. These indicators are summarized in Table 2.

Table 2 Listing of SPM advantages, disadvantages and reasons for use, as identified from the literature

	Indicators refer to	Literature source
Quantitative advantages		
Increase in revenue	R = Result	Malina and Selto (2001), Sim and Koh (2001), Davis and Albright (2002), de Waal (2002), Said et al. (2003), Braam and Nijssen (2004), Davis and Albright (2004), Neely et al. (2004) and Robinson (2004)
Increase in profit		Epstein et al. (2000), Davis and Albright (2002), de Waal (2002), Said et al. (2003), Braam and Nijssen (2004), Davis and Albright (2004), Neely et al. (2004) and Robinson (2004)
Reduction in costs		Sim and Koh (2001) and Neely et al. (2004)
Qualitative advantages		
More focus on the achievement of results	R = Result	Dumond (1994), Bititci et al. (2004), Lawrie et al. (2004), Neely et al. (2004) and Self (2004)
Higher innovativeness		Sim and Koh (2001), de Waal (2002) and Self (2004)
Better achievement of organisational goals		de Waal (2002), Hatch (2005) and Tapinos et al. (2005)
Higher quality of products and services		de Waal (2002) and Brown (2004)

(*continued*)

Table 2 (continued)

	Indicators refer to	Literature source
Closer collaboration and better knowledge sharing and information exchange between organisational units	O = Organization	Mooraj et al. (1999), Kald and Nilsson (2000), Neely et al. (2004), Lawrie et al. (2004), Papalexandris et al. (2004) and Robinson (2004)
Strengthened focus on what is important for the organization		Mooraj et al. (1999), Kald and Nilsson (2000), Baraldi and Monolo (2004), Neely et al. (2004) and Self (2004)
Better strategic alignment of organisational units		Malina and Selto (2001), Shulver and Antarkar (2001), Lovell et al. (2002), Neely et al. (2004), Lawson et al. (2005)
Higher operational efficiency		de Waal (2002), Neely et al. (2004) and Robinson (2004)
Improvement in the decision-making process		Dumond (1994), Mooraj et al. (1999) and Kald and Nilsson (2000)
More effective management control		Malina and Selto 2001 and Neely et al. (2004)
Stronger process orientation		Shulver and Antarkar (2001) and Neely et al. (2004)
Strengthened reputation of the organisation as a quality firm		de Waal (2002) and Self (2004)
Better strategic planning process		Lovell et al. (2002) and Tapinos et al. (2005)
Improvement in communication in the organization on the strategy	I = Information	Lovell et al. (2002), Baraldi and Monolo (2004), Heras (2004), Neely et al. (2004), Papalexandris et al. (2004), Robinson (2004) and Lawson et al. (2004)
Higher quality of performance information		Lawson et al. (2004), Neely et al. (2004), Robinson (2004), IOMA. Business Intelligence at Work (2005) and Tapinos et al. (2005)
Better understanding of organizational units of the strategy		Lovell et al. (2002), Heras (2004) and Neely et al. (2004)
More clarity of people about their contribution towards achievement of the strategy and organizational goals		Lawson et al. (2004), Neely et al. (2004) and Papalexandris et al. (2004)
Improvement of management quality	P = People	Malina and Selto (2001), de Waal (2002) and Neely et al. (2004)
Higher commitment of organizational members to the organization		Malina and Selto (2001), Neely et al. (2004) and Bititci et al. (2004)
More pro-activity of organizational units		Neely et al. (2004), Hatch (2005), Tapinos et al. (2005)
More clarity for organizational units about their roles and goals to be achieved		Lawson et al. (2004) and Neely et al. (2004)
Higher employee satisfaction		Sim and Koh (2001) and Papalexandris et al. (2004)

(*continued*)

Table 2 (continued)

	Indicators refer to	Literature source
Qualitative disadvantages		
It causes too much internal competition	S = System	Kald and Nilsson 2000 and Papalexandris et al. (2004)
There is too much financial information		Kald and Nilsson 2000 and IOMA, Business Intelligence at Work (2005)
It is too expensive and too bureaucratic		Braam and Nijssen (2004) and IOMA, Business Intelligence at Work (2005)
There are too many performance indicators		Dumond (1994), Kald and Nilsson (2000), Self (2004) and IOMA, Business Intelligence at Work (2005)
The performance information is too aggregated	I = Information	Kald and Nilsson (2000) and Neely et al. (2004)
There is not enough strategic information in the system		Kald and Nilsson (2000) and Sim and Koh (2001)
The performance indicators are too subjective and therefore unreliable		Kald and Nilsson (2000) and Malina and Selto (2001)
There is too much historical information		Kald and Nilsson (2000), IOMA, Business Intelligence at Work (2005)
Reason for use		
More accurate measurement of performance	C = Control	Robinson (2004)
Stronger accountability		Robinson (2004)
Need for a broader set of measures of performance		Robinson (2004)
Stronger individual accountability of employees		Robinson (2004)
Handling the increase in complexity of the organization		Tapinos et al. (2005)
Improve the performance of the organization		Lawson et al. (2004)
Obtain a better understandings of knowledge and skills of people		Lawson et al. (2004)
Better control and with that a better "obedience" of people		Lawson et al. (2004)
Tracking progress towards achievement of organizational goals		Lawson et al. (2004)
Being able to measure people, projects and strategy		Lawson et al. (2004)
Being able to measure performance at various organizational levels		Lawson et al. (2004)
Need to correlate measures and actions better		Lawson et al. (2004)
Linking rewards to performance		Lawson et al. (2004)
Enforcing and monitoring regulatory compliance		Lawson et al. (2004)

(*continued*)

Table 2 (continued)

	Indicators refer to	Literature source
Expectation of the stock market		Lawson et al. (2004)
Requirement of governmental regulations		Lawson et al. (2004)
Providing a better picture of customer and product profitability		Lawson et al. (2004)
Making responsibility accounting possible		Lawson et al. (2004)
Facilitate comparison with other, similar business units		Lawson et al. (2004)
Determination of the bonus of management and/or staff		Lawson et al. (2004)
Monitor whether the business is creating value for shareholders		Lawson et al. (2004)
More focus on the strategy	S = Strategy	Robinson (2004)
Better facilitation of cross-functional understanding		Robinson (2004)
Better goal setting		Robinson (2004)
Formalization of the strategic planning process		Robinson (2004)
Stronger commitment of top management		Robinson (2004)
Higher commitment to the strategy		Neely et al. (2004)
Better description of mission, strategy and goals		Neely et al. (2004)
Aligning employee behaviour with strategic objectives		Lawson et al. (2004)
Better communication of strategy to everyone in the organization		Lawson et al. (2004)
Aligning the organization to the strategy		Lawson et al. (2004)
Translating the strategy into operational terms		Lawson et al. (2004)
Need to make strategy everyone's job		Lawson et al. (2004)
Requirement of a business opportunity		Lawson et al. (2004)
Decision support at top management level		Lawson et al. (2004)
Decision support at operational level		Lawson et al. (2004)
Identity possible needs for changes in strategy		Lawson et al. (2004)
Facilitate implementation of business strategy		Lawson et al. (2004)
Provide information for external reporting		Lawson et al. (2004)
Enhance quality of the organization		Lawson et al. (2004)
Facilitate a process orientation		Lawson et al. (2004)

An analysis of these findings from the literature indicates that profit and non-profit organizations using SPM achieved better organizational results (significant increase in both financial and non-financial terms) than those organizations that are less SPM-driven. Organizations achieved for instance a significant increase in revenue and profit and succeeded in establishing a continuous forum for strategic communication. However, the implementation of SPM varies widely from organization to organization, but has a common feature, namely a focus on KPIs. The introduction of SPM and new style of fashionable management technique that came with it resulted not only into enhanced strategic awareness among the top and middle management but also into a significant shift in "governance policy" at board level. The creative thinking required to understand the need for change sharpened the vision for the future of the organization and the way it intended to fulfil its mission.

The development of corporate and department SPM made it possible to achieve quantifiable and qualitative breakthrough results. In general, the changes and increases in scorecards outcomes and financial performance have encouraged organizations to continue using SPM. However, various studies emphasised that the insignificant impact of SPM in the areas detected is not only a result of the inadequacies of SPM but also of those implementing it. Thus, it is understood that the factors that determine the success of SPM initiatives require commitment, effort and resource allocation at all organizational levels. A further conclusion from various sources is that the increase of complexity either expressed with the organizational size or with environmental turbulence increases the need for information, which can be provided by making effective utilisation of SPM. Further, the literature highlights the fact that there is relatively little evidence into performance impact of SPM and whether the SPM actually works. Therefore, it still requires an empirical investigation of more organizations that implemented SPM into a longer timescale over which this performance impact can be observed, because this would allow for a more detailed analysis. Finally, based on this literature SPM tools in general helped organizations to communicate the strategy of the business to the managers and their subordinates and determine what impact a potential change will have on the rest of the organization.

5 Research Approach: A Prior SPM Study

Recently, a major study on SPM among major business firms in the Netherlands was carried out (Kourtit 2007). This prior study provides now the basis for crafting a conceptual framework, which can be used for further research on experiences of advantages, disadvantages and reasons for use in high-tech SMEs. The same quantitative and qualitative methodological approach is undertaken for the present study in order to address the experiences of SPM in high-tech SMEs active in the three enterprise category (see Table 1) operating in the Netherlands, which are familiar with the firm's SPM and performance.

In the present study, a total of 52 members of a top management team - managers, CEOs and company owners – of 43 SMEs that had implemented SPM and a total of 15 members of a top management team – managers, CEOs and/or company owners – of 15 high-tech SMEs that were not using SPM at all, selected from the high-technology industries, participated in the research. The main research instrument was a self-established survey questionnaire based on the previous research and in-depth interviews.

Our sample comprised a new set of high-tech SMEs that were recently approached. Based on the answers to the interview questions, a comparison can now be made in terms of the performance of firms using SPM for a period of less than or more than 1 year.

Table 3 lists the average scores for the financial and non-financial advantages, disadvantages and reasons for implementing SPM, on a scale of 1 (not at all) to 5 (very strong) of high-tech SMEs – both with and without SPM – which are active in the enterprise categorization from Table 2. The higher the score in Table 3, the stronger the firms experienced the (dis)benefits and more important the reasons for implementing SPM are to steer the business firms into long lasting success.

Table 3 shows that the scores on advantages for small firms using SPM do actually differ significantly from other small firms not using SPM. These firms using SPM indicate that both financial as well as non-financial performance indicators (clear and determined KPIs) – e.g., increase in revenue, profit, communication, customers and employees satisfaction – really need attention to survive in a turbulent business environment.

The small firms not using SPM are only focused on very tight financial indicators and outputs, and less on the non-financial aspects. However, the disadvantages turn out to be hardly experienced by the small firms using SPM (μ is 1.7; $\sigma = 0,6$), while small firms that are not using SPM definitely experienced the disadvantages. This suggests that the use of SPM brings clear advantages which outweigh the disadvantages.

Furthermore, all firms using SPM (or without another fashionable management technique) want to improve continuously the performance of their organization and to achieve sustainable success to become and stay at a world-class level in everything they do. They all indicate that the identified reasons to steer the business are virtually equally important (μ is between 3.64 and 4.58). This suggests there is

Table 3 Average score of advantages, disadvantages and reasons for use with (out) SPM

Average scores per experienced SPM variable

Business firms	Financial advantages		Non-financial advantages		Disadvantages		Reasons for use	
	Average scores (μ)	Standard deviation (σ)	Average scores (μ)	Standard deviation (σ)	Average scores (μ)	Standard deviation (σ)	Average scores (μ)	Standard deviation (σ)
SMEs with SPM	3.56	0.8	3.49	0.7	1.72	0.6	4.58	0.8
SMEs without SPM	3.03	0.6	2.86	0.7	2.53	0.5	4.33	0.9

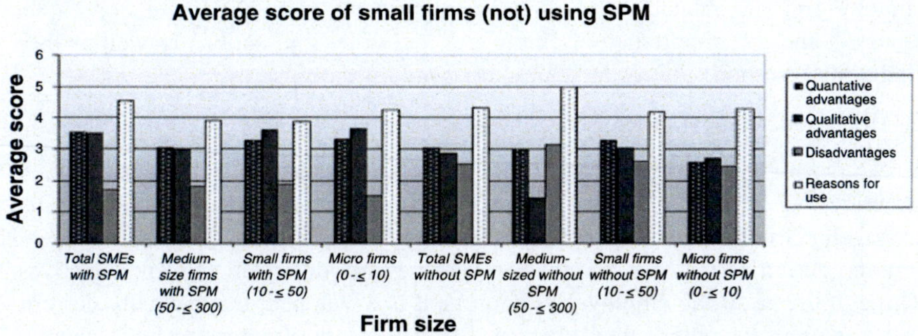

Fig. 2 Average score of advantages, disadvantages and reasons for use per firm size class

no particular reason that plays a dominant role in the decision to implement and use SPM (or another fashionable management technique). Further details on our results are recorded in Fig. 2.

Our sample of businesses was categorized in two groups, in relation to their size and use of SPM. The first group consists of 52 members of a top management team of 43 high-tech SMEs, which are engaged in the development, manufacturing, distribution of high-tech, bio- and agro-products, software and web design, technology transfer and consultancy (Commission of the European Communities 2003) and is using SPM for a period of less than or more than 1 year. The second group comprises in total 15 members of a top management team of 15 high-tech SMEs, which are engaged in the same sectors but not using SPM at all (or another management approach); they are also categorized. Figure 2 shows that the scores on both financial as non-financial advantages are strong for both the large firms and the high-tech SMEs using SPM which had practical experience with it, a finding which can be attributed to the introduction of SPM in practice, as described in the management and scientific literature. The introduction of SPM has had a positive impact on the financial and non-financial performance of firms, in particular of small firms with less than 10 employees followed by small firms with employees between 10 and 50. Through SPM, they have:

1. Strengthened their focus

In the first place they have strengthened their focus on what is important for the organization and the achievement of organizational results e.g. consciousness and involvement of personnel, better set-up priorities. Members of the business firms want to understand the KPIs and through determining clear objectives and KPIs they understand things better and what is important, which lead them to become more target-oriented and performance-driven. This is more a continuous process, which translates the strategies into focus, points out and clearly identifies individual activities (responsibility), and which also stimulates the participation of personnel. This leads to a better focus on transparency and clarity in the alignment of the strategy and the related KPIs (understanding how each KPI is related to the strategy) and the importance of employees' activities and actions as well as their (individual) responsibilities in achieving desired performances.

2. Higher operational efficiency

Next, a higher operational efficiency that is reflected in a better organizational structure (e.g., clarify processes through a decrease in ad hoc work) and flexible management to anticipate (external) changes can be achieved. They do not spend time on activities that will not lead to achievement of organizational objectives. Through this, a strong employee commitment is developed; they actually perform better in order to achieve their objectives (independent of the rewards). Managing on results helped them to focus more clearly on the position they take (operation-field) and where they ultimately want to be (success).

3. Better achievement of organizational goals

And in the third place, a better achievement of organizational goals is possible, because operational objectives have been translated into better and clearly measurable KPIs and undertaken actions, which are related to strategic objectives. There is a constant focus on the questions "*what do we want to do?*" "*what are we doing?*", "*what are the real priorities and real focus points*", *and is it feasible and how to deliver it*?

Furthermore, small firms using SPM experienced the financial advantages (e.g. revenue, profit) mainly through non-financial advantages. They indicate that through a better organizational structure e.g. better process and costs orientation, transparency, direct communication lines, and steering on structured methods and modes of outputs e.g. continuous focus on management and efficiency managers achieved the desired levels of performance and operational excellence, and realize what is promised in an efficient manner. And through enhanced strategic intensive periodical feedback, learning and evaluation (bottom-up) of what has been realized, objectives can be adapted, "it is a learning process". Finally, SMART (i.e. Specific, Measurable, Achievable, Realistic, and Timed) objectives and result-oriented appointments coupled with sustained improvement in decision-making through cutting off pointless discussions and reliable information – supported by informative reports – leads to more pro-activity, people react fast in developing actions and taking better "*GO-NO/GO*" decisions, "*Focus + Facts = Actions*", that contribute to organizational improvements. However, various external factors, e.g. economic, social, environmental, political, governmental, technological and competitive forces, influence also the financial advantages.

The scores on the advantages for small firms not using SPM are in general resulting from less experienced firms or are unclear (μ is 1.4 and 3.0), in particular small firms with employees less than 10. A possible explanation is that not all (important) organizational performances are determined and (well) measured to achieve sustainable success, in particular the "soft" performance indicators are not or poorly defined and not all indicators are identified correctly, and considered as important for these firms. This may cause firms to use their metrics ineffectively, because they believe that the best way to achieve a better focus on organizational performances is not to manage it! They do not plan consciously, because people have their own responsibility. If performance indicators that measure the execution of the strategy and the creation of value are not included in the SPM process, it will not be transparent whether strategic objectives and value creation are being achieved (Waal 2001). If organizations cannot define the overall performance of their organization, they cannot measure or manage it (Armstrong and Baron 1998). Clearly, the feedback of the results and measurement of various issues (hard and soft indicators) have to be clear, before an unambiguous statement can be made.

The disadvantages are rarely experienced by small firms using SPM (μ is 1.7 and 1.8). Their perception is "we can learn from our experiences, thus from historical information in the system". These firms indicate e.g. historical information in the system as a positive point of departure to determine their objectives, which are actually adjusted in the course of the time (process), and give them an image to steer the organization on it in order to be able to take corrective actions to keep the organization on track. They confirm that each action plan to achieve organizational objectives includes important strategic information (highly focused and result-oriented), so to know all the strategic assessments makes little sense to them. They said: "*How to steer the organization without a clear (definition of the) strategy and objectives of the organization*?" Thus, organizational objectives have to be specific and unambiguous by using structured business taxonomy, explained with simple terminology for the masses; understandable for organizational members. In other words "they must be well defined and precisely delimited by the start" (measurable and feasible). They content to have a dashboard scorecard with 5 or 6 performance indicators (mix of internal and external measures) to measure and focus on, but find it difficult sometimes to determine good quantitative objectives. Through regular communication and discussion of the "abstract context" on the indicators people remain focused. This suggests that the use of SPM brings clear advantages which are dominant over the disadvantages.

The scores on the disadvantages are strong for small firms not using SPM, in particular small firms with employees between 50 and 300, followed by small firms with employees between 10 and 50. They have to pay too much attention to various drivers of business performance that included too much historical and financial details (focusing excessively on the past and on the short run) without enough strategic information to choose a uniform and appropriate business direction and to consider what and how to improve, instead of prominent indicators that forecast future results. This can create frustration and unfocused decision making. Too much information and indicators can overload individuals and the provision of too many, or conflicting, performance measures may create an opposite reaction. This means that structurally the financial perspective dominates over the other, non-financial indicators (short-term strategy). This appears to give an unbalanced view of the total organization's performance, with a focus on mainly "hard indicators," e.g. revenue, profit, costs. These firms indicate that the financial perspective is an easy factor to measure and manage on, which may create a "tunnel vision" (narrow perspective) especially where problems, decisions and other issues are viewed from a "disciplinary" viewpoint; only from a "financial perspective". With a lack of openness to other organizational excellence outcomes (e.g., improve employee and customer satisfaction, improve organizational performance and productivity, innovativeness), which promotes a form of blindness given the complexity of the environmental factors, firms do not recognize the importance of non-financial measures of performance for both managing and evaluating their achievements, as financial figures alone did not identify the elements that may lead to good or poor future financial results. It is important to understand organizational excellence, which potentially leads to the success of a business in the future (better achievement of organizational goals).

Finally, all interviewees indicate that all the reasons for use of SPM (or other fashionable management techniques than SPM) to steer the business firm are important, in particular small firms not using SPM with employees between 50 and 300 (μ is 5.0), which showed also a strong score on the disadvantages. This suggests that they all want to improve continuously and optimize the performance of the organization (e.g. higher quality of products and services) which has now become standard, and to achieve sustainable success through a particular approach or mentality. Therefore, learning and innovation perspectives are very important for them, because it is intended to measure an organization's capacity to innovate, continuously improve and (cooperative) learn, and creative approaches to addressing business results.

Therefore, they have to be able e.g. to analyse how economic and social changes affect their businesses now and in the future; to anticipate on changing circumstances in the industry and to stay ahead of the extreme – often global – competition; to build a capacity for continuous adaptation of their organisations in order to achieve sustained high performance; translate strategy into action at each level within the organisation in order to bring the business strategy to successful life; to focus on 'doing the right things right' which implies that the link between information and successful management action in the business environment is essential; to have the right information at the right time to make the best decisions, to take the best actions for the benefit of the development of continuous and sustained organizational improvement and to enhance quality of the organization. In conclusion, they suggest that the fundamental reason for implementing a particular management technique is improvement of e.g. strategy and communication. Strategy and communication included both the need to communicate strategy to everyone and the need to align employee behaviour with strategic objectives. Thus, communication is essential for employee acceptance of SPM. They need to know the reason(s) behind the implementation of SPM and how it benefits both the organization and themselves.

6 Application of Common Factor Analysis

In this section we apply "Common factor analysis (CFA)" (Tucker et al. 1969; Rummel 1970; Ford et al. 1986) based on a "Maximum Likelihood Method" as a multidimensional analytical tool, to identify the main independent components: advantages, disadvantages and reasons for use. This approach avoids generation of an overwhelming amount of data. The results of this are discussed below.

6.1 *CFA Results of SPM Advantages for Small Firms Using SPM*

The application of CFA to the SPM advantages yielded four factors, namely Higher Result-Orientation of workforce (HROW), Better Organizational Structure (BOS), Better Information communication (BIC) and Higher Result-Orientation of Management (HROM), as shown in Table 4.

Table 4 Common factor analysis of the SPM advantages components

SPM advantages	Factor 1 (HROW)	Factor 2 (BOS)	Factor 3 (BIC)	Factor 4 (HROM)
Strengthened reputation of the organization as a quality firm (O)	0.884	0.209		−0.166
More innovativeness (R)	0.774	0.207	−0.106	
Higher quality of products and services (R)	0.748			0.236
Increase in revenue (R)	0.746			−0.114
Increase in profit (R)	0.711		0.168	
More pro-activity of organizational members (P)	0.697		0.368	
Reduction in costs (R)	0.474	−0.181	−0.248	0.253
Better understanding of organizational members of the strategy (I)	0.466			
More effective management control (O)		0.780		
Better focus on the achievement of results (R)	−0.130	0.737		
Stronger process orientation (O)	0.185	0.711	−0.316	−0.144
Better strategic alignment of organizational units (O)	0.137	0.682	0.158	
Higher quality of performance information (I)		0.603		
Improvement in the decision-making process (O)		0.570		0.229
Better strategic planning process (O)			0.759	0.228
More clarity for organizational members about their roles and goals to be achieved (P)			0.713	−0.129
Higher commitment of organizational members to the organization (P)	0.384		0.519	0.334
Improvement in communication in the organization on the strategy (I)	−0.207	0.492	0.514	
More focus on the achievement of goals (R)	0.295		0.458	0.295
Higher employee satisfaction (P)	0.327	0.205	0.453	
Better focus on what is important for the organisation (O)		0.213	0.430	−0.145
More clarity of people about their contribution towards achievement of the strategy and organizational goals (I)	0.343		0.402	
Higher operational efficiency (O)	0.191	0.117	0.185	0.705
Improvement of management quality (P)	0.176	0.241		0.604
Better knowledge sharing and information exchange between organisational units (O)	0.172	0.202		0.410

Factor 1, labeled as a higher result orientation of workforce (HROW), consists of variables which all have to do with a good communication, customer focus, higher result orientation and accountability of employees (shared governance become more efficient) on achieving organizational results, by using SPM. Firms want to manage effectively by clarifying the individual's responsibilities (defining specific goals and objectives) for employees which is part of the process of delegation, and developing accountabilities with employees through participation in decision-making and management processes to evaluate their needs to make better use of themselves and time, because within small firms the workforce is actually the "management" mainly due a simple and flat organizational structure (which encourage employees to take part in decision-making and management processes) with less or no organizational levels (removed middle management layers).

Thus, the level of participation of employees in decision-making and management processes is partly dependent on the degree to which an organization has many or few levels of management hierarchy. And because of the simpler organizational structures, fewer customers and flexibility and more adaptable to market changes, it seems plausible that result-orientation of small firms are higher on achieving organizational results than for large firms.

Furthermore, this factor can be explained by considering this advantage as a logical consequence or outcome measures of the other factors, which result in a higher orientation on organizational results. They make and communicate effectively a clear and understandable plan regarding the desired results aligned to the strategy (well-organized report of the results and how people perform). This is the result of the top-down and bottom-up method to make strategies tangible and understandable (better focus on defining objectives which furnishes also more quality), and to stimulate the participation of employees (responsibility and accountability) in the thinking-process throughout the entire organization in order to make the strategy owned and understandable.

The strengthened involvement and understanding of people of the strategy, coupled with the improvement in the quality (e.g., qualitative solutions lead to customers retention), motivated employees ("lower employees illness, less internal blockades and less complaints"), pro-activity (e.g., employees want to take a step further, anticipate fast on positive as well as negative situations, and are enthusiastic because thing are going well), better steering on projects (e.g., determined "*Critical to Quality*" parameters for improvement and secure) and more innovativeness (e.g., people are now thinking from different perspectives to achieve organizational objectives and many ideas and processes are carried out and developed, which give employees the space and time to realize these objectives and innovations without personal and financial consequences), considerably facilitates the achievement of organizational goals and results.

Finally, small firms also experienced an increase in revenue of approximately 10–15% on average, and a decline in cost of approximately 6% on average, (e.g., better insight and understanding into costs, failure reduction, less internal blockades and complaints, improved business activities), resulting in an increase in profit.

Factor 2, labeled as better organizational structure (BOS), consists of variables depicting advantages which are caused by SPM increasing focus, control, responsibility and alignment throughout the organization on the strategic goals to be achieved.

The business activities in small firms are more focused and structured on clear (financial) facts and targets (e.g. revenue, costs and profit consciousness of employees, clear measurable KPIs, conscious undertaken actions which are related to strategic objectives) more tightly planned, managed and monitored (clear rules and timelines, fast and simple to adjust) and objective decision-taking (fixed procedures, clear and determined values and norms, benefit-cost analysis, action and decision list), because of simple measurement tools (e.g., BSC, LEAN, Six Sigma), simpler organizational structures (e.g., enabled to delegate responsibility for operational activities and to focus more on planning and strategic functions, better flow of information, effective communication, better flow of authority and responsibility in the organization), greater flexibility of change (e.g., more ability to adept to changes in the environment), shorter communication-lines and higher interaction and consensus between various disciplines (e.g. implementation of extensive ICT-systems such as CRM, prick plate, clarity and openness, constructive and active meetings, a frequent review of progress performance regarding the strategy leads to a better alignment), daily face-to-face evaluation moments between manager(s) and their employees for improvements and their customers (e.g., clear appointments, delivery times), and fast reaction on pursuing and achieving desired strategic goals (e.g. to realize optimal sales results). In addition, this factor can be explained by clarity and accountability; a continuous strategic thinking-process which translates the strategy into operational results, identifies individual activities and responsibilities, and which also stimulates the participation of personnel or units and the alignments throughout the organization towards achieving the strategy.

SPM improve the business structure by systematic translating vision into operational strategy, which define tactics and processes to support the strategy and demonstrate measurable organizational and individual results; supported by regular and better quality of performance information. This creates more structure and insight for organizational members on goals to be achieved and their role in this and in (fixed) business processes (less ad hoc work) which leads to fast links and reaction within the management. A better translation of the strategy into clear focus points and individual activities (better cascading of the strategy), people better understand how their contribution fits within the total process (transparent added value of employees) or know why they are engaged in particular activities, it prevents a great deal of (behavioural) frustration and employees perform better. Thus, the outcome measures of BOS leads to a more transparency and clarity in the alignment of the strategy and the relevant KPIs (understanding how each KPI is related to the strategy), a better integral approach of distribution of people and production, the importance of employees' activities and actions as well as their specific responsibilities and accountabilities in achieving desired performances, a good planning and a more conscious strategic choice, coupled with actions (and rewards). This includes that there is a movement of "loose control" to "tight control" and frequently communication between members of the organization e.g. periodical feedback and evaluation, which created clarity and a better focus in the business (better focus leads to more effort of people and a better efficiency in sale actions), and a better steering and execution of business activities that help people to take a consistent business direction and to consider what and how to improve things and focus on issues that are important for the organization. This leads to tremendous results for instance increase in revenue, profit, innovativeness, and pro-activity which have also been improved and achieved in the organizations.

Factor 3, labeled as better information communication (BIC), consists of variables depicting the increased clarity and better communication on the strategy and business's performance to organizational members and their participation. A better and effective communication on the organizational strategy contributes to operationalizing the strategy from management further to individual objectives e.g. clear and transparent focus points. It creates more focused discussions on what is important for the organization, the business direction and segmentation(s), and on future performances and opportunities. BIC contributes to a better translation of the strategy into more concrete and tangible objectives and clear KPIs, with clear and well understood language (taxonomy). These firms want to make their members more aware and to provide a better understanding of the business direction and to encourage them to be more committed to their organization which contributes to a better achievement of business results. Employees are continuously informed (qualitative reports and communication-tools) about the business performance, the direction of which the organization intends to take (communicating clear targets), what their individual contribution is expected to be in the relation to the entire process (clear constructions) and how they fit within it (integral improvement) and their responsibility and accountability (this results in less external complaints and higher efficiency). All these factors make personnel more concerned about the overall success of the organization.

Firms make and communicate a clear and understandable plan regarding the desired results aligned to the strategy (well-organized report of the results and how people perform). This is the result of the top-down and bottom-up method to make strategies tangible and understandable, and to stimulate the participation of employees (accountability) in a integral process throughout the entire organization in order to make the strategy owned and understandable, and to perform more effectively. This leads to no-nonsense discussions on objectives and targets (result-oriented), people can refer to various policies, relate their work and to the strategy (know which direction the organization aims), and various issues are better to follow. Through this, a strong employee commitment is developed; they actually perform better in order to achieve their objectives and mainly independent of the rewards. Managing on results helped them to focus more clearly on the position they take in the operation-field and where they ultimately want to be (success).

Finally, Factor 4, labeled as higher result orientation of management (HROM), consists of variables, which have to do with improving the quality of management and processes on achieving organizational results, by using SPM. Managing real value drivers (quantitative as well as qualitative) behind the business with a continuous strategic planning process combined with learning- and thinking process for value-creation and achieving organizational results is a daily preoccupation for these firms. Thereby, it is important to communicate clear the key performance indicators (empowerment), which clear-up the accountability and responsibility of people and leads them to a better action-orientation and improvement of executing strategic planning and realising a competitive advantage. Through improved and intensive communication, knowledge share and exchange and cooperation have been improved between firm parts and members. Through this, members indicate earlier discrepancies in the indicators and understand the business strategy better, which reduce *Ad Hoc* work, reduce costs and increase profit. A better steering and execution, which in using specific concepts and taxonomy that everyone is expected to know, help people to take a consistent business direction and to consider what and how to improve things and a better focus on issues that are important for the organization and its shareholders.

6.2 CFA Results of the Disadvantages

The CFA of the SPM disadvantages yielded two factors, as depicted in Table 5.

Table 5 shows that Factor 1, badly aligned system (BAS), consists of variables showing that the implemented SPM system leads to a meaningless discussion and unclear focus on the business, which does not have the right fit with the organization. Despite that historical information in the system provides its users with appropriate information about their past performances; much historical information in the system is not always the best "measuring norm" for the future. However, historical information which gives a good overview of the achieved organizational performances is necessary during the implementation of SPM, which they can learn from the experiences (e.g. market prices, machine calculations, work hours and labour-intensive trade) in order to be able to take corrective actions to keep the organization on track. However, high-tech firms have not experienced too much historical information, because it is the starting point to define mission, strategy and short and long term goals and objectives of the organization. But that has to be adjusted with new and strategic information after the introduction year. Firms have

Table 5 Common factor analysis of the disadvantages high-tech SMEs

SPM disadvantages	Factor 1 (BAS)	Factor 2 (LIQ)
There is too much historical information in the system (S)	0.760	
There is not enough strategic information in the system (S)	0.685	
It is too expensive and too bureaucratic (S)	0.576	
The performance indicators are too subjective and therefore unreliable (S/I)	0.471	0.440
It causes too much internal competition (I)	−0.130	0.700
There are too many performance indicators (I)		0.598
The performance information is too aggregated (I)		0.493

to work on this minor experienced disadvantage; they have to use a combination of non-financial and financial measures to assess how well their operations are aligned with their business strategy to make it possible to measure strategy aspects. However, each action to achieve the determined objectives related to the strategy. This approach is a crucial component, because there is actually a strategy that goes beforehand at the choice of the performance indicators; and in managing the strategic direction and decisions of an organization and measuring its success: "*how would you like to organize without a strategy*?".

In addition, a quite technical organized system can be costly and time-consuming (i.e., depends on the size of a firm) for instance to develop "product information manual" for stakeholders and "manpower" which demands a lot of work hours and specialized employees (investing in specialized co-workers is a little risk and it can be costly on short term), but it yields opportunities and possibilities when there are good, clear and determined appointments without a lot of rules and bureaucracy.

Factor 2, low information quality (LIQ), consists of variables which depict the bad quality of the performance information generated by SPM system. The system contains too many performance indicators, which are too aggregated and do not give strategic information; and leads to too many reports, and finally bureaucracy. It becomes difficult for employees to focus on to many issues and results, and moreover to discus these in short time; important indicators will lose their value and context. However, improvements are certainly possible to make a selection of KPIs that are meaningful. The art is to get successful business information (findability and system usability) that serve both individual users and their organisation, which leads to "no-nonsense" discussions. The objectives are simple and good to understand, but the problem is the "translation" to other levels because of difficult terms and priorities statement. There are various terms and performance-indicators, which some of the employees do not always understand and concrete these in the frontline due to e.g. low-educational background. In addition, the system causes unwanted behaviour of employees as peer pressure escalates in internal competition and mutual strive. It leads to isolation of information for job security, low commitment to the organization, minimal collaboration improve the corporate, and other negative effects, which cause islands culture mainly due to miscommunication and poor alignment between collective and individual KPIs.

6.3 CFA Results of the Reasons for Use in High-Tech SMEs

The CFA of the reasons for use of SPM yielded two factors in high-tech SMEs, as depicted in Table 6.

Table 6 shows that Factor 1, focus on strategy (FoS), consists of reasons for use that have to do with creating a focus on formulating, deploying, communicating, implementing and understanding the strategy throughout the organization. A better two-way exchange of information (top-down and bottom-up), systemic direction of communication (structurally more attention for communicating results) with

Table 6 Common factor analysis of the reasons for use high-tech SMEs

High-tech SMEs: reasons for use	Factor 1 (FoS)	Factor 2 (FoC)
Better communicating of strategy to everyone in the organization (S)	0.911	
Aligning employee behaviour with strategic objectives (S)	0.749	
Stronger accountability (C)	0.739	
Stronger individual accountability of employees (C)	0.734	0.132
Translating the strategy into operational terms (S)	0.638	0.273
Higher commitment to the strategy (S)	0.610	0.167
Improve the performance of the organization (C)	0.557	0.260
Linking rewards to performance (C)	0.548	
To measure better the performance of organisation parts (C)	0.493	0.269
To obtain a better understanding in knowledge and skills of people (C)	0.484	−0.130
Better control and with that better "obedience" (C)	0.416	
Handling the increase in complexity of the organization (C)	−0.248	0.922
To describe the mission, strategy and targets of the organisation (S)	0.190	0.617
Enhance quality of the organization (S)	0.185	0.428
More focus on the strategy (S)	0.216	0.417

limited number of concretely and objectively defined KPIs allows the cascading of the strategy to the operational level to be more successfully carried out. Through that employees do not lose their focus on the business (people are not overwhelmed with information). Consequently, a greater awareness develops and more consistent attention is given to matters that relate to quality performance.

Factor 2, focus on control (FoC), consists of reasons for use that have to do with a better control of the organization and strategic objectives. SPM is used to deploy accountabilities and responsibilities on all levels in the organization and subsequently measure and control the performance of these levels. Firms want to be the "best-in-class" and achieve better organisational goals, therefore it is important that they have to create a better focus on the strategy (Factor 2, FoS), future performances, market and business changes, and new challenges and opportunities through e.g. a better communication and translate of the strategy to the operational level. Managing actively changes which take place in order to respond to new opportunities, is a transform of "re-active" into "proactive" approach to anticipate on changes and to avoid threats, indicate earlier discrepancies in organization and understand the business better.

7 Results of the Correlation Analysis

It is necessary that the factors are not subjected to the principle of multicollinearity, nor with a strong correlation, to analyze the relationships between the factors (Hoerl 1959, 1962; Hoerl and Kennard 1970, 1976). This is because strongly correlated factors would explain the same phenomenon. Although there is no clear limit in the literature for the strength of a correlation, an informal rule of thumb suggests it has to be around 0.6 ($X1 = \lambda X2$). Table 7 gives the correlation matrix.

Table 7 shows that the correlations indicate a linear positive as well as a negative relationship between two variables that are not stronger than (−) 0.6, which are mainly autonomous features. But the variables will almost always have some correlation with one another. These results make it possible, initially, to launch a multiple regression analysis without fearing an obvious problem of multicollinearity between factors.

8 Results of Multiple Regression Analysis

Using a multiple regression analysis, a structural model can be created from the SPM advantages, disadvantages and reasons for use factors (Fig. 3). This model is constructed to identify the various relationships between the factors. In this respect, several hypotheses can be made, such as: (1) specific reasons for using SPM will yield specific advantages (positive relationship) and disadvantages (negative relationship); (2) specific SPM advantages will create specific disadvantages (negative relationship); and (3) specific SPM advantages will cause specific other SPM advantages (positive relationship). Figure 3 depicts the results of the multiple regression analysis.

Figure 3 shows there are several significant relations between the SPM advantages, disadvantages and reasons for use factors. Results indicate that the reasons for use factors have significant positive relations with three of the advantages factors, namely Better Organizational Structure (BOS), Better Information and Communication (BIC), and Higher Result Orientation Management (HROM), and a significant relation with one disadvantage factor, namely Low Information Quality (LIQ). However, the reasons for use do not have a direct relation with the advantage Higher Result Orientation Workforce (HROW). This can be explained by considering this advantage as a logical consequence of the other advantages: a better organizational structure, which is drive by the organization's strategy to help firms translate the strategy into operations and reach their objectives, and better information and communication will result in a higher orientation on results by the workforce and subsequently achieving higher organizational results.

Moreover, a better organizational structure improves organizational efficiency or effectiveness, reflects efficient (information) flows and internal business processes, and identifies lines of authority and splits responsibilities (less ad hoc work). Next, results indicate that there is a significant relationship between the reasons for use Focus on Control (FoC) and the disadvantage Low information Quality (LiQ). A possible explanation is that firms during the introduction of SPM still feel traditional financially oriented and therefore still too much and emphasis on financial indicators and measures. The provision of too many performance measures can effect the quality of information and overload individuals. Employees do not always understanding and are not always conscious of these indicators.

Table 7 Component correlation matrix of factors high-tech SMEs

Factors	Mean (μ)	Standard deviation	HRO	BSC	HPQ	HOQ	BAS	LIQ	FoC	FoS
Achieved organizational results Workforce (HROW)	3.5	0.8	1							
Better organizational structure (BOS)	3.5	0.8	0.351	1.000						
Better information communication (BIC)	3.5	0.7	0.449	0.386	1.000					
Achieved organizational results Management (HROM)	3.5	0.8	0.409	0.331	0.515	1.000				
Badly aligned system (BAS)	1.6	0.8	−0.340	−0.423	−0.462	−0.213	1.000			
Low information quality (LIQ)	1.8	0.7	−0.007	0.203	0.269	0.437	0.087	1,000		
Focus on control (FoC)	4.6	0.8	−0.191	−0.028	−0.118	0.026	0.073	0.440	1.000	
Focus on strategy (FoS)	4.5	0.7	0.141	0.065	0.047	−0.098	−0.123	0.131	0.476	1

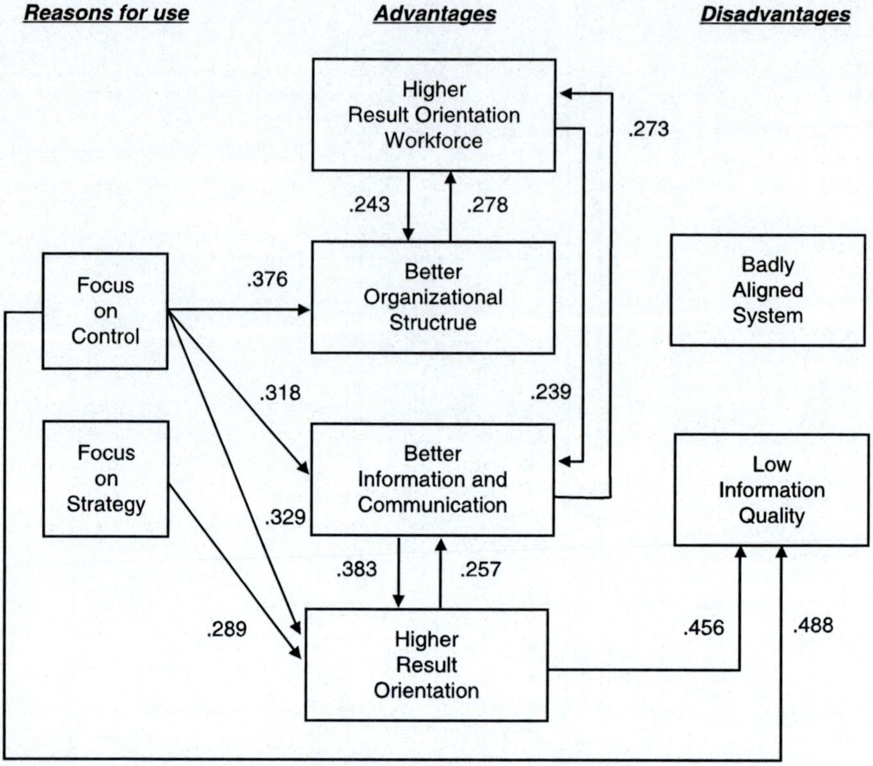

Fig. 3 Structural relations in high-tech SMEs

Further, results indicate also that there is a significant relationship between the advantage High Result Orientation of Management and the disadvantage Low information Quality (LiQ), a possible explanation is information that is needed to build quality-oriented decisions and actions. For many employees too many performance indicators, in particularly financial measures, are too aggregated and not appropriately and accurately related to outcomes they directly influence, and organizational goals to provide useful feedback on their decisions and actions. Finally, the disadvantages, which did not occur very often anyway (see Table 5), do not 'automatically' stem from specific reasons for use or are inescapably linked to specific advantages. If these disadvantages are experienced in firms, they occur stand-alone.

9 Summary and Evaluation

The research described in this paper focused on answering the research questions first introduced in Chap. 3 as follows: *To what extent have small high-tech business firms practiced SPM? What are the advantages, disadvantages and reasons behind the implementation of SPM in these firms in practice? How do the reasons for SPM influence advantages and disadvantages*? At this point, the study undertook to solve the research questions by exploring the understanding of and knowledge about SPM and experiences in practice by the Dutch high technology business firms. The evidence in this research supports a well-developed understanding of and extensive knowledge about SPM by the Dutch high technology firms. These firms have incorporated a number of unique financial as well as non-financial performance measures to enforce their quality and service initiatives in terms of e.g. employees, customer focus and improvements points, related to the strategy for achieving organizational sustainability and long-term success, and for maximizing opportunities for gaining and enhancing a competitive advantage.

They strengthened involvement and understanding of employees (cost and added value consciousness of personnel) in the decision-taking and determining and formulating organizational and individual objectives which are fast and simple to adjust. In particular small-sized firms, with less than 10 employees followed by small firms with employees between 10 and 50, have a more distinct focus on firm and employee performance than the large-sized firms, because of simple measurement tools, simpler organizational structures, shorter (and informal) communication-lines and higher interaction and consensus between various disciplines (e.g. daily face-to-face evaluation moments between manager(s) and their employees for improvements) and fast reaction on achieving organizational results (e.g. to realize optimal sales results).

Further, based on a literature study and practical research at 43 Dutch high-tech firms, four main advantages, two main disadvantages and two main reasons for use were discovered, which are to be expected from using SPM. The practical implication of this research is that implementing and using SPM in crafting and executing of strategy and increase the contribution of employees to strategic plans in a daily preoccupation yields specific benefits for an organization in gaining and sustaining their competitive advantages in terms of products and services. Overall, the findings and their possible implications give insight into the firms' experiences and enables management to adopt a useful approach to increasing their potential to progress (local) firms' health and create a strong and professional human capital pool that sustains the firm's competitive advantage.

This research contains a few limitations. Although there have been statistically significant improvements and correlations after using SPM within the business firms, findings are not generalisable beyond the context of this study in which the research was conducted, because the sample size is limited to 43 Dutch high-tech firms. On the other hand, the study approach facilitated the understanding of SPM using and its experiences in various Dutch high technology business firms. Further research might replicate this study in a large sample of similar organizations for further research on SPM. A large sample of similar organizations would increase the overall validity of the results. Another limitation is that this research is not longitudinal. Longitudinal studies would better examine and observe (without manipulating) the developments and shifts in the relations between SPM advantages, disadvantages at different points of time, and reasons for use.

9.1 Future Research

Future research is needed into environmental factors (external driving forces) such as spatial dimensions (e.g. accessibility, agglomeration principles) or localized concentrations of economic activity. This interest is warranted because of the far-reaching influence that localized concentrations of economic activity (development of creative clusters) and business performance have on regional and national modern economies.

The future research will be able to answer these research issues address the location challenges identified in the future research.

References

Andersen B, Henriksen B, Aarseth W (2006) Holistic performance management: an integrated framework. Int J Product Perform Manage 55(1):61–78

Armstrong M, Baron A (1998) Performance management, the new realities. Institute of Personnel and Development, London

Atkinson AA, Balakrishnan R, Booth P, Cote JM, Groot T, Malmi T, Roberts H, Uliana E, Wu A (1997) New directions in management accounting research. J Manage Acc Res 9:70–108

Baraldi S, Monolo G (2004) Performance measurement in Italian hospitals: the role of the Balanced Scorecard. In: Neely A, Kennerly M, Waters A (eds) Performance measurement and management: public and private. Centre for Business Performance, Cranfield University, Cranfield pp 75–82

Berry AJ, Broadbent J, Otley DT (1995) Management control: theories, issues and practices, 1st edn. Macmillan Press, London

Berry MMJ, Taggart JH (1998) Combining technology and coporate strategy in small high tech firms, Research Ploicy 26:883–895

Bititci U, Mendibil K, Nudurupati S, Turner T, Garengo P (2004) The interplay between performance measurement, organizational culture and management styles. In: Neely A, Kennerly M, Waters A (eds) Performance measurement and management: public and private. Centre for Business Performance, Cranfield University, Cranfield pp 107–114

Bommer M, Jalajas D (1996) Technological innovation support by domestic and multinational high-tech enterprises in Canada. Technology Studies 3(1):50–73

Bourne M, Franco M, Wilkes J (2003) Corporate performance management. Measuring Bus Excell 7:15–21

Braam GJ, Nijssen EJ (2004) Performance effects of using the balanced scorecard: a note on the Dutch experience. Long Range Plann 37(4):335–349

Brignall TJ (2002) The unbalanced scorecard: a social and environmental critique. In: Neely, A., Walters, A., Austin, R. (Eds.), Performance Measurement and Management: Research and Action. Centre for Business Performance, Cranfield School of Management, Cranfield, UK, 17–19 July, pp. 85–92

Brown A (2004) Implementing a system of performance management in England's primary school. In: Neely A, Kennerly M, Waters A (eds) Performance measurement and management: public and private. Centre for Business Performance, Cranfield University, Cranfield, pp 155–162

CBS (2011) *Vestigingen en bedrijven naar economische activiteit*, grootte en rechtsvorm en regio (*corop*) (SBI 2008 4- and 5 digit)

Chow CW, Ganulin D, Haddad K, Williamson J (1998) The balanced scorecard: a potent tool for energizing and focusing healthcare organization management. J Healthc Manage 43(3): 263–280

Communities C o t E (2003) European Competiveness Report. Brussels.

Cooper SY, Park JS (2008) The Impact of 'Incubator' Organization on Opportunity Recognition and Technology Innovation in New, Entrepreneurial High-technology Ventures, International Small Business Journal, 26(1):27–51

Cummings G, Worley CG (2005) Organization Development and Change, 8th Ed., Thomson South-Western, USA

Davis S, Albright T (2002) Relationship between high quality implementation procedures and improved financial performance for new performance measurement systems, FSR Forum 4 (September): 22–31

Davis S, Albright T (2004) An investigation of the effect of balanced scorecard implementation on financial performance. Manage Acc Res 15(2):135–153

de Haas M, Kleingeld A (1999) Multilevel design of performance measurement systems: enhancing strategic dialogue throughout the organization. Manage Acc Res 10:233–261

de Waal, A.A. de (2001) Power of Performance Management: How Leading Companies Create Sustained Value. John Wiley & Sons, New York

de Waal AA (2002) The role of behavioural factors in the successful implementation and use of performance management systems. PhD thesis, Vrije Universiteit, Amsterdam

de Waal AA, de and Counet H (2006) Lessons learned from the balanced scorecard, Research into problems encountered and lessons learned from implementation and use of performance management systems. In: A. Neely, M. Kennerley and A. Walters (ed.), *Performance Measurement and Management: Public and Private*. Cranfield School of Management, UK

de Waal AA (2007) Strategic performance management, a managerial and behavioural approach. Palgrave MacMillan, London

Dumond EJ (1994) Making best use of performance-measures and information. Int J Oper Prod Manage 14(9):16–32

Epstein M, Kumar P, Westbrook R (2000) The drivers of customer and corporate profitability: modeling, measuring, and managing the causal relationships. Adv Manage Acc 9(1):43–72

Fernández WD, Martin MA, Gregor SD, Stern SE Vitale MR (2006) "A Multi-Paradigm Approach to Grounded Theory," Information Systems Foundations Workshop: Constructing and Criticising, School of Accounting and Bussiness Information Systems, College of Business and Economics, The Australian National University, Canberra, Australia

Ford JK, MacCallum RC, Tait M (1986) The application of exploratory factor analysis in applied psychology: a critical review and analysis. Pers Psvchology B 2:291–314

Groves G, Valsamakis V (1998) Supplier–customer relationships and company performance. Int J Logistics Manage 9(2):51–64

Glaeser EL (2005) Review of Richard Florida's the rise of the creative class, Regional Science and Urban Economics 35(5): 593–596

Heras MA (2004) Performance measurement and quality systems: results of qualitative research carried out in companies that had won the Catalan quality award. In: Neely A, Kennerly M, Waters A (eds) Performance measurement and management: public and private. Centre for Business Performance, Cranfield University, Cranfield, pp 459–466

Ho KS, Mckay RB (2002) Innovative performance measurement; balance scorecard-tow perspectives. The CPA Journal 72(3):20–25

Ho SK, Chan YL (2002) Performance measurement and the implementation of balanced scorecards in municipal governments. The Journal of Government Financial Management 51:8–19

Hoerl AE (1959) Optimal solution of many variables equations. Chem Eng Prog 55:69–78

Hoerl AE (1962) Application of ridge analysis to regression problems. Chem Eng Prog 58:54–59

Hoerl AE, Kennard RW (1976) Ridge regression: iterative estimation of the biasing parameter. Commun Stat A 5:77–88

Hoque Z (2003) Total quality management and the balanced scorecard approach: a critical analysis of their potential relationships and directions for research. Critic Perspect Acc 14(5):553–566

Hoque Z, James W (2000) Linking balanced scorecard measures to size and market factors: impact on organizational performance. J Manage Acc Res 12(1):1–18

Hronec SM (1993) Vital signs: using quality, time, and cost performance measurements to chart your company's future. AMACOM, New York

IOMA. Business Intelligence at Work (2005) Two studies reveal how firms are improving their budgeting and planning. performance reporting: majority of companies need to fix their balanced scorecards IOMA's Financial Analysis, Planning & Reporting 2005 Yearbook. IOMA, Newark, pp 4–5

Jones-Evans D, Klofsten M (1997) *Technology, Innovation, and Enterprise—The European Experience*. London, England: Macmillan Publishers

Kanji GK (2005) Sustainable growth and business excellence. Total Quality Management and Business Excellence, Vol. 16, pp 1069–1078

Kald M, Nilsson F (2000) Performance measurement at Nordic companies. Eur Manage J 14(1): 113–27

Kaplan RS, Norton DP (1992) The balanced scorecard measures that drive performance. Harvard Bus Rev 70(1):71–79

Kaplan RS, Norton DP (1996a) Using the balanced scorecard as a strategic management system. Harvard Business Review (January/February): 75–85

Kaplan RS, Norton DP (1996b) The balanced scorecard – translating strategy into action. Harvard Business School Press, Boston

Kaplan PS, Norton DP (2001a) The Strategy-Focused Organization: How Balanced Scorecard Companies Thrive in The New Business Environment, Harvard Business School Press. Boston

Kaplan RS, Norton DP (2001b) Transforming the balanced scorecard from performance measurement to strategic management: Part 1. Acc Horiz 15(1):87–104

Kourtit K (2007) Strategic performance management: reasons for implementation, and advantages, disadvantages and experiences in practice. Master thesis, VU University Amsterdam

Lawrie G, Cobbold I, Issa K (2004) The design of a strategic management system in an industrial private sector organisation. In: Neely A, Kennerly M, Waters A (eds) Performance measurement and management: public and private. Centre for Business Performance, Cranfield University, Cranfield, pp 579–586

Lawson R, Stratton W, Hatch T (2003) The benefits of a scorecard system. CMA Manage 77(4):24–26

Lawson R, Stratton W, Hatch T (2004) Automating the balanced scorecard. CMA Manage 77(9):39–43

Lawson R, Stratton W, Hatch T (2005) Achieving strategy with scorecarding. Journal of Corporate Accounting & Finance 16(3):63–68

Lee J-N, Miranda SM, Kim Y-M (2004) IT Outsourcing Strategies: Universalistic, Contingency, and Configurational Explanations of Success. Information Systems Research 15(2):110–131

Lingle JH, Schiemann WA (1996) From balanced scorecard to strategic gauges: is measurement worth it? Manage Rev 5(3):56–61

Lingle JH, Schiemann WA (1999) Bullseye! Hitting your strategic target through high impact measurement. Free Press, New York, NY

Lipe MG Salterio SE (2000) The balanced scorecard: judgmental effects of common and unique performance measures, Accounting Review 75(3):283–298

Lovell B, Radnor Z, Henderson J (2002) A pragmatic assessment of the balanced scorecard: an evaluation of a new performance system for use in a NHS multi agency setting in the UK. University of Bradford Working Paper Series 2(13):339–346

Lynch R, Cross K (1995) Measure up! Yardsticks for continuous improvement. Basil Blackwell, Cambridge

Malina MA, Selto FH (2001) Communicating and controlling strategy: an empirical study of the effectiveness of the balanced scorecard. J Manage Acc Res 13:47–90

Marr B, Schiuma G, Neely A (2004) The dynamics of value creation: mapping your intellectual performance drivers. J Intellect Cap 5(2):312–325

Mcgranahan M Timothy W (2007) http://ideas.repec.org/a/taf/regstd/v41y2007i2p197-216.html. Recasting the Creative Class to Examine Growth Processes in Rural and Urban Counties, http://ideas.repec.org/s/taf/regstd.html Regional Studies, 41(2):197–216

McDonald D, Smith A (1995) A proven connection: performance management and business results. Compensat Benefit Rev 27(1):6, viewed 19 June 2000 EBSCOhost File: Academic Search Elite, item: AN9502270385

Mohrman A, Morhman S (1995) Performance management is Running the Business. Compensation & Benefits Review 27(4):69–75

Mooraj S, Oyon D, Hostettler D (1999) The balanced scorecard: a necessary good or an unnecessary evil? Eur Manage J 17:481–91

Millett B (1998) Performance management: a strategic human resource function, in K Parry and D Smith (eds.), Human Resource Management: Contemporary Challenges and Future Directions, pp 95–114, USQ Press, Toowoomba

Neely A, Kennerley M, Martinez V (2004) Does the balanced scorecard work: an empirical investigation. In: Neely A, Kennerly M, Waters A (eds) Performance measurement and management: public and private. Centre for Business Performance, Cranfield University, Cranfield, pp 763–770

Niven PR (2002) Balanced scorecard: step-by-step maximizing performance and maintaining results. Wiley & Sons, New York

Norreklit H (2000) The balanced on the balanced scorecard: a critical analysis of some of the assumptions. Manage Acc Res 11(1):65–88

Oakey R (2007) Clustering and the R&D management of high-technology small firms: in theory and practice. R&D Management 37:237–248

Papalexandris A, Ioannou G, Prastacos GP (2004) Implementing the balanced scorecard in Greece: a software firm's experience. Long Range Plann 37(4):347–362

Rigby D (2001) Management tools and techniques: a survey. Calif Manage Rev 43(2):139–160

Robinson SP (2004) The adoption of the balanced scorecard: performance measurement motives, measures and impact. In: Neely A (ed) Performance measurement: past, present and future. Centre for Business Performance, Cranfield School of Management, Cranfield University, Cranfield, pp 883–890

Rummel RJ (1970) Applied factor analysis. Northwestern University Press, Evanston, IL

Said AA, HassabElnaby HR, Wier B (2003) An empirical investigation of the performance consequences of nonfinancial measures. J Manage Acc Res 15:193–223

Scheipers G, Ameels A, Bruggeman W (2004) The Cascading process for multi-level balanced scorecard design: the case of a Belgian wastewater management company. In: Neely A (ed) Performance measurement: past, present and future. Centre for Business Performance, Cranfield School of Management, Cranfield University, Cranfield, pp 1209–1216

Self J (2004) Metrics and management: applying the results of the balanced scorecard. Performance Measurement Metrics 5(3):101–105

Shulver M, Antarkar N (2001) The balanced scorecard as a communication protocol for managing across intra-organizational borders. Proceedings of the 12th annual conference of the Production and Operations Management Society, Orlando, FL

Silk S (1998) Automating the balanced scorecard. Manage Acc 79(11):38–44

Sim KL, Koh HC (2001) Balanced scorecard: a rising trend in strategic performance measurement. Measuring Bus Excell 5(2):18–28

Simons R (2000) Performance measurement and control systems for implementing strategy: text & cases, Prentice Hall, Upper Saddle River, NJ

Spence J (2004) Targeting, Accountability and Youth Work Practice. Practice. A Journal of the British Association of Social Workers 16(4):261–272

Sureshchandar G, Leisten R (2005) Holistic scorecard: Strategic performance measurement and management in the software industry, Measuring Business Excellence 9(2):12–29

Tapinos E, Dyson RG, Meadows M (2005) The impact of performance measurement in strategic planning. Int J Prod Perform Manage 54(5/6):370–384

Tether BS, Storey DJ (1998) Smaller firms and Europe's high technology sectors: a framework for analysis and some statistical evidence. Research Policy, Elsevier 26(9):947-971

Trumbach C, Payne D, Kongthon A (2006) Technology mining for small firms: Knowledge prospecting for competitive advantage. Technological Forecasting & Social Change 73: 937–949

Torraco RJ, Swanson RA (1995) The strategic roles of human resource development. Hum Resour Plann 18(4):10–21

Tucker LR, Koopman RF, Linn RL (1969) Evaluation of factor analytic research procedures by means of correlation matrices. Psvchometrika 34(4):421–459

Zairi M, Jarrar Y (2000) Becoming world class through a culture of measurement. In: Neely A (ed) Performance measurement: past, present and future. Centre for Business Performance, Cranfield School of Management, Cranfield University, Cranfield, pp 688–694

Zeng DD, Zhao JL (2005) Effective role resolution in workflow management. INFORMS Journal on Computing 17(3):374–380

Part III
Evidence-Based Analysis: Non-European Studies

Modelling Endogenous Regional Employment Performance in Non-metropolitan Australia: What Is the Role of Human Capital, Social Capital and Creative Capital?

Robert J. Stimson, Alistair Robson, and Tung-Kai Shyy

1 Introduction

Over the years many studies have been conducted in Australia investigating regional spatial differentials in regional economic performance, including investigations of the inter-relationships between regional economic growth, population growth, structural shifts in employment distribution across industry sectors, industrial diversification, levels of income, and the location of regions in the national settlement system. It has been relatively common for variables relating to human capital to be included in such studies, but rarely has research incorporated a consideration of social capital and creative capital. There are difficulties in pursuing such research because there are no national data sets that explicitly provide operational measures of human capital, social capital and creative capital. As such, it is necessary to use data derived from the Census of Population and Housing to form proxy measures relating to those constructs.

Building on our previous research (Stimson et al. 2008, 2009a), the modelling discussed in this chapter explicitly does that in an exploratory investigation of what roles variables relating to human capital, social capital and creative capital might play within a broader modelling framework in which we seek to identify which variables relating to a broader range of factors might explain the spatial variation in endogenous regional employment growth performance over the decade 1996–2006 across Australia's non-metropolitan regions. The spatial units of analysis used are Local Government Areas (LGAs) outside the Australia's five major capital city metropolitan Statistical Divisions.

R.J. Stimson (✉) and T.-K. Shyy
School of Geography, Planning and Environmental Management,
The University of Queensland, Brisbane, QLD 4072, Australia
e-mail: r.stimson@uq.edu.au

A. Robson
The Institute for Social Science Research, The University of Queensland,
Building 31B, St Lucia, QLD 4072, Australia

K. Kourtit et al. (eds.), *Drivers of Innovation, Entrepreneurship and Regional Dynamics*, Advances in Spatial Science, DOI 10.1007/978-3-642-17940-2_9,

The chapter begins with an overview of the findings from existing research investigating differentials in regional economic development and performance in Australia. The paper then discusses the data sources used to compile a battery of 32 independent variables relating to factors that the literature suggests might influence regional economic development and growth that are incorporated within an exploratory analytical modelling framework that was developed by the to investigate the potential determinants spatial variation in regional performance, and in particular to identify the degree to which the variables measuring aspects of human capital, social capital and creative capital are significant explanatory factors. The results of that modelling are discussed. The chapter then concludes with a brief discussion of the potential policy implications of this research.

2 Existing Research on Differential Regional Economic Performance in Australia

Understanding the factors influencing regional economic development and growth (or decline) and isolating those which might explain the differentials that exist across the regions of a national economy is an important issue. It has been attracting the attention of regional scientists for a long time.

It is not our purpose here to rehearse the arguments in the copious literature international literature on regional development and growth theory and how that has evolved over the last half century and what might be the factors that explain spatial differentials in regional economic performance. A summary of that is provided by Stimson and Stough (2009) and Stimson et al. (2009b). The reader is also referred book edited by Capello and Nijkamp (2009) which provides a comprehensive coverage of research to the study of regional economic development theory and its application.

However, it is evident that with the emergence of the "new growth theory" there is an increasing interest in the role of endogenous factors and process in regional economic development and growth. That incorporates a consideration, *inter alia*, of explanatory factors such as the following:

- *Regional resource endowments* (see, e.g., Blakely 1994)
- *Industrial structure, specialisation and diversification* (Henderson et al. 1995; Gordon and McCann 2000)
- *Urbanisation, population growth, market size, and agglomeration* economies (Patten 1991; Duranton and Puga 2000; Maier 2001; Taylor et al. 2002; Rosenthal and Strange 2001; van Oort and Atzema 2004)
- *Innovation and entrepreneurship* (Grossman and Helpman 1991; Hubbard and Hall 1998; Acs et al. 2002; Nijkamp and Stough 2004; Döring and Schnellenback 2006)
- *Knowledge spillovers and human capital* (Romer 1986, 1990; Lucas 1988; Krugman 1991; OECD 1998, 2001; Hanushek and Kimko 2000; Goetz and Rapasingla 2001; Faggian and McCann 2009)

- *Social capital* (Coleman 1990; Bolton 1992, 1999; Fukuyama 1995)
- *Creative capital* (Florida 1995, 2002, 2005)
- *Leadership and institutional factors* (North 1990; Blakely 1994; de Santis and Stough 1999; Clingermayer and Feiock 2001; Vazquez 2002; Stimson et al. 2009b)

In developing operational models of regional development and growth, for some of those factors it is relatively easy to specify variables that are widely used surrogates measuring an aspect of aspects of a factor using, for example, national data collections such as census data that is readily available at different levels of spatial scale. But for others of those factors there are not readily available national data sets that may be used to derive variables that are satisfactory surrogate measures of a factor.

There is a relatively rich literature that has investigated aspects of regional economic performance in Australia, some of which explicitly seeks to examine the influence of endogenous factors in explaining differentials in regional economic performance. What follows is an overview of that research.

The Bureau of Transport and Regional Economics (2004a: p. ix) has argued that, in Australia, the distribution of a region's economic activity across industry sectors, the mix of occupations, and the skills mix of jobs, all combine to determine the level and distribution of income in a region and the resilience of its economy and its ability to grow. Its research suggests that how regional industrial mix changes over time is an important consideration for regional economic performance and competitiveness. In particular, this is in consideration of the context of how that performance reflects shifts in the structure of the national economy and how the shifts produce an increase or decline in gross value added for regions and their industry sectors. In the Australian context in particular, a region's industry structure, its occupation mix and its human capital structure, will be affected not only by the size of a region's economy, but also by its level of remoteness in the context of the nation's settlement system. Over the years there have been a number of studies that have investigated the nature of regional industry structure and regional economic performance (see, e.g., Department of Home Affairs and the Environment 1982; Beer and Maude 1995; Beer 1999) and the effects of structural change on employment growth (see, e.g., Productivity Commission 1998; Bradley and Gans 1998; Garnaut et al. 2001; Lawson and Dwyer 2002).

The role of human capital – its development and levels – has received specific attention by The Bureau of Transport and Regional Economics (2004b: p. 4) which notes that education, skills and qualifications are not evenly distributed across populations and regions. Earlier Quiggin (1999) had pointed out that higher levels of education and skills have benefits, including higher employability, increased earnings and higher job quality. Education is thus seen as a significant contributor to economic growth and regional performance. There may also be non-monetary benefits, including better health knowledge and health status, transmission of cultural values and reduced criminal behaviour.

Research by Baum et al. (1999) has demonstrated that variation in levels of human capital is highly significant in differentiating between the performance of statistical local areas (SLAs) across Australia's cities and towns. When linked to other aspects of labour force engagement, income distribution, and employment in the "symbolic analyst" occupations, human capital is shown to be crucial in Baum et al.'s (1999) development of a typology of community opportunity/vulnerability. Furthermore, the Bureau of Transport and Regional Economics (2004b: p. 1) has stated that

> . . . education, skills and qualifications are increasingly considered to be central to regional productivity, employment opportunities, personal development and the resilience of communities. (p. 1)

Thus, the

> . . . acquisition, production, distribution and use of knowledge are valued as a contributor to individual and social well-being and economic prosperity. (p. 1)

In another study, Stimson et al. (2004) analysed the patterns of human capital and industry and occupation diversification/specialisation across the settlement system in Australia's states and the Northern Territory using data from the 2001 Census of Population and Housing. They investigated the relationship between the population size of urban centres and localities (UCLs) and their performance on three derived indices of human capital and on three indices measuring the degree of diversification/specialisation in employment across industry sectors and across occupation categories, and in the distribution of people across levels of educational qualifications/skills.

The findings of studies such as those referred to above seem to indicate that in Australia a rise in education levels nationally is being reflected widely across all regions nationally over the decade 1991–2001. This is even though the level of educational attainment does continue to decline with increasing remoteness and also with the size of a region or place (Bureau of Transport and Regional Economics 2004b). It is among the Indigenous population that that decline is most marked. In the late 1990s and early 2000s there seemed to be emerging something of an over-supply of skills in some of the more populated regions. Additionally, there has been greater absorption of skills in the regional areas of Australia, and that was reflected through the increasing shares of some employment in some of the professions across some of the regional areas. But there is a continuing dichotomy at a general level between the capital cities and regional urban areas in their human capital performance.

Research by the Bureau of Transport and Regional Economics (2004b: p. ix) suggests that a degree of equalisation was occurring in educational levels across Australia's regions. They found that there was some evidence of convergence in skills and qualification outcomes for the more populated regions. However, there also some evidence that this diminishes with increasing remoteness. That work reveals how professionals in other fields – such as teaching – are becoming more evenly distributed regionally. Comparatively, some professions – such as computing professionals – are highly concentrated in the major cities and some of the larger regional centres. That study shows the following:

> ….Overall, regions with higher 1991 shares of university educated residents tended to experience higher income growth over the 1991 to 2001 decade (p. 42). The regions with high shares of Managers, administrators and professionals – the highly skilled workers – showed no clear link with income growth. The research findings suggest that 'the relationship between higher education and income growth does not persist indefinitely'. (p. 42)

The Stimson et al. (2004) research shows that, across the nation's urban system, there does seem to be a positive relationship between the size of an urban place and its level of performance on the human capital indices derived by that research. It also found that there is a tendency for occupation diversity and skills density to increase with increasing size of a place. But the relationships for the latter are of a considerably lesser magnitude than for the former. Additionally, it is evident that some larger places do not have high levels of human capital performance, and that some of them have specialised occupation and skills structures. Comparatively, some smaller places have strong human capital performance and reasonably diverse occupational and skill structures.

Taking a different approach, research by Stimson et al. (2008) have uses a spatial econometric model to analyse the patterns of endogenous employment growth performance across the non-metropolitan regions in the five mainland states. It identifies those factors that explain spatial variations in regional economic performance over the decade 1991–2001. Using a battery of variables derived from census data, those authors model the explanatory power of variables relating to both static and dynamic measures of regional industrial structure and specialisation/diversification; population size and growth; labour force participation and unemployment; human capital and income; occupational structure; and the effects of proximity to the state capital city which is the metropolitan area and proximity to the coast. The findings indicated the following:

1. Population growth was a particularly factor with a strong positive impact in all five mainland States, and population size at the beginning of the decade had an important positive impact in all States except New South Wales.
2. The level of industry sector specialization, and the change in it over time, both had an important positive effect in all States except New South Wales, and the structural change index and the dynamics in it over time were important positive factors in Queensland, South Australia and Western Australia.
3. The level of income at the end of the decade 1991–2001 had a significant factor only in Queensland.
4. Only in New South Wales did the unemployment variables have an impact.
5. There was a mixed picture regarding the impacts of the degree of concentration of jobs in broad industry sectors, and the effects were most apparent in New South Wales and in Western Australia; but there was neither a consistent directional effect for these variables, nor for the change over time in their location quotients.
6. There was a particularly marked impact from the incidence of people with university and technical qualifications in New South Wales and Queensland,

but to a lesser degree in Victoria and the other States; however, the incidence of people with university qualifications at the beginning of the decade had a negative impact, but the change in that incidence over the decade was a positive impact factor in New South Wales and Western Australia, where the change in incidence of people with technical qualifications also had a positive impact.

7. There was no really discernable pattern in the impact of the incidence of jobs in the Reich (1991) broad occupation categories, except in Victoria where the effects were negative, while in South Australia, Western Australia and Queensland the effects were positive when such a variable is significant.
8. Only in Western Australia and South Australia was it evident that the proxy variables relating to location on or near the coast or proximity to the metropolitan area have a significant effect, and that effect is negative.

From a review of the Australian literature investigating differentials in regional economic development and performance in Australia, Stimson (2007) suggested four general conclusions that might be drawn. Those were:

1. Diversification is important but not a panacea for economic growth, and a

> ... regional push for greater industrial diversity might better reflect preferences for slow and steady growth over a boom and bust economy. (Bureau of Transport and Regional Economics 2004a: p. 46)

2. Where people live can certainly affect the diversity of job opportunities with respect to the mix of industry, occupation and skill requirements, and it can affect overall levels of labour market participation and engagement. Thus, "place does matter" (Stimson et al. 2004, Ch 4).
3. The level of human capital is a factor differentiating the performance of places. Thus, "people skills do matter" (Stimson et al. 2004, Ch 10: p. 4).
4. Population size does matter, as seen in the degree to which large urban places have more diversified job opportunities across both industry sectors and occupational categories, and larger places have higher levels of human capital performance:

> .... Large urban places also tend to offer employment opportunities for a greater range of educational attainment skill levels. This is not surprising and most likely it represents a self-selection outcome as the best people seek the better performing places so the direction of causation becomes blurred. (Stimson et al. 2004, Ch 10: p. 5)

3 Investigating the Role of Human, Social and Creative Capital in Differential Endogenous Regional Growth Performance

We now turn to discuss new research which explicitly attention is directed to ascertaining the degree to which variables that are surrogate measures of human capital, social capital and creative capital play a role within what is a battery of variables referring to a broader set of factors that are hypothesised to be potential explaining factors accounting for spatial variations in the endogenous regional employment growth/performance of non-metropolitan regions in Australia over the decade 1996–2006.

3.1 Data Issues

The five yearly Census of Population and Housing conducted in Australia by the Australian Bureau of Statistics (ABS) provides researchers with a wealth of aggregated data at various levels of geographic scale that may be used in spatial econometric modelling. In particular it provides cross-sectional information and time-series data at one particular level of spatial scale, namely the Statistical Local Area (SLA), which tend to be Local Government Areas (LGAs) or parts of them.

The census data provides statistics on the demographic and socio-economic characteristics of people, the labour force, households and dwellings. This information relates to place of enumeration (where a person is on census night) and not to place of employment (which would be an ideal for labour market related analysis and modelling). Additionally, researchers are restricted to use the specific categories that are defined in the census and which relate to attributes such as individual and household income, the industry and occupation of people working, and the levels of educational attainment and skills of people.

Thus, there are significant constraints that researchers face in using data from the census of population and housing to derive measures for variables relating to factors such as human, social and creative capital. There are also constraints in the data which might be used in operational analytical models to investigate the role of those constructs in explaining differentials in the pattern of regional economic development and growth.

Typically in using census data, researchers are restricted to use categories of the demographic and socio-economic attributes of people as defined by the Australian Statistician. Often those categories may only be used to identify inadequate surrogate measures as variables depicting those constructs that might be hypothesised to be of importance with respect to a dynamic outcome state – that is, the *dependent variable* – across specified geographic regions. It might involve the use of a surrogate variable measure derived as a summative combination of a number of categories for an attribute or across a number of attributes. Researchers might also select a battery of variables derived from range of data from the census that attribute categories which relate to constructs which are hypothesised to influence the outcome state. Those variables represent the *explanatory variables* which may be used in a model – such as a multiple regression model – to investigate the relationship between the dependent variable and the explanatory variable. In addition, it is common for the battery of explanatory variables to subject a data reduction tool, such as principal components analysis, in order to derive a smaller number of synthetic variables which represent a small number of summary dimensions relating to constructs which are hypothesised to be *explanatory factors* and to use those measures as the explanatory variables in a model.

From our analysis of the literature on endogenous growth, and based on our previous research reported in Stimson et al. (2008, 2009a), particular variables were derived from census data to represent surrogate measures for factors which, *a priori*, were viewed as having a significant contribution to endogenous regional performance. Those variables were derived from data available in the last three censuses (1996, 2001, and 2006).

3.2 The Dependent Variable

There are difficulties encountered in the use of data from the census to derive a satisfactory measure of the outcome state that is the *dependent variable* in a model investigating spatial differentials in regional economic performance. Here we need a variable measuring regional economic growth or performance over a period of time. We followed the method proposed by Stimson et al. (2005, 2009b) and which has been as used in a previous investigation into the determinants *endogenous regional employment growth* across Australia's regions over the decade 1991–2006 conducted by Stimson et al. (2008, 2000). That methodology used the *regional (or differential) shift component* in a shift-share analysis of employment change, standardised by the size of the labour force at the beginning of the period, as a surrogate measure of endogenous regional employment growth.

3.3 The Explanatory Variables

For the categories of 32 variables used as the *explanatory variables* in the analysis it was important to include a mixture of both cross-sectional and dynamic measures of regional characteristics that might impact on endogenous regional performance. As the purpose of this paper is to give explicit attention to investigating the potential roles of human capital, social capital and creative capital as factors that might account for spatial variations in patterns of regional endogenous growth performance across non-metropolitan LGAs in Australia, surrogate measures for those factors were deliberately included in the set of variables used in the analysis reported here.

The categories of factors that were deemed to be important to include in the model and the resultant operational variables that were derived mostly from census data are the variables are listed in Table 1.[1] The variables are surrogate measures

[1]Here we do not provide detailed descriptions of all of the variables listed in Table 1 that are used in the modelling undertaken and which is presented in this chapter.

Table 1 Variables used in the modelling

Variable label	Variable description
Dependent variable	
REG_SHIFT	Regional shift (1996–2006)/labour force (1996)
Independent (explanatory) variables	
SPEC_96	Specialization index for 1996 (Herfindahl–Hirschman index)
SPEC_CH	Change in specialization index from 1996 to 2006 (Herfindahl–Hirschman index)
SCI	Structural change index (1996–2006)
SCI_CH	Change in the structural change index (from 1996–2001 to 2001–2006)
L_INC_96	Median individual income – 1996 annual (log) (real)
L_INC_CH	Change in median individual income – 1996–2006 annual (log) (real)
UNEMP_96	Unemployment rate in 1996 (%)
UNEMP_CH	Change in unemployment rate from 1996 to 2006 (pps)
L_POP_96	Log of population (1996)
L_POP_CH	Change in log of population (1996–2006)
LQ_MAN_96	Location quotient for the manufacturing industry in 1996
LQ_INF_96	Location quotient for the information media and telecommunications industry in 1996
LQ_FIN_96	Location quotient for the financial and insurance services industry in 1996
LQ_PRO_96	Location quotient for the professional, scientific and technical services industry in 1996
LQ_MAN_CH	Change in the location quotient for the manufacturing industry, 1996–2006
LQ_INF_CH	Change in the location quotient for the information media and telecommunications industry, 1996–2006
LQ_FIN_CH	Change in the location quotient for the financial and insurance services industry, 1996–2006
LQ_PRO_CH	Change in the location quotient for the professional, scientific and technical services Industry, 1996–2006
POSTGRAD_96	*Proportion of labour force with a postgraduate degree of higher in 1996*
BACHELOR_96	*Proportion of labour force with a bachelor degree of higher in 1996*
TECHQUALS_96	*Proportion of labour force with technical qualifications in 1996*
POSTGRAD_CH	*Change in the proportion of labour force with a postgraduate degree of higher, from 1996 to 2006*
BACHELOR_CH	*Change in the proportion of labour force with a bachelor degree of higher, from 1996 to 2006*
TECHQUALS_CH	*Change in the proportion of labour force with technical qualifications, from 1996 to 2006*
SYMBA_96	Proportion of symbolic analysts (Managers + Professionals) in employment in 1996
SYMBA_CH	Change in the proportion of symbolic analysts (Managers + Professionals) in employment from 1996 to 2006
VOLUNTEER_06	*Proportion of volunteers in working age population (15–64) in 2006*
CREATIVE_06	*Proportion of total employment in creative industries in 2006*
A_COAST	Border is adjacent to coastline (No = 0; Yes = 1)
P_METRO	Border is adjacent to metropolitan statistical division (No = 0; Yes = 1)
D_URBAN	Classified as urban under Australian classification of local government system (1 = Yes; 0 = No)
D_REMOTE	Classified as remote under Australian classification of local governments system (1 = Yes; 0 = No)

Note: The variables that are surrogate measures for human capital, social capital and creative capital are in italics relating to factors that the regional science literature has suggested might be factors that might explain variation in endogenous regional employment performance. The factors are: industrial structure, including industry specialisation and structural change; population size and growth; labour force participation; human capital (skills); income distribution; occupational shifts; social capital; creative capital. Some variables listed in Table 1 are cross-sectional measures at the beginning or end of the period 1996–2006 while others are change-over-time dynamic variables. In addition four locational proxies were included.

As we have an explicit interest in this chapter to investigate the degree to which human capital, social capital and creative capital might play a role as explanatory factors in the modelling, a full description of those variables follows:

1. The *human capital* measures were based on the highest level of qualification at the time of the census. Two main categories of human capital measures are used: first category of variables measured the number of people with a particular level of qualification as a proportion of the working age population (15–64). The levels of qualification selected were: Postgraduate degree (POSTGRAD_96); Bachelor degree (BACHELOR_96); and, those with technical qualifications[2] (TECHQUALS_96). The second category of variables was the change in these human capital measures from 1996 to 2006 (that is: POSTGRAD_CH; BACHELOR_CH; and, TECHQUALS_CH).
2. It was not until the 2006 census that any information was collected that might be used as a surrogate measure for *social capital*. The only meaningful variable the people who report being involved in volunteering, which is a rather surrogate measure of social capital. The variable used is VOLUNTEER_06, which measures the proportion of Volunteers in the working age population (age groups 15–64) in 2006.
3. Identifying and constructing a measure of *creative capital* was also a challenge. Part of the reason for this is the difficulty in gaining employment data at the disaggregated industry level to discern "creative" industries at the spatially disaggregated level of the LGA. Two-digit employment by industry data from the working population profile of the census LGA was based on the place-of-work count method rather than the place of enumeration count method. Identifying which two-digit industries considered to be "creative" was somewhat subjective, and those included were: The industries are: Construction services; Information media and telecommunications/Not defined; Publishing (except internet and music publishing); Motion picture and sound recording activities; Broadcasting (except internet); Internet publishing and broadcasting; Library and other information services; Computer system design and related services; Tertiary education; Arts and recreation services\Not defined; Creative and performing arts activities The number of jobs in those industries was then divided by the total number of jobs to identify the proportion of jobs in creative industries in each LGA (CREATIVE_06).

[2]This includes the categories of: Advanced Diploma and Diploma, Certificate nfd, Certificate I and II, Certificate III and IV.

3.4 Spatial Units of Analysis

The focus here is on investigating non-metropolitan regional across Australia. It was thus decided to focus on spatial units that relate to the third tier of government, namely LGAs. But for various technical reasons – and in particularly for future spatial autocorrelation analysis – it was necessary to actually use three different units of geography, including in some remote parts of the nation Statistical Local areas (SLAs) and/or Statistical Sub Divisions (SSDs).[3]

A total of 493 spatial units drawn from the above three units of geography are used as the spatial framework for the analysis reported here. For the sake of simplicity the term LGA is used throughout the chapter to refer to the spatial units used in the analysis of non-metropolitan regions discussed in what follows.

3.5 The Analytical Model

Our objective here was to investigate the potential determinants of spatial variations in endogenous regional employment performance over the decade 1996–2006 across non-metropolitan regions in Australia. Within that context we wished to specifically identify the degree to which variables that are surrogate measures of human capital, social capital and creative capital might be explanatory factors. We employed the same modelling approach used in our earlier study (Stimson et al. 2008).[4]

As discussed above, the *dependent variable* of interest is endogenous regional employment (the REG_SHIFT variable) derived from a shift share analysis, which

[3]A problem with using LGAs in some parts of Australia (particularly in Queensland and the Northern Territory) is that a few LGAs are non-contiguous. Mostly those tend to be Indigenous communities. To solve this problem, those LGAs were removed in Queensland. In the Northern Territory, the solution to the non-contiguous LGA problem is to use the spatial scale of the Statistical Sub-Division (SSD). Statistical Local Areas (SLAs) – which are based on the boundaries of incorporated bodies of local government where these exist – often are in fact LGAs or may be aggregated into LGAs. A LGA is an SLA if it fits entirely within an SSD and is broadly similar in size, economic significance and user needs for statistics to other LGAs in Australia. An LGA will be composed of two or more SLAs when the above conditions are not met. This can occur if an LGA is divided by the boundary of one or more SSDs or where the LGA is substantially different in size, economic significance and user needs for statistics to other LGAs. Unincorporated SLAs are defined for unincorporated on-shore area(s) and/or off-shore island(s) in an SSD or are defined for that part of an unincorporated area which is considered of sufficient economic significance as to warrant the formation of a separate SLA (ABS 2006: p. 11). There are, however large parts of Australia that are not administered by incorporated local government bodies. Those unincorporated areas tend to be in remote mainly inland areas that are largely uninhabited, and where they are the habitation tends to be concentrated with Indigenous peoples. For such areas, an SLA is an unincorporated area. Finally, in some cases, such as in the remote parts of the Northern Territory, the Statistical Sub Division (SSD), which comprise one or more SLAs, had to be used.

[4]Note that in the earlier study by Stimson et al. (2008) separate modelling was undertaken for the five mainland states of Australia. Here the modelling was performed for non-metropolitan LGAs across the whole nation.

was expected to exhibit a high degree of spatial variability across the non-metropolitan regions.

The initial analysis undertaken was to investigate the simple ecological relationships between this regional endogenous growth/decline performance measure and the surrogate measures for human capital, social capital and creative capital that are listed among the explanatory variables in Table 1. That involved the calculation of simple product moment correlation coefficients. However, that simple ecological correlation analysis did not inform us as to whether or not any one or all of those the variables measuring human capital, social capital and creative capital plays a significant role in explaining the differential patterns of spatial performance of the non-metropolitan regions across Australia as measured by the REG_SHIFT variable of regional endogenous growth/decline.

In order to do that the full set of potential explanatory variables listed in Table 1 were used in a replication of the modelling undertaken in the previous research by Stimson et al. (2008) which had used a two-stage regression modelling approach:

1. First a *general model* Ordinary Least Squares (OLS) model was run.
2. Second, that was followed by running a *specific model* to address the issue of multi-colinearity, a *stepwise backward iterative elimination method* was used to determine a model that exhibited the minimum number of statistically significant variables, but maximised the explanatory power of the model. The threshold for eliminating a variable was the variable which had the highest *p*-value. This procedure was completed when the variable had a *p*-value of 0.05. This technique has encountered some criticism, such as missing the optimal model, treating *p*-values too literally, picking models that are smaller than desirable, etc. (Faraway 2002: p. 126).
3. Third, and most importantly, we then proceeded to incorporate a spatially weighted regression approach to account for the effects of spatial autocorrelation that was evident. The neighbourhood matrix approach was used. Both a *spatial error model* and a *spatial lag model* were run, and the *Moran's I test* and the *Lagrange Multiplier Test Statistics* for spatial autocorrelation were calculated.

The modelling approach adopted enabled us to explicitly identify the degree to which the surrogate measures of human capital, social capital and creative capital do or do not have a statistically significant role in helping explain the spatial variation in the endogenous growth/decline performance of non-metropolitan regions across Australia over the decade 1996–2006.[5]

[5]Note that in the modelling it was decided to remove some influential observations (spatial units) from the general model. The technique used was the Cooks Distance Method. The threshold for removing an observation selected was 0.5 (that is, there was a 99.5% confidence that no other observation influenced the model). That resulted in five spatial units being removed from the analysis, all in either the Northern territory or in Western Australia. The observations removed from the general model were Daly (Northern Territory, with a cook score of 11.6); Murchison (Western Australia, with a Cook score of 1.1); Ravensthorpe (Western Australia, with a cook score of 1.0); and, Palmerston Eat Arm (Northern Territory, with a cook score of 0.5).

4 Results

The results of the application of the methodology discussed above to model the potential determinants of spatial variability in endogenous regional employment performance over the decade 1996–2006 in non-metropolitan LGAs across Australia to explicitly investigate, and explicitly to investigate the explanatory role of human capital, social capital and creative capital are discussed in what follows.

4.1 Mapping Patterns of Regional Performance

An important task was to be able map the geographic patterns, across non-metropolitan Australia, of the dependent variable (REG_SHIFT). The method used was to produce maps that display the pattern for a variable by using shaded symbols which differentiate the spatial units of analysis according to population size categories of LGAs, and which classifies the level of a region's performance on that variable.

The pattern of regional performance on the REG_SHIFT variable is mapped in Fig. 1. Note that scores in his measure may generate a negative or positive score for a LGA because of the regional (differential) shift component derived from a shift-share analysis is a residual that may be positive or negative after allowing for the national shift and the industry mix shift effects:

1. A *positive score* indicates a LGA has endogenous processes or factors within the region that are creating employment growth over and above that which is occurring as a result of the national and the industry shifts that are occurring. It is indicative that a region has particular local attributes that are advantageous in creating comparative and competitive advantage for the region.
2. In contrast, a *negative score* indicates a LGA has endogenous processes or factors within the region that are conducive to declining employment. While the region may have been experiencing aggregate growth in employment, after taking account of the impact of national and industry shift effects there is a residual negative endogenous effect which has been detrimental to the job generation processes within the region.

Figure 1 shows that across Australia there is a preponderance – N = 298 (or 60%) – of non-metropolitan LGAs with negative scores that had experienced negative endogenous processes over the decade 1996–2006. Conversely, a minority – N = 195 (or 40%) – of non-metropolitan LGAs had experienced positive endogenous employment growth.

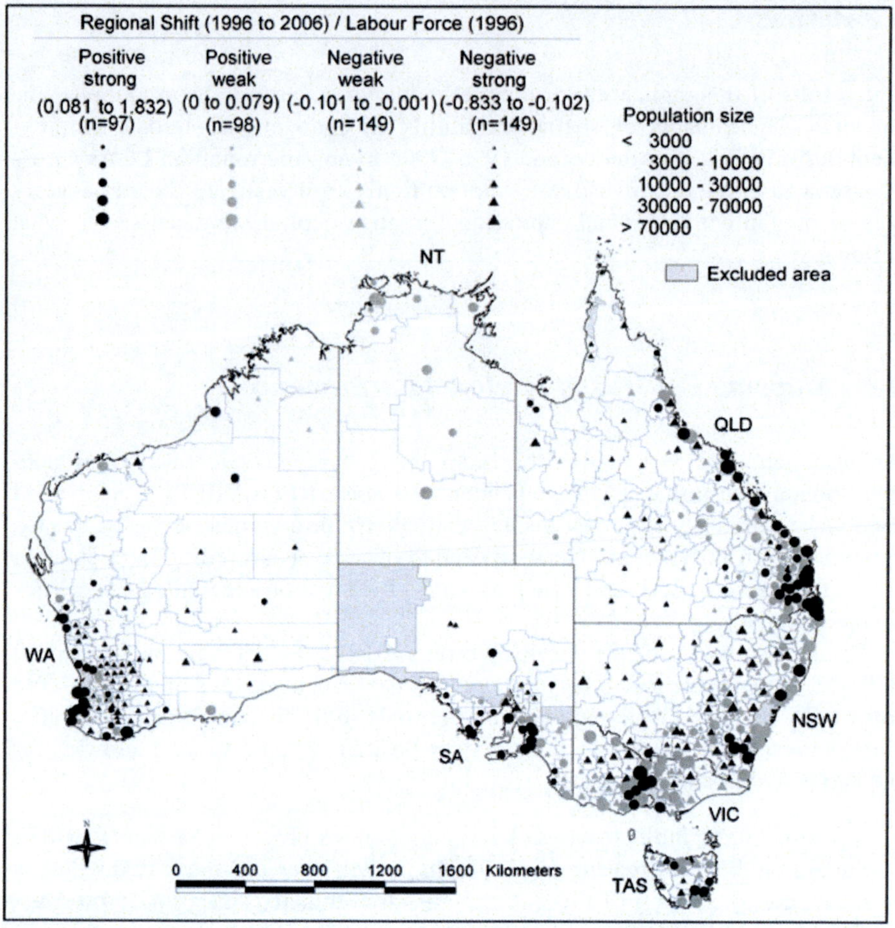

Fig. 1 Pattern of endogenous regional employment performance, 1966–2006.
Source: The authors

From a visual inspection of the patterns in Fig. 1 it would seem that:

- The incidence of negative endogenous regional performance is most heavily concentrated in LGAs with smaller populations that have urban places that are well down the settlement hierarchy
- The incidence of positive endogenous regional performance is more associated with larger population LGAs
- A greater incidence of negative endogenous regional employment performance is evident across the inland areas of Australia, and in particular in smaller population regions and remote areas; however, positive endogenous regional employment growth is associated with a number of remote locations, and in particular with mining towns

- Some of the regions across the wheat-sheep belt of inland Australia – the rural heartlands – that have larger population urban service centre have a positive endogenous regional employment growth
- There is a mixture of both positive and negative endogenous regional employment performance along the coastal areas of Australia, and it appears that positive growth is more associated with regions with larger population urban centres
- Close proximity to a metropolitan state capital city is associated with positive endogenous employment growth performance

The modelling that will be discussed later will test the degree to which some of those apparent locational attributes are or are not significant explanatory factors in differentiating between non-metropolitan LGAs in their performance on this REG_SHIFT outcome dependent variable.

4.2 Factors Explaining Spatial Variations in Endogenous Regional Employment Performance, with Particular Reference to the Roles of Human, Social and Creative Capital

4.2.1 Simple Correlation Analysis

The ecological correlations between the REG_SHIFT variable and surrogate variable measures for human capital, social capital and creative capital measures are presented in Table 2:

- The strongest *positive* relationship between the endogenous regional employment growth variable (REG_SHIFT) and these measures was with the BACHELOR_CH variable, followed by CREATIVE_06; TECHQUALS_CH; TECHQUALS_96; POSTGRAD_96; and POSTGRAD_CH
- The strongest negative relationship between endogenous employment growth and these measures was with the BACHELOR_96 variable, followed by VOLUNTEER_06

Table 2 Correlations between REG-SHIFT variable and human capital, social capital and creative capital measures

Variable	Correlation coefficient	Rank
POSTGRAD_96	0.07	5
BACHELOR_96	−0.09	8
TECHQUALS_96	0.12	4
POSTGRAD_CH	0.05	6
BACHELOR_CH	0.34	1
TECHQUALS_CH	0.18	3
VOLUNTEER_06	−0.07	7
CREATIVE_06	0.26	2

Source: The authors

4.2.2 Regression Modelling

We now turn to consider the results of the regression modelling conducted on the dataset.

An OLS General Model

Initially an OLS regression approach was used to investigate which of the 32 independent variables listed Table 1 might be significant in explaining the spatial variation across non-metropolitan regions using the total data matrix. That generated a *general model* solution with an adjusted $R^2 = 0.89$ (that is, the variables explained 89% of the variance in the dependent variable).[6] In that general model, several statistically significant relationships with the dependent variable (REG_-SHIFT) were found:

- *Positive* relationships: SPEC_CH, SCI_CH, L_POP_CH, LQ_MAN_CH, LQ_MAN_CH, POSTGRAD_96, POSTGRAD_CH, BACHELOR_CH, TECH-QUALS_CH, SYMBA_96
- *Negative* relationships: UNEMP_96, UNEMP_CH, LQ_PRO_96, BACHE-LOR_96, SYMBA_CH, VOLUNTEER_06

However, it was evident that only two of the variables that are measures of human capital, social capital and creative capital used in the model were found to be significant in explaining variations in endogenous employment performance over the decade 1996–2006 across the non-metropolitan regions of Australia:

1. One was the BACHELOR_CH variable, which had a positive impact. This indicated that an increase in the proportion of workers with a bachelor level degree over the decade 1996–2006 was likely to enhance endogenous regional employment performance. This is supportive of the notion that attraction of higher levels of human capital can enhance regional growth performance.
2. The other was the BACHELOR_96 variable, which had a negative impact. This indicates that regions with a higher proportion of bachelor level qualified workers at the beginning of the period in 1996 were more likely to have a lower level of endogenous regional employment performance over the period 1996–2006. This result that might seem to be somewhat counter-intuitive against that theory of human capital enhances regional growth.

[6]For reasons of space we do not provide a table showing the results one of the OLS general model solution.

An OLS Backward Step-Wise Specific Model

To address the high level of multi-colinearity that exists between many of the independent variables in the general model, a stepwise backward iterative elimination method was used to determine a *specific model* that exhibited the minimum number of statistically significant variables, but maximised the explanatory power of the model. The threshold for eliminating a variable was whether it had the highest *p*-value of 0.05. The results of this model are shown in Table 3.

In that specific model:

- The variables found to have a *positive* statistically significant impact on the endogenous growth variable at a 95% confidence level were: SPEC_CH; SCI; SCI_CH; L_POP_CH; LQ_FIN_CH; BACHELOR_CH; TECHQUALS_CH; and, SYMBA_96
- The variables found to have a *negative* statistically significant impact on the endogenous growth variable at a 95% confidence level were: UNEMP_96; UNEMP_CH; LQ_PRO_96; BACHELOR_96; SYMBA_CH; and, VOLUNTEER_06

Thus four of the variables that are measures of human capital and social capital that are significant in explaining the spatial differentials in endogenous

Table 3 Specific model results

Coefficients	Estimate	Std. error	t value	Pr(>|t|)
(Intercept)	−0.18	0.02	−8.13	0.00***
SPEC_CH	0.54	0.06	9.31	0.00***
SCI	0.00	0.00	2.00	0.05*
SCI_CH	0.00	0.00	2.32	0.02*
UNEMP_96	−0.01	0.00	−6.08	0.00***
UNEMP_CH	−0.02	0.00	−9.60	0.00***
L_POP_CH	2.24	0.05	44.44	0.00***
LQ_PRO_96	−0.04	0.02	−2.40	0.02*
LQ_FIN_CH	0.04	0.02	2.08	0.04*
BACHELOR_96	*−0.84*	*0.21*	*−4.08*	*0.00****
BACHELOR_CH	*1.66*	*0.25*	*6.67*	*0.00****
TECHQUALS_CH	*0.90*	*0.12*	*7.50*	*0.00****
SYMBA_96	0.40	0.06	7.25	0.00***
SYMBA_CH	−0.51	0.10	−5.34	0.00***
VOLUNTEER_06	*−0.11*	*0.05*	*−2.17*	*0.03**

Note: The variables that are surrogate measures of human capital, social capital and creative capital are in italics
Residual standard error: 0.06 on 475 degrees of freedom
Multiple R-squared: 0.8949, adjusted R-squared: 0.8918
F-statistic: 289.00 on 14.00 and 475 DF, p-value: <2.20E-16
Significance codes: *** 0.001, ** 0.01, * 0.05
Source: The authors

employment growth/decline performance over the decade 1996–2006 across the non-metropolitan regions on Australia:

1. The BACHELOR_CH variable has a positive impact. This indicates that an increase in an incidence of workers with bachelor level qualifications over the decade 1996–2006 was likely to have a positive effecting endogenous employment growth in a non-metropolitan region. This is what would be expected from the human capital theory of regional development.
2. However, the BACHELOR_96 variable has a negative impact. This indicates that a region with a higher incidence of workers with bachelor level qualifications at the beginning of the study period (1996) was more likely to experience a lower level of endogenous performance. This is somewhat counter intuitive with the theory of human capital effects on regional development.
3. The TECHQUALS_CH variable has a positive impact. This indicates that those non-metropolitan regions that had experienced a greater increase in the incidence of workers with technical qualifications over the decade 1996–2006 were more likely to have experienced stronger endogenous employment growth performance.
4. The VOLUNTEER_06 variable has a negative impact. This indicates that those non-metropolitan regions with a higher incidence of volunteering in 2006 were more likely to have experienced a lower level of endogenous regional employment performance over the decade 1996–2006. This finding questions the notion that a high level of social capital might have an enhancing effect on endogenous regional growth; from this result it might have a negative effect at least in non-metropolitan regions of Australia.

It is noteworthy that the creative capital variable (CREATIVE_06) was not included in the specific model and thus is not a significant factor in explaining spatial differentials in endogenous regional employment over the decade 1996–2006.

Addressing the Spatial Autocorrelation Problem: A Spatial Error Model and a Spatial Lag Model

The Moran's I test was run to test for spatial autocorrelation in the specific model.[7] From this we discovered that the probability of spatial autocorrelation in the specific model was statistically significant at the 99.9% confidence level (with a p-value of less than 0.01). Furthermore, the Moran's I statistic was positive, which indicates that nearby LGAs have similar rates. That indicates global spatial clustering.

There are two options to adjust for spatial autocorrelation effects, namely: the *spatial error model*, and the *spatial lag model*. The results from applying those

[7]For the specific model, the Moran I statistic standard deviate = 3.6756, p-value = 0.0001187.
Alternative hypothesis: greater.
Observed Moran's I: 0.0973747747.
Expectation: −0.0070965101.
Variance: 0.0008078735.

Table 4 Spatial error model: specific model coefficients

	Estimate	Std. error	z value	Pr(>|z|)
(Intercept)	−0.19	0.02	−8.09	0.00***
SPEC_CH	0.52	0.06	9.27	0.00***
SCI	0.00	0.00	2.54	0.01*
SCI_CH	0.00	0.00	1.28	0.20
UNEMP_96	−0.01	0.00	−5.57	0.00***
UNEMP_CH	−0.02	0.00	−9.37	0.00***
L_POP_CH	2.27	0.05	43.45	0.00***
LQ_PRO_96	−0.05	0.02	−2.91	0.00***
LQ_FIN_CH	0.03	0.02	1.73	0.08
BACHELOR_96	−0.82	0.20	−4.03	0.00***
BACHELOR_CH	1.69	0.24	6.97	0.00***
TECHQUALS_CH	0.90	0.12	7.68	0.00***
SYMBA_96	0.40	0.06	7.18	0.00***
SYMBA_CH	−0.51	0.09	−5.57	0.00***
VOLUNTEER_06	−0.10	0.05	−1.82	0.07

Lambda: 0.053845, LR test value: 13.18, p-value: 0.00028299
Asymptotic standard error: 0.012582, z-value: 4.2794, p-value: 1.8739e-05
Wald statistic: 18.313, p-value: 1.8739e-05
Log likelihood: 685.758 for error model
ML residual variance (sigma squared): 0.0035109, (sigma: 0.059253)
Number of observations: 490
Number of parameters estimated: 17
AIC: −1337.5, (AIC for lm: −1326.3)
Significance codes: *** 0.001, ** 0.01, * 0.05
Source: The authors

modelling approaches are shown for the *specific model* (step-wise approach) in Tables 4 and 5. It is evident that there some differences in these results compared to that for the OLS specific model.

As seen in Table 4, for the *spatial error* specific model the SCI_CH, LQ_FIN_CH, and VOLUNTEER_06 variables are no longer statistically significant in explaining spatial variations in endogenous regional employment performance over the decade 1996–2006 across the non-metropolitan regional of Australia. But the LQ_PRO_CH variable becomes a more significant explanatory factor in the spatial error model.

Turning to the *spatial lag specific model* results in Table 5, all the variables from the OLS specific model remain significant explanatory variables, but in thr spatial lag model solution the SCI, LQ_PRO_96 and VOLUNTEER_06 variables are of greater significance, while the UNEMP_96 and BACHELOR_96 variable are of lesser significance.

In comparing the results of the *spatial error specific model* and the *spatial lag specific model*, we see the following:

- The SCI variable has lesser explanatory significance in the spatial error model than in the spatial lag model
- The UNEMP_96 and BACHRLOR_96 variables are of lesser explanatory significance in the spatial lag model than in the spatial error model

Table 5 Spatial lag model: specific model coefficients

	Estimate	Std. error	z value	Pr(>|z|)
(Intercept)	−0.19	0.02	−8.09	0.00***
SPEC_CH	−0.18	0.02	−8.24	0.00***
SCI	0.54	0.06	9.44	0.00***
SCI_CH	0.00	0.00	2.04	0.04*
UNEMP_96	0.00	0.00	2.35	0.02*
UNEMP_CH	−0.01	0.00	−6.16	0.00***
L_POP_CH	−0.02	0.00	−9.76	0.00***
LQ_PRO_96	2.25	0.06	40.12	0.00***
LQ_FIN_CH	−0.04	0.02	−2.41	0.02*
BACHELOR_96	0.04	0.02	2.08	0.04*
BACHELOR_CH	−0.84	0.20	−4.15	0.00***
TECHQUALS_CH	1.66	0.25	6.78	0.00***
SYMBA_96	0.90	0.12	7.60	0.00***
SYMBA_CH	0.40	0.05	7.34	0.00***
VOLUNTEER_06	−0.51	0.09	−5.43	0.00***

Rho: −0.0015433, LR test value: 0.089998, p-value: 0.76418
Asymptotic standard error: 0.0050134, z-value: −0.30783, p-value: 0.75822
Wald statistic: 0.094756, p-value: 0.75822
Log likelihood: 679.213 for lag model
ML residual variance (sigma squared): 0.0036604, (sigma: 0.060501)
Number of observations: 490
Number of parameters estimated: 17
AIC: −1324.4, (AIC for lm: −1326.3)
LM test for residual autocorrelation
Test value: 13.682, p-value: 0.00021653
Significance codes: *** 0.001, ** 0.01, * 0.05
Source: The authors

- In the spatial error model the SCI_CH, LQ_FIN_CH and VOLUNTEER_06 variables are not significant explanatory factors whereas they are in the spatial lag model
- The L_POP_CH variable is significant in both models but in different directions, it being positive in the spatial error model and negative in the spatial lag model
- The BACHELOR_96 and the BACHELOR_CH variables are significant in both models, but in different directions, with the BACHELOR_96 having a negative effect in the spatial error model and a positive effect in the spatial lag model, while the BACHELOR_CH variable has a positive effect in the spatial error model and a negative effect in the spatial lag model
- The SYMBA_CH variable is significant in both models but the direction of influence is different, it being negative in the spatial error model and positive in the spatial lag model

It is evident that of some of the variables that are surrogate measures of human capital do play significant explanatory roles in the spatially-weighted regression modelling approaches. In both the spatial error and the spatial model solutions the change in the incidence of workers with technical qualifications over the decade 1996–2006 has a significant positive impact on endogenous regional employment performance. The incidence of workers with bachelor qualifications at the beginning of the decade has a positive impact in the spatial lag model while it is negative in the spatial error model. And the change over the decade in the incidence of workers with a bachelor degree has a negative effect in the spatial lag model solution while in the spatial error model it has a positive impact on endogenous regional employment performance.

With respect to the variable measuring social capital, in both the spatial error and the spatial lag model solutions the incidence of people engaged in volunteering in 2006 had a negative impact on endogenous regional employment performance. We note again that the variable measuring creative capital is not a significant explanatory factor influencing regional endogenous employment performance over the decade 1996–2006 across the LGAs in non-metropolitan Australia.

These differences referred to above between the spatial error model and the spatial lag models in the significance of variables – and for some of them in the direction of their influence on the dependent variable – is an issue of interest and perhaps of concern, and it makes it important to be able to ascertain which of the models might be more valuable or the "preferred" model for furnishing explanation of variation in the dependent variable (REG_SHIFT).

Thus, in order to determine which of these two models might represent the "better" approach for modelling the determinants of spatial variation in endogenous regional employment performance over the decade 1996–2006 across non-metropolitan LGAs across Australia. To assist with this, *Lagrange multiplier diagnostics* for spatial dependence were run.[8]

It would seem that the *spatial error model* is the preferred model to use. This is because the probability of the spatial autocorrelation (both with the normal language multiplier and the robust version) being present in the error term is statistically significant at the 99.9% level of confidence (with *p*-values of less than 0.01), whilst for the lag model it is not (with *p*-values of 0.77 for the normal Lagrange Multiplier and 0.12 for the robust version).

5 Policy Implications and Conclusions

The review of the literature on regional economic development and differentials in regional performance in Australia conducted by Stimson (2007) had suggested that an important question for policy makers to address is the degree to which the differentiation that exists across regions can be addressed by people-based as

[8]The Lagrange Multiplier results are:

- LM spatial error model = 11.7, df = 1, p-value = <0.01.
- LM spatial lag model = 0.09, df = 1, p-value = 0.77.
- RLM spatial error model = 14.0, df = 1, p-value = <0.01.
- RLM spatial lag model = 2.4, df = 1, p-value = 0.12.

against place-based policies and programs or by a mixture of both approaches. That had also been proposed in earlier studies by O'Connor et al. (2001) and Baum et al. (1999).

The argument may be summarised as follows:

1. People based approaches certainly enhance human capital development, and thus post-secondary education and training become critical. "And enhancing geographical access to those education and training services also became important" (Stimson et al. 2004: p. 108).
2. The overwhelming evidence is that investment in human capital development as a people-based policy is associated with advantageous place-based outcomes, as well as advantages for people (Stimson et al. 2004: p. 108).
3. Place-based interventions are typically oriented towards selective industry assistance, payroll tax exemptions, land deals, and the like. But such measures can have detrimental impacts on the economic welfare of populations in particular regions (Industry Commission 1993, 1996). Additionally, industry assistance packages in reality have limited potential for State and local government regional development policy to impact on regional economic activity in the longer-term (Giesecke and Maddern 1997: p. 17).
4. Strategies for regional development "need to be built upon local comparative advantage, and capitalize on region-specific resources, knowledge and location" (BTRE 2004a: p. 45).

In his overview Stimson (2007) pointed out that:

> . . .the nature of Australia's space economy is changing rapidly and the processes of change are impacting people and places in differential ways. One of the challenges will be to develop and implement regional policies and strategies that build successful regions and places into even more success.

In an earlier paper, Stimson had stated that:

> . . .inevitably that will require greater selectivity, but would be more likely to result in enhanced national performance and improved competitiveness, giving a better return on limited government resources. (p. 35)

Stimson (2007) also noted that:

> . . .it is inevitable that tensions result from differential levels of regional performance and that will continue. The challenge is how best might Australia, with its three-tier system of government respond with appropriate people-based and place-based policies for all the nation's regions.

What might be added as a result of the research as a result of the new modelling reported in this chapter?

It is clear from the modelling results discussed in this chapter that a number of the variables that are surrogate measures of human capital, social capital and creative capital do play some role as potential explanatory factors accounting for the spatial variation in endogenous regional employment performance over the decade 1996–2006 across non-metropolitan regions in Australia; but that is not necessarily a pervasive powerful explanatory role. For example, from the "preferred" *spatial error* model solution results, four of those surrogate variables – BACHELOR_CH, BACHLLOR_06, TECHQUALS_CH and VOLUNTEER_06 – are statistically significant at the 95% confidence level. The direction of that influence was positive in the case of BACHELOR_CH and TECHQUALS_CH; but it was negative in the case of the BACHELOR_96 and for the VOLUNTEER_06 variable. Thus, there is a mixed impact of the human capital variables, being both positive and negative in their impact on endogenous regional employment performance. The social capital variable measure derived from the 2006 census that was used in the modelling shows that there might be a negative relationship between the incidence of social capital and endogenous regional employment performance. However, the incidence of employment in creative industries (CREATIVE_06) is not a significant factor influencing endogenous regional employment performance.

What the modelling does indicate is that there were also a number of other factors that are significant in explaining differentials in endogenous regional employment performance over the decade 1996–2006 across Australia's non-metropolitan LGAs. Positive impacts were evident from the variables that purport to measure the following: change in industrial specialisation and the structural change index; population growth; and the incidence of workers in the symbolic analyst occupations at the beginning of the period. In contrast, it seems that there was a negative impact on endogenous regional employment performance as a result of a region having the following: a higher incidence of unemployment at the beginning of the decade and where there was an increase in the incidence of unemployment over the decade; an increase in the incidence of jobs in the professional, scientific and technical services industries; and an increase in the incidence of workers with symbolic analysis occupations over the decade.

In many respects these findings are similar to those uncovered in the previous analysis by the authors (Stimson et al. 2008) which focused on explaining differentials in patterns of endogenous employment performance across non-metropolitan regions separately within each of the five mainland states of Australia over the decade 1991–2001.

There are, however, two caveats need to be stated regarding the modelling results discussed in this chapter:

1. First, the use of the LGA as the predominant spatial unit for analysis is not particularly satisfactory. It would be much better if future analysis was to use functional labour market areas that are being demarcated in research that is being undertaken at the time of writing that will produce a new national geography that will greatly enhance the spatial analysis of labour market performance in Australia.
2. Reliance on census data to derive the surrogate measures of social capital and of creative capital is most restrictive, and the variables thus used – VOLUNTEER_06 and CREATIVE_06 – are far from satisfactory measures for those constructs.

Nonetheless, the research reported here is a first if exploratory attempt to explicitly investigate the roles of human capital, social capital and creative capital as explanatory factors in the variation that exists in endogenous regional employment performance over the decade 1996–2006 across regions in Australia – albeit restricted to non-metropolitan LGAs. The results highlight the apparently limited explanatory roles of those factors using variables that are surrogate variable derived from census data, and indeed the directional influence of some of those variables revealed in the modelling may be somewhat counter-intuitive.

Acknowledgements The research conducted by the authors that is discussed in this paper has been supported by Australian Research Council Discovery Program grant #DP0558722.

References

ABS (2006) Australian Standard Geographical Classification (ASGC), catalogue no. 1216.0, July 2006, Canberra, Australia

Acs ZJ, Anselin L, Varga A (2002) Patents and innovation counts as measures of regional production of new knowledge. Res Policy 31:1069–1085

Baum S, Stimson R, O'Connor K, Mullins P, Davis R (1999) Community opportunity and vulnerability in Australia's cities and towns: characteristics, patterns and implications. University of Queensland Press for the Australian Housing and Urban Research Institute, Brisbane

Beer A (1999) Regional cities within Australia's evolving urban system, 1991–96. Australas J Reg Stud 5(3):329–348

Beer A, Maude A (1995) Regional cities within the Australian urban system, 1961 to 1991. Urban Pol Res 13(3):135–148

Blakely EJ (1994) Planning local economic development: theory and practice, 2nd edn. Sage, Thousand Oaks, CA

Bolton R (1992) 'Place prosperity' vs 'people prosperity'. Urban Stud 29(2):185–203

Bolton R (1999) Place surplus, exit, voice, and loyalty. Uddevalla Symposium on Evaluation of Regional Policies. Fiskebackkil, Sweden, June

Bradley R, Gans J (1998) Growth in Australian cities. Econ Rec 74(226):266–278

Bureau of Transport and Regional Economics (2004a) Focus on regions no. 1: industry structure. Information Paper 49, BTRE, Department of Transport and Regional Services, Canberra

Bureau of Transport and Regional Economics (2004b) Focus on regions no. 2: education, skills and qualifications. Information Paper 51, BTRE, Department of Transport and Regional Services, Canberra

Capello R, Nijkamp PJ (eds) (2009) Regional growth and development theories in the XXI century: theoretical achievements and future challenges. Edward Elgar, Cheltenham

Clingermayer JC, Feiock RC (2001) Institutional constraints and policy choice: an exploration of local government. State University of New York Press, New York

Coleman JS (1990) The foundations of social theory. Harvard University Press, Cambridge, MA

Commission I (1993) Impediments to regional industry adjustment. Australian Government Publishing Service, Canberra

Commission I (1996) State, territory and local government assistance to industry (draft report). Industry Commission, Melbourne

Commission P (1998) Aspects of structural change in Australia. Research Report, Ausinfo, Canberra

De Santis M, Stough RR (1999) Fast adjusting urban regions, leadership and regional economic development. Region et Development 10:37–56

Department of Home Affairs and the Environment (1982) Australian urban environmental indicators. Australian Government Publishing Service, Canberra
Döring Th, Schnellenback J (2006) What do we know about geographical knowledge spillover and economic growth? Reg Stud 40(3):375–395
Duranton G, Puga D (2000) Diversity and specialisation in cities: why, where and when does it matter? Urban Stud 37(3):533–555
Faggian AA, McCann P (2009) Human capital and regional development. In: Capello R, Nijkamp PJ (eds) Regional growth and development theories in the XXI century: theoretical achievements and future challenges. Edward Elgar, Cheltenham, pp 133–151
Faraway J (2002) Practical Regression and Anova using R, cran.r-project.org/doc/contrib/Faraway-PRA.pdf. Accessed 13 June 2008, 2002
Florida R (1995) Toward the learning region. Futures 27(5):527–537
Florida R (2002) The rise of the creative class: and how it's transforming work, leisure, community and everyday life. Basic Books, New York
Florida R (2005) Cities and the creative class. Routledge, London
Fukuyama F (1995) Trust: the social virtues and creation of prosperity. Free Press, New York
Garnaut J, Connell P, Lindsay R, Rodriguez V (2001) County Australia: influences on employment and population growth, ABARE Research Report, 2001.1, Canberra
Giesecke AD, Maddern JR (1997) Regional government economic policy: assessing the policy instruments. Australas J Reg Stud 3(1):3–18
Goetz SJ, Rapasingla A (2001) The returns to higher education: estimates for the contiguous states, paper to regional science association international, North American annual meeting, Charleston, SC, November
Gordon IR, McCann P (2000) Industrial clusters, complexes, agglomeration and/or social networks. Urban Stud 37(3):513–532
Grossman GM, Helpman E (1991) Innovation and growth in the global economy. MIT, Cambridge, MA
Hanushek EA, Kimko DD (2000) Schooling, labour-force quality, and the growth of nations. Am Econ Rev 90(5):1184–1208
Henderson JV, Kuncoro A, Turner M (1995) Industrial development in cities. J Polit Econ 103:1067–1090
Hubbard T, Hall PH (eds) (1998) The entrepreneurial city. Wiley, Chichester, pp 77–99
Krugman P (1991) Geography and trade. MIT, Cambridge, MA
Lawson J, Dwyer J (2002) Labour market adjustment in regional Australia. Reserve Bank of Australia, Sydney
Lucas R (1988) On the mechanics of economic development. J Monetary Econ 22:3–42
Maier G (2001) History, spatial structure, and regional growth: lessons for policy making. In: Johansson B, Karlsson Ch, Stough RR (eds) Theories of endogenous regional growth. Springer, Heidelberg, pp 255–277
Nijkamp P, Stough R (eds) (2004) Entrepreneurship and regional economic development. Edward Elgar, Cheltenham
North D (1990) Institutions, institutional change and economic performance. Cambridge University Press, Cambridge
O'Connor K, Stimson R, Daly M (2001) Australia's changing economic geography and society dividing. Oxford University Press, Melbourne
OECD (1998) Human capital investment: an international comparison. Organisation for Economic Cooperation and Development, Paris
OECD (2001) The well-being of nations: the role of human and social capital. Organisation for Economic Cooperation and Development, Paris
Patten C (1991) The competitiveness of small firms. Cambridge University Press, Cambridge
Quiggin J (1999) Human capital theory and education policy in Australia. Aust Econ Rev 32(2):130–144

Reich R (1991) The work of nations: preparing ourselves for 21st century capitalism. Vintage Books, New York
Romer P (1986) Increasing returns and long-run growth. J Polit Econ 94(5):1002–1037
Romer P (1990) Endogenous technological change. J Polit Econ 98:S71–S102
Rosenthal SS, Strange WC (2001) The determinants of agglomeration. J Urban Econ 59:191–229
Stimson RJ (2002) Success and stress: regional performance across Australia, Keynote Address, Inspiring Regional Australia National Conference, Bunbury, October
Stimson RJ (2007) Endogenous factors in regional performance: evidence from Australia. In: Johansson I (ed) Institutions for knowledge generation and knowledge flows: building innovative capabilities for regions, Proceedings of the Uddevalla symposium 2007, Research Reports 2007:4. University West, Sweden, pp 671–700
Stimson RJ, Stough RR (2009) Regional economic development: methodology and analysis. In: Rowe JE (ed) Theories of local economic development: linking theory to practice. Ashgate, Aldershot, pp 169–192, Ch 9
Stimson R, Baum S, van Gellecum Y (2004) A typology of economic and human capital performance across Australia's large and medium size regional towns. Australas J Reg Stud 10(3):367–382
Stimson R, Stough R, Salazar M (2005) Leadership and institutional factors in endogenous regional economic development. Investigaciones Regionales 7:23–52
Stimson RJ, Robson A, Shyy T-K (2009) Modeling regional endogenous growth: an application to the non-metropolitan regions of Australia. Ann Reg Sci 43:379–398
Stimson RJ, Robson A, Shyy TK (2009a) Measuring regional; endogenous growth. In: Capello R, Nijkamp PJ (eds) Regional growth and development theories in the XXI century: theoretical achievements and future challenges. Edward Elgar, Cheltenham, pp 354–372, Ch 18
Stimson RJ, Stough RR, (with Maria Salazar) (2009b) Leadership and institutions in regional endogenous development. Edward Elgar, Cheltenham, p 151
Taylor PJ, Catalano G, Gane N (2002) A geography of global change: services and cities 2000–01, Globalisation and World Cities Study Group and Network. GaWC Res Bull 77:1–9
Van Oort FG, Atzema OALC (2004) On the conceptualisation of agglomeration economies: the case of new firms in formation. Ann Reg Sci 38:1–28
Vazquez BA (2002) Endogenous development: networking, innovation, institutions and cities. Routledge, London

Domestic Innovation and Chinese Regional Growth, 1991–2004

William Latham and Hong Yin

1 Introduction

Many other papers, notably those of Jefferson et al. (1996, 1997, 2002, 2003) have examined foreign investment and innovation in China. Other papers listed in the Appendix to this paper give the history and development of examinations of the modern Chinese innovation system. The contribution of this paper is its focus on the contributions of Chinese ***domestic*** innovation to economic growth at the provincial level.[1] Such an examination is a necessary part of an assessment of whether or not policies that promote domestic R&D, such as China's Science and Technology Policy, could be productive for China's regions.[2] The return to domestic innovation at the provincial level is estimated using a value-added Cobb-Douglas production function. The measure of the effect of innovation (patenting activity) is valued-added industrial output. The data are a balanced panel for 30 provinces for the period 1991–2004. The estimation results indicate that technology plays a positive role in China's provincial growth, but the contribution from technology (and thus from China's Science and Technology Policy) is small.

The effects of inter-regional innovative-knowledge spillovers on value added industrial output are also examined. Econometric evidence of positive inter-regional knowledge spillovers is found, however, the magnitude of these spillover effects is even smaller than that of the own technology effect.

Section 2 describes the basic model specifications. Section 3 describes data sources and problems. Section 4 presents the results of estimating the models

[1]We have elsewhere (Latham and Yin 2009) described the nature of China's domestic innovation system.

[2]We are grateful for the comments of an anonymous referee for helping us to clarify that our work on the effects of innovation on growth cannot reveal whether or not the Science and Technology policies in place during the whole period have actually had any effect because we do not have data for the counterfactual. Our results show only that, if S&T policies promote innovation and innovation has an effect on grow, then S&T policy may be effective in promoting growth.

W. Latham (✉) and H. Yin
Department of Economics, University of Delaware, Newark, DE 19716, USA
e-mail: latham@udel.edu

K. Kourtit et al. (eds.), *Drivers of Innovation, Entrepreneurship and Regional Dynamics*, Advances in Spatial Science, DOI 10.1007/978-3-642-17940-2_10,

with the data. Section 5 describes alternative model specifications and estimation results. Section 6 presents the impact of knowledge spillovers on value-added industrial output and Sect. 7 presents conclusions and suggestions for further research. In an appendix we provide an extended list of references on Chinese patenting for researchers interested in pursuing this subject.

2 Basic Model Specification

Griliches (1990) showed the power of using aggregate production functions to model innovation and the knowledge economy. While alternative formulations have been found useful, the basic framework continues to be a powerful analytic tool. Our analysis is consistent with this tradition. Assuming a conventional Cobb-Douglas production function, the basic model specification is:

$$\log(Y_{it}) = \alpha_i + a(t) + \beta_1 \log(C_{it}) + \beta_2 \log(L_{it}) + \beta_3 \log(P_{it}) + \varepsilon_{it}, \quad (1)$$

where Y_{it} is the value added industrial output in the region i at time t; a is the rate of exogenous technical progress; C_{it} and L_{it} are capital and labor inputs in the region i at time t; P_{it} is the technology input in the region i at time t, which is proxied alternatively by either contemporaneous patent applications or patent stocks. Fixed effects of regional specific characteristics are controlled by α_i. In this equation the elasticity of valued added output to technology is measured by the coefficient β_3.

3 Data Sources and Problems

Our investigation of this topic would certainly be improved by using more disaggregated data, especially in the regional dimension, over a longer time period. However, the data are quite limited as we discuss in the following paragraphs. In the end we are able to obtain meaningful results with the available data.

Fourteen years (1991–2004) of industrial data by province and by domestic and foreign-owned firms are available and are collected directly from the various issues of the China Statistical Yearbook (NBS 1992–2005). These data are based on reporting by all the independent accounting units by regions.

Output, Y_{it}, is constructed as the sum of value-added industrial output by domestic firms by region. Capital, C_{it}, is constructed as regional total assets of domestic firms. Total assets are a reasonable measure of capital input compared to fixed assets because total assets are the net values of funds used plus the fixed assets. Both value-added industrial output and total assets are reported in nominal terms and are adjusted to constant 2,000 yuan by an ex-factory producer price index.

It is very difficult to find an accurate measure of labor input, L_{it}, since no effective measure of working hours is reported in the statistical yearbooks.

We have no alternative to using the official number of manufacturing employees in domestic firms for labor input, although China's employment data are considered to be deeply flawed (Banister 2005; Wu 2001), compared to the other official data.[3] Before the industrial restructuring of 1994–1999, unproductive working hours were common in state-owned enterprises (SOE) due to shirking, lack of jobs, shortages of energy and/or political reasons. Thus manufacturing employment figures were highly inflated. After 1995 there were massive lay-offs in the SOEs: total manufacturing employment in SOE's declined from 66.1 million in 1995 to 37.5 million in 2002. Figure 1 presents total manufacturing employment along with value-added industrial output for the years 1991–2004. The structural break in manufacturing employment is very clear: manufacturing employment has declined continuously since 1995, while valued added output has a continuous upward trend over the years. Because of the large measurement errors and the structural change in labor inputs, the precision of our estimation results will be greatly affected.

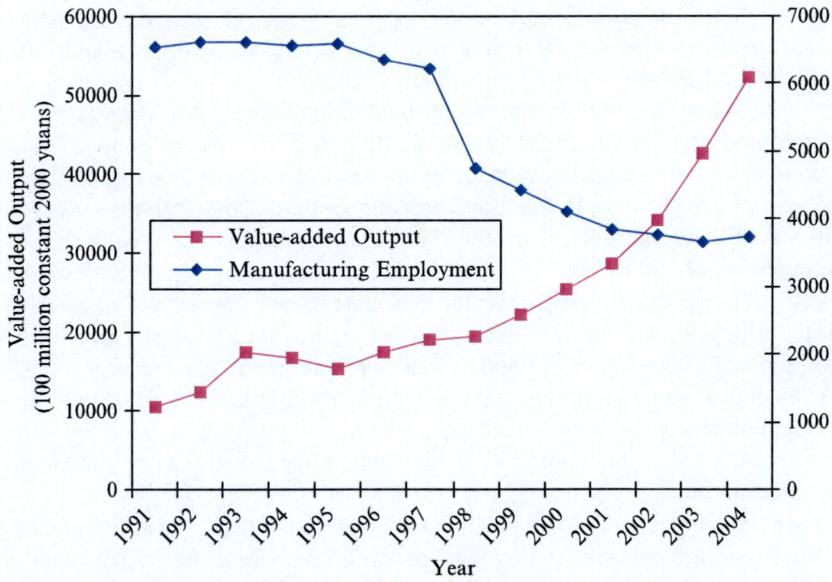

Fig. 1 Aggregated value-added industrial output and manufacturing employment (staffs and workers) in China from 1991 to 2004. Note: Data are based on the total manufacturing employees from 30 provinces reported in the various issues of China Statistical Yearbook (Tibet is excluded)

[3]We attempted to adjust the employment data for changes in human capital as measured by years of education per worker in the provinces. However, data on the education levels of employees by province are only available from 1996 to 2000, so the education levels of employees for the years 1991–1995 and for the years 2001–2004 had to extrapolated. The equations were then estimated with labor input adjusted by educational levels, however, the estimation results were not significantly improved.

Technology input, P_{it}, is measured both by both contemporaneous patent applications and by a measure of patent stocks. The patent data and the construction of patent stocks are described in this and the following paragraphs. Chinese patent data are available both on-line and on CD-ROMs. There are at least two official databases distributed by the State Intellectual Property Offices (SIPO): (1) CNPAT ABSDAT, which is in Chinese, and (2) CNPAT ACCESS, which is in English and has been distributed worldwide. However, the covered periods of these two databases vary slightly: CNPAT ABSDAT is the most comprehensive one which covers patent applications beginning in April 1985 when the first Chinese patent was filed. In contrast, CNPAT ACCESS began only in November 1985. Thus the CNPAT ABSDAT database is slightly more comprehensive and is the database used for this study.

There are two official on-line versions of the chosen database (CNPAT ABSDAT) that are maintained by the SIPO and which have the most complete patent documents up to the present.[4] Careful comparisons of search results retrieved from these two on-line databases and to the patent data published by SIPO revealed that the on-line databases are identical and comparable to the published patent data. Thus we used one of the official on-line databases (www.sipo.gov.cn) to retrieve the patent data for this study. All the patent data were retrieved between March 10, 2006 and April 30, 2006.

Patent applications are used as a proxy for innovation output in this study. There are two reasons for us to use patent applications rather than patent grants. First, there are potentially long lags between a patent's application and its grant: it might take 3–5 years for a patent to be examined and granted (and some patents may not be granted at all). Accordingly, if patent grants were used, the most dynamic and interesting period of 2000–2004 would be excluded in the analysis.

Second, it has generally been observed that patents are applied for relatively early in the lifecycle of a research project. Most studies find that there is a very strong relationship between R&D and patent applications at the cross-sectional level: the median R-square is around 0.9 (Griliches 1990). This relationship is close to contemporaneous with some small lags which are difficult to be estimated (Hausman et al. 1984). Thus most studies use patent applications as an indicator for innovation output.

There are three types of Chinese patents: (1) invention patents, (2) utility model patents and (3) design patents. An invention patent is comparable to a utility patent in the USA. A utility model patent is a "petty" patent, not recognized in the USA. A design patent is for improvement in aesthetic features rather than technical features. For the purpose of analyzing China's true technological capabilities, only invention patents are used in this study. Hereafter, the term patent refers to an invention patent application.

Because patent documents are not accessible to the public until 18 months after an application has been filed, the period of patent filings in this study is restricted to the period of April 1, 1985 to December 31, 2004. It should be noted that the reported

[4]The two websites of on-line databases are www.sipo.gov.cn and www.cnipr.com.

numbers of patents from 2004 might be slightly biased downward because some patent filings were not yet published when the data for this paper were collected due to the 18-month restriction before publication. In addition, there can be multiple patentees on a single patent located in different provinces. In this case, both provinces are recorded since the address of first patentees is not separated from those of other patentees in the patent documents. We carefully compared the results of a multiple-provinces search with those of a single-province search and found that the statistical error caused by double counting is very small, on average only about 2%.

There are several additional potential problems related to Chinese patent data. First, China's patent law went through a significant change in 1992 and was further revised in 2000. Consequently, we may expect that those changes might have a considerable impact on both domestic and foreign filings. Our analysis should be considered within this context. Second, Chinese patents include filings from both domestic and foreign patentees, and the majority of invention patents are actually filed by foreigners. We treat domestic patents as those patents with patentees' addresses from the 31 provinces and independent municipal cities of China. This raises the question of patents applied for by joint ventures with foreign firms. Compared to purely domestic firms, firms with foreign partners may be more competitive and may have more intensive innovation activities. However, unfortunately it is impossible to separate patents filed by joint ventures from other domestic filings in our data, as patentees from joint ventures are classified as having origins in China. With respect to the impact of foreign firms' R&D and patenting activity, the evolution of technological development in China has been greatly influenced by and has benefited from its increasing exposure to world-class technologies from foreign firms. China's domestic innovation activities have been stimulated and pushed forward by their foreign competitors. In recent years, more and more R&D centers of multinational corporations have moved to China. It is not clear how many patents filed by multinational corporations are actually generated in these offshore R&D centers in China. Surely, intensive and high-quality innovative activities in these R&D centers will generate spillover effects on domestic innovation activity. Unfortunately, in our data it is impossible to separate these spillover effects of foreign inventions from domestic firms' own innovations efforts.

Table 1 presents summary statistics for the variables. The correlation matrix of the variables in logs is in Table 2. As expected, among the three inputs, labor is least

Table 1 Summary statistics of variables in levels (1991–2004)

Variables	Observations	Mean	Std. Dev.	Min.	Max.
OUTPUT (Y) (value-added industrial output in 100 million constant 2,000 yuan)	420	793.97	933.63	2.12	6,763.23
CAPITAL (C) (total assets in 100 million constant 2,000 yuan)	420	3,572.27	3,451.51	15.03	20,805.20
LABOR (L) (manufacturing employees in 10,000 persons)	420	175.87	125.37	1.30	547.60
PATENTS (P) (patent applications)	420	612.08	973.76	2	6,847

Table 2 Correlation matrix of variables in logs

Variables	OUTPUT	CAPITAL	LABOR	PATENTS
OUTPUT	1.000			
CAPITAL	0.980	1.000		
LABOR	0.829	0.812	1.000	
PATENTS	0.916	0.915	0.716	1.000

correlated with value-added output: the correlation coefficient between the output and labor is only about 0.83. In contrast, the correlation coefficient is 0.98 and 0.92 between the output and capital and between the output and patents, respectively.

4 Estimation Results

Equation (1) is first estimated with both a fixed effects estimator and a random effects estimator. Table 3 presents the robust estimation results and the Hausman specification tests for comparisons of the fixed effects and random effects estimators. In columns (2) and (5), the contemporaneous patent applications are used as the technology input. In columns (3) and (6), the patent stocks are used as the technology input. The equation without the technology input are estimated and reported in columns (1) and (4).

First, we notice that the estimated coefficients of fixed effects models and random effects models are quite different. Although the random effects estimations seem to be better, with higher R-squared values and better-estimated coefficients, the Hausman tests reported in the first three columns reject all the random effects estimators. In the following analysis, only the estimation results of fixed effects estimators are reported [column (1) to column (3)].

In column (1) without the technology input, the estimated elasticity of value-added output to capital is only 0.194, which is much smaller than the one usually found in the literature. The estimated elasticity to labor is 0.305. The overall fit of the model is improved when the technology input is included: the R-squared increases from 0.833 in column (1) to 0.871 in column (2) and 0.85 in column (3), respectively. The estimated elasticity of value-added output to technology is 0.26 for the contemporaneous patent applications and is 0.385 for the patent stocks. The higher coefficient for the patent stocks is not surprising, as the magnitude of cumulated patent stocks is much larger than that of contemporaneous patent applications.[5] The elasticity of labor drops significantly to 0.135 when the patent stocks are used. This result is common in the literature: an increase in the elasticity to technology is at the expense of a declining elasticity to labor.

[5]The results are robust to the use of different depreciation rates in the construction of the patent stocks.

Table 3 Estimation results of the knowledge production function at the chinese provincial level (1991–2004); (1)

Dependent variable: LOG OUTPUT						
Independent variables[a]	Fixed effects			Random effects		
	(1)	(2)	(3)	(4)	(5)	(6)
LOG CAPITAL	0.194* (0.051)	0.168* (0.077)	0.167* (0.079)	0.629** (0.000)	0.415** (0.000)	0.468** (0.000)
LOG LABOR	0.305** (0.000)	0.324** (0.000)	0.135* (0.070)	0.420** (0.000)	0.374** (0.000)	0.268** (0.000)
LOG PATENTS		0.261** (0.000)			0.274** (0.000)	
LOG PATENT STOCKS			0.385** (0.000)			0.322** (0.000)
TIME TREND	0.094** (0.000)	0.066** (0.000)	0.033** (0.027)	0.057** (0.000)	0.042** (0.000)	0.019* (0.071)
R-squared[b]	0.833	0.871	0.850	0.958	0.959	0.953
Observations	420	420	420	420	420	420
Hausman specification test:						
Chi-square statistics	24.320 (0.000)	17.850 (0.007)	45.490 (0.000)			

Notes: The P-values are reported in parentheses. All estimation results are based on robust standard errors

*Significant at the 0.10 level. **Significant at the 0.05 level

[a]PATENTS refers to the contemporaneous patent applications. PATENT STOCKS is constructed from a perpetual inventory model using by a 7% depreciation rate

[b]The R-squared is within R-squared for the fixed effects estimators and overall R-squared for the random effects estimators

[c]The Chi-square statistics are for the Hausman specification test for comparing the fixed effects models and random effects models [column (1) vs. column (4), column (2) vs. column (5), and column (3) vs. column (6)]

Although the overall fit of models in columns (2) and (3) is good, the large measurement errors in the labor input may have biased the estimates of technology input upward: the elasticities to technology are even larger than those to capital input. The structure break in the manufacturing employment is obviously not captured by the model specifications. Further, it is found that the estimation results are sensitive to the price index used to deflate the capital input.[6] In addition, there might be an omitted variable problem: the R-squared increases significantly when the technology input is included in the estimations. Using a single time trend in the model may be also inappropriate as the exogenous technological change is unlikely to be linear over the years. Because of these problems, the model specification of (1) is modified to improve the estimation results in the next section.

5 Alternative Specifications

As there are large labor shocks during the years 1995–1999, we consider using a set of year dummies to capture these changes. The modified equation with year dummies is:

$$\log(Y_{it}) = \alpha_i + \sum_{t=1}^{14} a_t + \beta_1 \log(C_{it}) + \beta_2 \log(L_{it}) + \beta_3 \log(P_{it}) + \varepsilon_{it}, \tag{2}$$

where a_t is a set of year dummies from 1991 to 2004. Estimation results of fixed effects models are reported in Table 4. The estimation results with contemporaneous patent

Table 4 Estimation results of the modified knowledge production function at the Chinese provincial level (1991–2004); (2)

Dependent variable: LOG OUTPUT			
Independent variables	Fixed effects		
	(1)	(2)	(3)
LOG CAPITAL	0.449** (0.000)	0.419** (0.000)	0.424** (0.000)
LOG LABOR	0.326** (0.000)	0.304** (0.000)	0.237** (0.000)
LOG PATENT		0.099** (0.000)	
LOG PATENT STOCKS			0.235** (0.000)
YEAR DUMMY	YES	YES	YES
R-squared	0.913	0.911	0.919
Observations	420	420	420

Notes: The P-values are reported in parentheses. All estimation results are based on robust standard errors. PATENTS refers to contemporaneous patent applications. PATENT STOCKS is constructed from a perpetual inventory model using a 7% depreciation rate. The R-squared is within R-squared for the fixed effects estimators. Diagnostics tests reveal both heteroscedasticity and serial correlation. Therefore, the estimated results reported are heteroscedasticity and autocorrelation consistent (HAC) robust standard errors
**Significant at 0.05 level

[6]The capital input (total assets) is also deflated by the fixed-asset price index and the equation is re-estimated. The precision of estimated coefficients of all variables drops sharply.

applications are presented in column (2); results with patent stocks are listed in column (3); and results without patent variables are reported in column (1).

5.1 *Revised Results for Effects of Technology*

Compared to the models reported in Table 3, the overall fit of the regressions has improved with an R-squared of about 0.91 for all the three models. The estimated elasticities have also increased: without the technology input the estimated elasticities to capital and labor are 0.449 and 0.326, respectively. Those elasticities are in line with the ones found in the literature (Movshuk 2004; Wu 1996). Wu (1996) reports that the elasticities of gross industrial output to capital and labor are 0.54 and 0.23, respectively, at the Chinese provincial level for the years 1985–1990.

Movshuk (2004) estimates a similar Cobb-Douglas production function for Chinese domestic firms for the years 1988–2000 and finds that the elasticities of gross industrial output to capital and labor are 0.14 and 0.63, respectively.

The estimated elasticity of output to technology is 0.099 for contemporaneous patent applications, which implies that a 1% increase in a region's patent applications results in a 0.099% increase in that region's valued-added industrial output, other things being equal. In comparison, the estimated elasticity to technology is 0.235 for the patent-stocks model. The impact of patent stocks is much larger as expected: a 1% increase in a region's patent stocks increases the value-added output by 0.235%. However, the magnitude of technology's contribution to the value-added output is small, either with the patent applications or patent stocks. Similar studies conducted at the level of European regions and US states usually find that the elasticity coefficient of technology inputs is closer to that of capital input.

5.2 *Time Effects*

To see the effects of time in the analysis, the coefficients of the year dummies are plotted in Fig. 2. The structural break due to the industrial restructuring of 1994–1999 is clear: the coefficients of the year dummies started to drop in 1994 and were particularly low in 1995. The effects are most striking when patent stocks are used in the equation: there was practically no economy-wide exogenous technical progress during the period 1995–1999. In contrast, the effects of technical progress increase linearly for the years 2000–2004. Those results seem to suggest that using year dummies in the equation is a better choice to capture the effects of structural changes.

As a robustness check, capital input (total assets) is further deflated by the fixed asset price index and the equation is re-estimated. The estimation results are similar to those reported in Table 4. The results are also robust to the depreciation rates of patent stocks. Thus it can be concluded that the results in Table 4 are robust and (2)

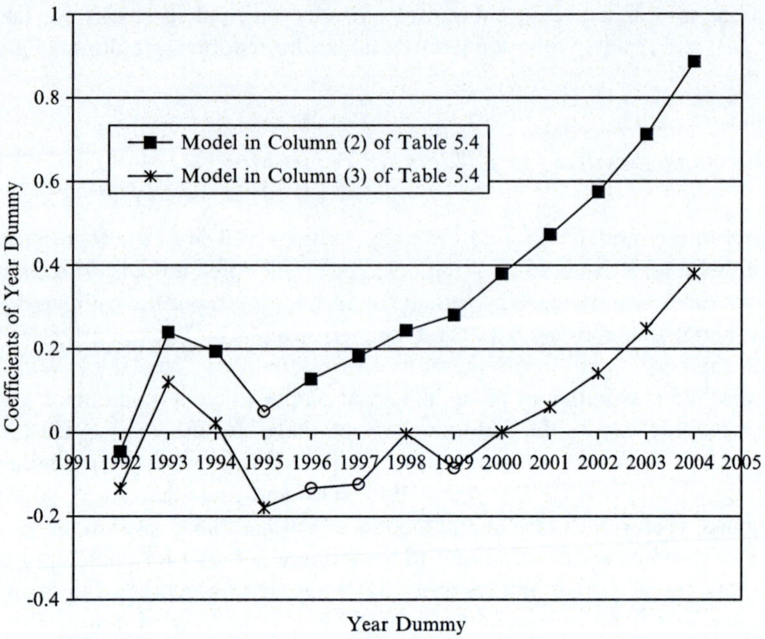

Fig. 2 Estimated coefficients of the year dummies of (2). Note: The data points marked with "o" mean that the estimated coefficients are not significant at the 10% significance level

is more appropriate for estimating the knowledge production function at the Chinese provincial level. In the following analysis, only the results using patent stocks with a 7% depreciation rate are reported.

5.3 Effects in Three Macro Regions

The empirical results of our prior work (Latham and Yin 2008) point out that there are enormous regional differences in technological development (patenting activity) among the three major macro-regions of China. Here, the contribution of regional variations in technology to industrial growth is further explored.

Location dummies for the EAST, CENTRAL, and WEST regions, are created and are interacted with the technology input (the patent stocks). The estimated equation is:

$$\log(Y_{it}) = \alpha_i + \sum_{t=1}^{14} a_t + \beta_1 \log(C_{it}) + \beta_2 \log(L_{it}) + \sum_{j=1}^{3} \beta_j \log P_{it} + \varepsilon_{it}, \quad (3)$$

where β_j is the elasticity of value-added output to technology input in the macro-region j.

Table 5 Effects of locations and the year dummies on the estimations of the knowledge production function at the Chinese provincial level (1991–2004); (3) and (4)

Dependent variable: LOG OUTPUT		
Independent variables	Fixed effects	
	(1)	(2)
LOG CAPITAL	0.414** (0.000)	0.450** (0.000)
LOG LABOR	0.263** (0.000)	0.293** (0.000)
LOG PATENTSTOCKS_EAST	0.233** (0.000)	
LOG PATENTSTOCKS_CENTRAL	0.216** (0.001)	
LOG PATENTSTOCKS_WEST	0.155** (0.006)	
LOG PATENTSTOCKS_91_94		0.041** (0.038)
LOG PATENTSTOCKS_95_99		0.013 (0.515)
LOG PATENTSTOCKS_00_04		0.039** (0.018)
YEAR DUMMY	YES	YES
R-squared	0.920	0.915
Observations	420	420

Notes: The P-values are reported in parentheses. All estimation results are based on robust standard errors. PATENTSTOCKS_EAST (CENTRAL and WEST) is the interaction term between PATENT STOCKS and the macro region dummy, EAST (CENTRAL and WEST). PATENTSTOCKS_91_94 (95_99 and 00_04) is the interaction term between PATENT STOCKS and the time dummy D91_94 (D95_99 and D00_04). PATENT STOCKS is constructed from a perpetual inventory model using a 7% depreciation rate. The R-squared is within R-squared for the fixed effects estimators

**Significant at 0.05 level

Estimations with separate slopes of patent stocks are reported in column (1) of Table 5. The elasticity coefficient of technology in the eastern region is 0.23, compared to 0.21 in the central region and 0.155 in the western region. These estimates suggest that the separate treatment of three macro-regions is appropriate. The evidence points out how the western region lags behind in terms of technology's contribution to valued-added output. With respect to the role of technology in industrial growth, the differences among the three macro-regions are relatively small, and are significantly smaller than the differences in the effect of R&D on patenting found in our prior work (Latham and Yin 2008). This result suggests that regional variations in the adoption of new technology are smaller than differences in the production of new technology. Given the small elasticities of output to technology input, the disconnect between innovations and commercialization of new technology seems to be a common problem across the regions.

5.4 *The Effects of Industrial Reforms*

The estimated coefficients of the year dummies only capture certain time-specific effects of industrial reforms on economy-wide rates of technical progress. In this section, the effect of industrial reform on technology's contribution to value-added output is further examined. Dummies are created for: (1) the pre-reform period of

1991–1994; (2) the reform period of 1995–1999; and (3) the post-reform period of 2000–2004. The three time dummies are interacted with the technology input (patent stocks), so the separate slopes of patent stocks of three periods can be estimated. The estimated equation is:

$$\log(Y_{it}) = \alpha_i + \sum_{t=1}^{14} a_t + \beta_1 \log(C_{it}) + \beta_2 \log(L_{it}) + \sum_{T=1}^{3} \beta_T \log(P_{it}) + \varepsilon_{it}, \quad (4)$$

where β_T is the elasticity to patent stocks for the period T. The results are presented in column (2) of Table 5.

The elasticity coefficient of patent stocks is 0.041 for the pre-reform period of 1991–1994 and is 0.039 for the post-reform period of 2000–2004. In contrast, the coefficient of patent stocks for the period 1995–1999 is not only smaller (0.013) but also insignificant, which implies that there is no technology's contribution to the valued-added output at the provincial level during the industrial reform period. The empirical results here again support the findings in our previous work (Latham and Yin 2008) and point out that the effects of industrial reform on China's technological development during the period 1994–1999 are very negative: there is no economy-wide technical progress and technology's contribution to industry growth is nonexistent.

We notice that the estimated coefficients of patent stocks for the three separate periods decrease significantly, while the coefficients of capital and labor increase. This is not surprising: the fixed effects estimators only use within variations of the data. The within-variations of the patent stocks decline substantially when 14 years are divided into three sub-periods. Consequently, the precision of estimated coefficients of patent stocks declines sharply.

6 The Effects of Technology Spillovers

In this section, the impact of technology spillovers to value-added industrial output is investigated. Equation (2) is extended by including a spillover variable. This spillover variable is the weighted patent stocks from other regions, which are described in an appendix. The estimated knowledge-spillover production function is:

$$\log(Y_{it}) = \alpha_i + \sum_{t=1}^{14} a_t + \beta_1 \log(C_{it}) + \beta_2 \log(L_{it}) + \beta_3 \log(P_{it}) + \beta_4 W \log(S_{it}) + \varepsilon_{it}, \quad (5)$$

where W is the weight matrix and S_{it} is the technology stocks (patent stocks) from other regions. The spillover variable is represent by Wlog (S_{it}). Two different

Table 6 Effects of technology spillovers on the estimation results of the knowledge production function at the Chinese provincial level (1991–2004); (5)

Dependent variable: LOG OUTPUT		
Independent variables	Fixed effects	
	(1)	(2)
LOG CAPITAL	0.425** (0.000)	0.417** (0.000)
LOG LABOR	0.224** (0.000)	0.232** (0.000)
LOG PATENT STOCKS	0.225** (0.000)	0.227** (0.000)
Spillover variable		
LOG SPILLOVER (PATENT STOCKS) (Contiguity-weighted)	0.098** (0.038)	
LOG SPILLOVER (PATENT STOCKS) (Gravity-weighted)		0.134 (0.164)
YEAR DUMMY	YES	YES
R-squared	0.918	0.917
Observations	420	420

Notes: The P-values are reported in parentheses. All estimation results are based on robust standard errors. PATENT STOCKS is constructed from a perpetual inventory model using a 7% depreciation rate. SPILLOVER (PATENT STOCKS) refers to the knowledge stocks available in the other regions, proxied by the patent stocks in the other regions. The results are similar with patent stocks using a 0% depreciation rate and a 12% depreciation rate. The R-squared is within R-squared for the fixed effects estimators
**Significant at 0.05 level

weights are used to construct the spillover variable: a contiguity weight and a gravity-weight.[7]

The robust estimation results of (5) are presented in Table 6.[8] Results with the contiguity-weighted spillover variable are reported in column (1) and results with the gravity-weighted variable are listed in column (2). The coefficient of the contiguity-weighted spillover variable is 0.098, which implies that 1% increase in the patent stocks from the neighboring regions will lead to a 0.098% increase in the region's valued-add output. In comparison, the coefficient of the gravity-weighted spillover variable is 0.134, but it is only significant at 0.16 significance level.

The estimated results seem to suggest that geographical proximity is very important in the inter-regional technology spillovers: solid technology spillovers are only found with the contiguity-weighted spillover variable and the magnitude of technology spillover is much smaller than that of own technology. This implies that the inter-regional technology linkage only exists between the bordering provinces and even that linkage is not strong.

[7]The equation with an unweighted spillover variable is also estimated and no evidence of spillover effects is found at all.

[8]Robustness checks for the results reported in Table 6 are conducted. The estimation results are insensitive to either the price index used to deflate the capital input or the depreciation rates of patent stocks.

7 Conclusion

We find that the production function including innovation fits the Chinese provincial level data. The elasticities of value added industrial output to capital and labor are 0.425 and 0.224, respectively. The elasticity of value-added industrial output to the region's own technology is 0.099 for the contemporaneous patent applications and 0.235 for the patent stocks. These estimates indicate that technology plays a positive role in industrial growth at the provincial level; however, the contribution of technology is far too small which indicates that China's economic growth is largely driven by the factor inputs. The results here seem to support the views that the linkages between innovation activity and commercialization of new technology are weak within Chinese domestic firms (Sun 2002). Domestic firms apparently have difficulties in exploiting and adopting the new technologies. This naturally raises the questions about the current technology policy in China: does current S&T policy emphasize too much on the generation of new technology, compared to the adoption of new technology? For long-term sustainable economic growth, how to facilitate and encourage the adoption of new technology should be the main concerns for China's policymakers.

The results also indicate that the inter-regional technology spillovers are positive but relatively small and weak, compared to the European regions and the states in the USA. The evidence here confirms the low developmental stage of China's industry as the ability to adopt and diffuse the new technology is weak across the Chinese provinces.

The estimated results further confirm that the impact of industrial reforms during the period of 1994–1999 on China's technological development is negative, as there seems to be neither exogenous technical progress nor technology's contribution to the value-added industrial output at all in those years.

References

Banister J (2005a) Manufacturing employment in China. Mon Labor Rev 128:11–29

Griliches Z (1990) Patent statistics as economic indicators: a survey. J Econ Lit XXVIII:1661–1707

Hausman J, Griliches Z, Hall BB (1984) Econometric models for count data with an application to the patent-R&D relationship. Econometrica 52(4):909–938

Latham W, Yin H (2008) R&D investment and domestic patenting at the provincial level in China: 1998–2004, University of Delaware Department of Economics Working Paper

Movshuk O (2004a) Restructuring, productivity and technical efficiency in China's iron and steel industry, 1988–2000. J Asian Econ 15:135–151

NBS (1992b) China statistical yearbook. National Bureau of Statistics, People's Republic of China

SIPO (2002) White paper on the intellectual property rights protection in China in 2002, State Intellectual Property Office, People's Republic of China. Retrieved 4 March 2005, from http://www.sipo.gov.cn/sipo_english/ndbg/nb/ndbg2003/default.htm

SIPO (2003) White paper on the intellectual property rights protection in China in 2003, State Intellectual Property Office, People's Republic of China. Retrieved 4 March 2005, from http://www.sipo.gov.cn/sipo_english/ndbg/nb/ndbg2004/default.htm

Sun Y (2002a) Source of innovation in China's manufacturing sector: imported or developed in-house. Environ Plann 34:1059–1072

Wu Y (1996a) Productive performance in Chinese enterprise: an empirical study. St. Martin's Press, New York

Wu XH (2001a) China's comparative labor productivity performance in manufacturing, 1952–1997: catching up or falling behind. China Econ Rev 12:162–189

Appendix: Extended List of References on Chinese Patenting

Allison JR, Lin L (1999) The evolution of Chinese attitude toward property rights in invention and discovery. SSRN Electronic Paper Collection (http://papers.ssrn.com/abstract=22539)

Banister J (2005b) Manufacturing employment in China. Mon Labor Rev 128:11–29

Bhattashi D, Li S, Martin W (eds) (2004) China and the WTO. Oxford/World Bank, UK/US

Bosworth D, Yang D (2000) Intellectual property law, technology flow and licensing opportunities in the People's Republic of China. Int Bus Rev 9(4):453–77

Broadman HG (2002) A litmus test for China's accession to the WTO: reform of its state-owned enterprises. In: Ostry S, Alexandorff AS, Gomez R (eds) China and the long march to globe trade. Routledge Curzon, London, pp 47–65

CAS (2005) Chinese Academy of Sciences, People's Republic of China. Retrieved 2 Oct 2005 from http://english.cas.cn/eng2003/page/about_03.htm

Cauley J, Cornes R, Sandler T (1999) Stakeholder incentives and reforms in China's state-owned enterprises. China Econ Rev 10(2):191–206

CERNET (www.cernet.edu.cn) (2005a) Science and technology system. Ministry of Education, People's Republic of China. Retrieved 2 Oct 2005, from http://www.edu.cn/Homepage/english/R&D/laws

CERNET (www.cernet.edu.cn) (2005b) Science and technology policy. Ministry of Education, People's Republic of China. Retrieved 2 Oct 2005, from http://www.edu.cn/Homepage/english/R&D/laws

Chen C, Shih H (2005) High-tech industries in China. Edward Elgar, Cheltenham, UK

Cheung K, Lin P (2003) Spillover effects of FDI on innovation in China: evidence from provincial data. SSRN Electronic Paper Collection (http://papers.ssrn.com/abstract=419020)

Chow G (2001) China's economic reform and policies at the beginning of the twenty-first century. China Econ Rev 11(4):427–431

Demurger S, Sachs JD, Woo WT, Bao S, Chang G, Mellinger A (2002) Geography, economic policy and regional development in China. Harvard Institute of Economic Research, Discussion Paper No.1950, March 2002

Fai FM (2005) Using intellectual property data to analyze China's growing technological capabilities. World Patent Inf 27:49–61

Gabriele A (2002) S&T policies and technical progress in China's industry. Rev Int Polit Econ 9 (2):333–373

Ho SM (1997) Technology transfer to China during the 1980s: how effective? Some evidence from Jiangsu. Pac Aff 70:85–107

Jefferson GH, Hu AGZ (2002) FDI impact of spillover: evidence from China's electronic and textile industries. World Econ 25(8):1063–1076

Jefferson GH, Rawski TG, Zheng Y (1996) Chinese industrial productivity: trends, measurement issues, and recent developments. J Comp Econ 23:146–180

Jefferson GH, Rawski TG, Zheng Y (1997) Innovation in Chinese manufacturing enterprises: a preliminary analysis of survey data. MOCT-MOST 7:101–120

Jefferson GH, Hu AGZ, Guan X, Qian J (2003) R&D and technology transfer: firm-level evidence from Chinese industry. William Davison Institute Working Paper 582, June 2003

Jefferson GH, Hu AGZ, Guan X, Yu X (2003b) Ownership, performance, and innovation in China's large and medium-size industrial enterprise sector. China Econ Rev 14:89–113

La Croix SJ, Konan DE (2002) Intellectual property rights in China: the changing political economy of Chinese-American interests. World Econ 26:759–88

Latham W, Yin H (2008) R&D investment and domestic patenting at the provincial level in China: 1998–2004, University of Delaware, Department of Economics Working Paper

Latham W, Yin H (2009) Technological development in China 1985–2004: evidence from domestic patents. Economie et Société. série W: Dynamique technologique et organization, June

Lin P (2001) Protection of intellectual property right in China. In: Kueh YY, Zheng WM (eds) Globalization and Sino-US economic and trade relationship. Social Press, Beijing, pp 372–397

Liu X, White S (2001) Comparing innovation systems: a framework and application to China's transitional context. Res Policy 30:1091–1114

Lo D (1999) Reappraising the performance of China's state-owned enterprises, 1980–96. Camb J Econ 23(6):693–718

Lu Q (2000) China's leap into the information age. Oxford University Press, New York

Maskus KE (2004) Intellectual property rights in the WTO accession package. In: Bhattashi D, Li S, Martin W (eds) China and the WTO. Oxford University Press, Oxford, pp 49–68

Movshuk O (2004b) Restructuring, productivity and technical efficiency in China's iron and steel industry, 1988-2000. J Asian Econ 15:135–51

NBS (1992–2005) China statistical yearbook. National Bureau of Statistics, People's Republic of China

NBS (2003) Statistical bulletin on the input of science and technology 2002. National Bureau of Statistics, People's Republic of China. Retrieved 20 Oct 2005, from http://www.stats.gov.cn/english/newrelease/statisticalreports/t20031111_119097.htm

Nolan P (2002) China and the WTO: the challenge for China's large-scale industry. In: Holbig H, Ash R (eds) China's accession to the world trade organization. Routledge Curzon, New York, pp 43–74

O'Keeffe M (2005) Cross comparison of US, EU, JP and Korean companies patenting activity in Japan and in the Peoples Republic of China. World Patent Inf 27(2):125–134

Pangestu M (2004) China's trade policy: post WTO accession commitments and experiments with regionalism. In: Garnaut R, Song L (eds) China: is rapid growth sustainable? The Australian National University, Asia Pacific Press, Canberra, pp 232–245

Patent Law of People's Republic of China (2000) Retrieved 2 Oct 2005, from http://www.sipo.gov.cn/sipo_English/flfs/zlflfg/t20020327_33872.htm

PatVal-EU Project (2005) The value of european patents: evidence from a survey of European inventors, Contract HPV2-CT-2001-00013

Phillip K, Chen B (2004) A panel-data sensitivity analysis of regional growth in China. SSRN Electronic Paper Collection (http://papers.ssrn.com/abstract=599662)

SIPO (2002) White paper on the intellectual property rights protection in China in 2002. State Intellectual Property Office, People's Republic of China. Retrieved 4 Mar 2005, from http://www.sipo.gov.cn/sipo_english/ndbg/nb/ndbg2003/default.htm

SIPO (2003) White paper on the intellectual property rights protection in China in 2003. State Intellectual Property Office, People's Republic of China. Retrieved 4 Mar 2005, from http://www.sipo.gov.cn/sipo_english/ndbg/nb/ndbg2004/default.htm

Sun Y (2000) Spatial distribution of patents in China. Reg Stud 34:441–454

Sun Y (2002b) Source of innovation in China's manufacturing sector: imported or developed in-house. Environ Plann 34:1059–1072

Sun Y (2003) Determinants of foreign patents in China. World Patent Inf 25(1):27–27

Sutherland D (2003) China's large enterprises and the challenge of late industrialization. Routledge Curzon, London

Walsh KA (1999) US commercial technology transfers to the People's Republic of China. Office of Strategic Industries and Economic Security, Bureau of Export Administration and DFI International, Washington, DC

Wang J, Tong X (2005) Sustaining urban growth through innovative capacity: Beijing and Shanghai in comparison. Word Bank Policy Research Working Paper 3545, March 2005

WIPO (2006) Exceptional growth from north east Asia in record year for international patent filings. Press Release 436. The Word Intellectual Property Organization. Retrieved 21 Mar 2007, from http://www.wipo.int/

Wu Y (1996b) Productive performance in Chinese enterprise: an empirical study. St. Martin's Press, New York

Wu XH (2001b) China's Comparative labor productivity performance in manufacturing, 1952–1997: catching up or falling behind. China Econ Rev 12:162–189

Wu Y (2004) Productivity and sustainable growth. In: Garnaut R, Song L (eds) China: is rapid growth sustainable? The Australia National University, Asia Pacific Press, Canberra, pp 181–196

Wu C, Liu Y (2004) Use of IPC and various retrieval systems to research paten activities of US organization in the People's Republic of China. World Patent Inf 26:225–33

Xu X (2002) Have Chinese provinces become integrated under reforms? China Econ Rev 13 (2–3):116–133

Yueh L (2006) Patent laws and innovation in China. University of Oxford, Department of Economics, Economics Series Working Papers 271

The Spatial Dynamics of China's High-Tech Industry: An Exploratory Policy Analysis

Junbo Yu, Peter Nijkamp, and Junyang Yuan

1 Introduction

In recent years, increasing attention has been paid to the development of China's high-tech industry, not only because it reflects China's domestic achievement in transforming its industrial structure, but also because it represents a challenge, from a global perspective, to the preeminence of the technological competitiveness of developed countries (Chen and Shi 2005; Gilboy 2004; Jefferson 2005; Liu and Buck 2007). However, existing studies largely concentrate on describing and analyzing the temporal dynamics of China's high-tech industry against a global competition scenario rather than drilling down and exploring more details of the issue from a complementary point of view (OECD 2007, 2008). As a result, apart from a convinced impression that China's production and R&D investment is soaring in high-tech industry, very little is known about the spatial pattern of China's high-tech industry, the underlying factors that drive the formation and evolution of this pattern, its profound socioeconomic influences on crucial issues like regional competitiveness and disparity, and their implications for policymaking. The objective of this paper, in response, is to address the gap in studies on China's high-tech industry from a spatial perspective.

Studies analyzing the experience of developed countries or regions have progressed to the point of concluding that much of the traditional location theory

J. Yu (✉)
Regional Research Institute, 886 Chestnut Ridge Road, PO Box 6825, Morgantown, WV 26505–6825, USA
e-mail: junbo_yu@mail.jlu.edu.cn

P. Nijkamp
Department of Spatial Economics, VU University Amsterdam, De Boelelaan 1105, 1081 HV Amsterdam, The Netherlands
e-mail: p.nijkamp@vu.nl

J. Yuan
School of Public Policy, George Mason University, PMB 3704, 4450 Rivanna Ln, Fairfax, VA 22030, USA

K. Kourtit et al. (eds.), *Drivers of Innovation, Entrepreneurship and Regional Dynamics*, Advances in Spatial Science, DOI 10.1007/978-3-642-17940-2_11,

is applicable to explain the spatial pattern of high-tech industry (Norton 2000). For instance, high-tech firms tend to agglomerate and collocate, after historical accidents and exogenous forces created the initial critical mass in one location; they are to be attracted to this place by traditional location factors that are generally referred to as "cost-of-doing-business-measures"; then the spinoff mechanism and the incentive to exploit agglomeration economies will increasingly reinforce the existing location pattern (Cortright and Mayer 2001; DeVol et al. 1999; Maggioni 2002; Polese and Shearmur 2006). Nevertheless, there are also some novel elements identified and this distinguishes the clustering process of high-tech industries, for example, the capabilities of local economic agents to absorb and utilize spillovers from high-tech agglomerations are not only bounded by pure physical distance, but they depend also strongly on territorialized and indigenous endowments such as entrepreneurship, trust and creativity (Capello 2009; Florida 2005); contrary to common wisdom, high-tech regions tend to specialize more exclusively in a few products or technologies and thus differ significantly from one another (Cortright and Mayer 2001; Keeble and Wilkinson 1999, 2000); other factors such as the access to a trained/educated workforce and close proximity to excellent educational facilities and research institutions weigh more heavily in terms of attracting and retaining high-tech firms (Audretsch and Feldman 1996; Feldman and Audretsch 1999; Capello 1999). In general, the high-tech industry assumes a geographic feature in the form of specialized locational clusters (Scott 2006). It resembles traditional industries in concentration behavior, but the glue to hold such concentrations differs, while its degree of specialization is also much higher (Cutrini 2010).

Drawing upon the preceding observation, our attempt to investigate the spatial dynamics of China's high-tech industry is directed to first examine its pattern in agglomeration and specialization. Similarities and differences in comparison with the experience of developed countries from these two perspectives will be identified and highlighted. Next, in search of explanations to the similarities and differences, a tentative explanatory framework is constructed and applied to the Chinese context in high-tech industry development. And lastly, the validity of our explanatory framework is assessed in the concluding section, while its implications for policy-making are also discussed.

2 Exploratory Spatial Analysis

The high-tech industry is normally distinguished by an industry's relatively higher research and development (R&D) intensity and proportion of professionals and scientists or skilled workers in the workforce (Goetz and Rupasingha 2002; OECD 2006). The definition of high-tech industries in China is consistent to the one used in OECD countries (Appendix, Table 1). However, associated industrial statistics based on this definition only came into being after 1994, when the most recent adjustment took place in the Chinese industrial classification (IC) system aiming at a better accordance with the International Standard Industrial Classification (ISIC) system.

Therefore, official statistics about China's high-tech industries under the current IC system have only been available since 1995. Our subsequent analysis finds in general its sources in the China Statistical Yearbook on High Technology Industry (2002–2008), an annually publication released by the National Bureau of Statistics (NBS) of China since 2002. Thus, it can only report data for most statistics from 1995 to 2007 (13 years). Our analysis begins with a sketch of high-tech industry development history before the year 1995, when the official statistics became available, and proceeds by organizing and mapping relevant industrial data to reveal and analyze the agglomeration and specialization pattern of China's high-tech industry after 1995.

2.1 Industrial Geography Before 1995

Although there are no statistics for an explicitly defined high-tech industry in China before 1995, some sub-industries well-known for their high-tech features, for example, the spacecraft and aircraft industry, have existed and developed at various scales in different regions ever since the late 1950s, primarily driven by military and defensive purposes. As a result, they were intentionally designed to disperse across the country so as to increase their survival rates as a whole during wars – some located in far interior places like Sichuan (southwest), some situated in close proximity to ideological comrade countries such as the Soviet Union and North Korea (northeast) (Chen and Shi 2005; Simon 1989).

The social and economic reform since 1978 has initiated an incremental and moderate restructuring of China' high-tech industry. On the one hand, in response to the surging domestic demand from civil sectors, foreign investment and technology in sub-industries like electronic and telecommunication and medical and pharmaceutical sectors were introduced into authorized regions and these thus proliferated the components of the industry; on the other hand, in face of the inability of domestic firms in these emerging sub-industries, the Chinese government aspired to build indigenous capabilities by fostering Chinese firms to cooperate, and then to compete, with foreign firms under the assistance of policy intervention (Sigurdson 2004; Sutherland 2007; Ning 2007). As a result of this twofold concern, the following eight regions were officially selected as the priority areas for high-tech center development in China since 1988: the Shanghai–Suzhou–Wuxi–Changzhou region of east China; the Pearl River Delta in Guangdong province; the Qi–Lu region of Shandong; the Guanzhong district Shaanxi; the "Five Districts and One Corridor" region in Hunan; the South Fujian High-tech Corridor; the Beijing–Tianjin–Tangshan region of north China, and the Songjiang–Liaohe region of northeast China (Chen and Shi 2005; Zhou and Sun 2006).

2.2 Spatial Dynamics (1995–2007)

To demonstrate the spatial high-tech patterns, three maps have been aggregated in Fig. 1 to compare the distribution of high-tech regions as planned by the

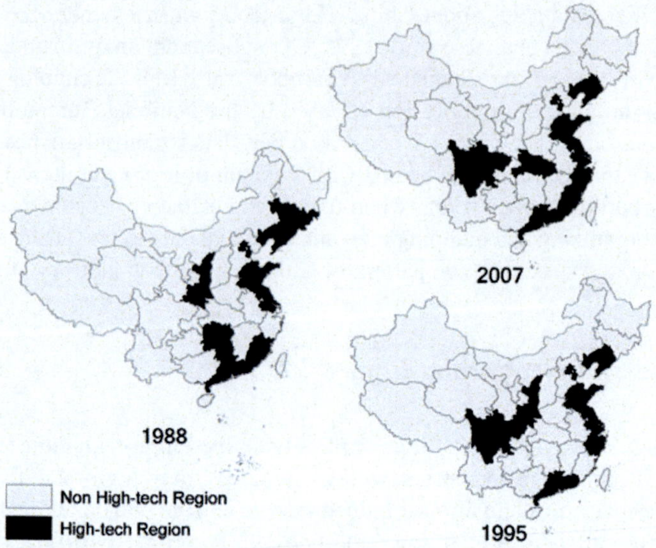

Fig. 1 High-tech regions in China, 1988, 1995 and 2007

government in 1988 and the real concentration of high-tech industry across provinces in 1995 and 2007. As can be seen, after less than a decade of development (1988–1995), discrepancies have already appeared between the reality and the government design. Judging from the gross output value of the high-tech products in 1995,[1] Sichuan has replaced Hunan and turned into China's interior high-tech center along with Shaanxi province; a continuous coastal high-tech band has apparently emerged which should be attributed to the rise of Zhejiang province; on the contrary, high-tech industries hardly thrived in Jilin province, where the whole manufacturing industry plunged due to its historical overdependence on the state economy and thereby the institutional reluctance to reform (Pearson 2007). Unlike the contrast between 1988 and 1995, the high-tech industry gross output value in 2007 reveals that the "interior center" in Sichuan and Shaanxi and the "band" region constituted by coastal provinces are still far ahead of other regions in high-tech production capacity. Therefore, our first impression on the spatial dynamics of China's high-tech industry is that, after a cluster reformation period since 1978 and especially between 1988 and 1995, the spatial distribution of the industry production capacity has concentrated in one interior center and a costal band, and thereafter entered a stage of path dependence, implying stabilization and constant self-reinforcement.

[1]All values in this paper have been deflated into 1996 prices based on a relevant associated price index.

2.3 Agglomeration

After an overview of the spatial dynamics, our research next proceeds by investigating the agglomeration and specialization features of China's high-tech industry in retrospect, while it will contrast the results with experiences from developed countries in pursuit of the uniqueness of the Chinese case. With respect to the identification of agglomerations, two approaches have proven to be effective in the literature: first, employing a variety of direct measures associated with the magnitude of industrial economic activities and mapping them to highlight regional gaps and the agglomeration locus; second, when a mapping technique is unavailable or regional gaps are not visually significant, computed indirect measures such as location quotients are often adopted (Cruz and Teixeira 2009). In this paper, we start with the first approach, and complement this with the second one, if necessary.

In Figs. 2–4, the spatial-temporal dynamics of establishment and comprehensive production capacity in China's high-tech industry in 1995 and 2007 are mapped. As shown in Fig. 2, the spatial distribution of the first tertile provinces in terms of high-tech establishments remains largely the same in 1995 and 2007, where an interior center at Sichuan province and a band constituted by coastal provinces conspicuously outnumber other regions[2]. Meanwhile, as the arithmetic average of high-tech establishment numbers across provinces increased from 1,168 in the year 1995 to 1,503 in the year 2007, the gap between the average of the first tertile provinces and the average of the remaining provinces expanded from 987 in the year 1995 to 1,476 in the year 2007. The existing literature alleges that establishment numbers, which in the Chinese context include incumbents and firms newly founded through foreign direct investment (FDI), domestic entrance and spinoffs, not only provide a

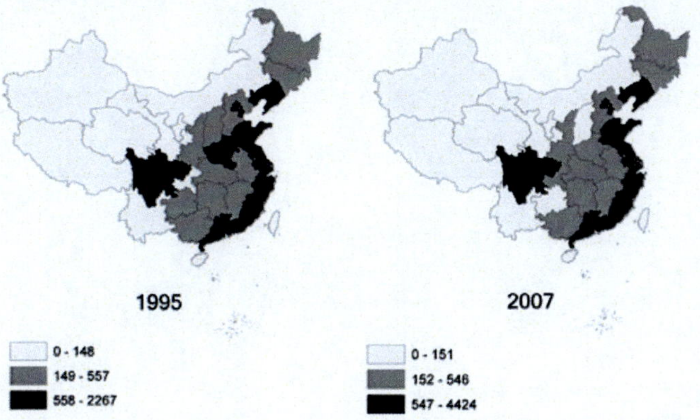

Fig. 2 High-tech establishment numbers, 1995 and 2007

[2]The underlying rationale of grouping all 31 provinces with the three quantiles conforms to the NBS norm of conducting economic-geographic analysis in China (see Demurger et al. 2002).

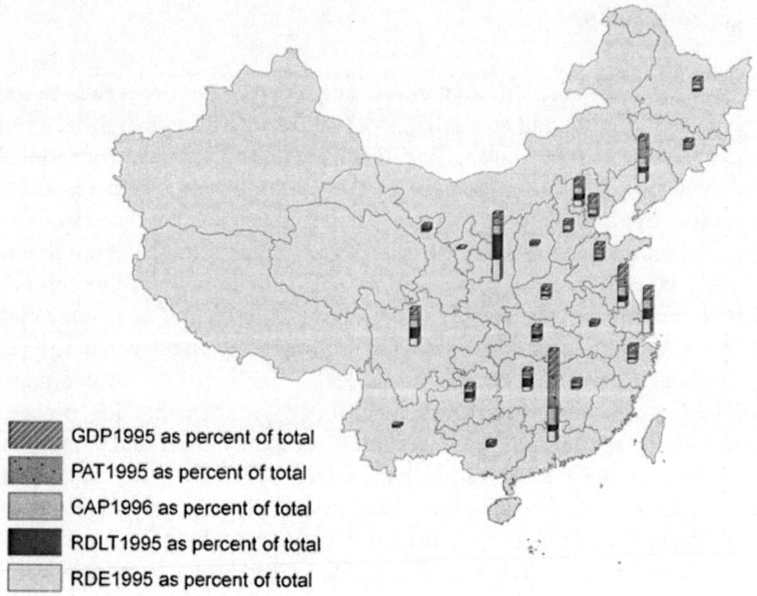

Fig. 3 Regional comprehensive high-tech production capacity, 1995

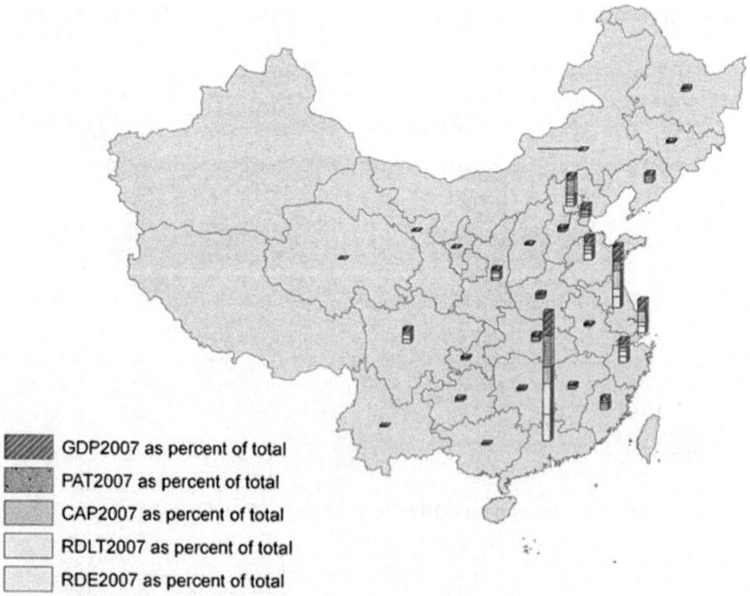

Fig. 4 Regional comprehensive high-tech production capacity, 2007

measure for agglomeration (Lu 2010), but also indicate the potential to agglomerate, since investors and spinoffs are both prone to follow their precedents regarding location choice (Ran et al. 2007; Cheng 2008; Lyons 1995). In light of this observation, it can be argued that, while high-tech firms are persistently concentrated in the previously defined "interior center" and "coastal band", the extent of this agglomeration has been intensified between 1995 and 2007, and this is expected to deepen afterwards as a result of the path-dependency mode in firm location choice.

In addition, the relative comprehensive high-tech production capacity for each province, by measuring their individual proportion in the national total along an array of conventional industrial input–output measures, is mapped in Figs. 3 and 4, respectively, for the year 1995 and 2007. As a supplement to the establishment measure, the term "relative comprehensive production capacity" is defined, which serves to verify the previously identified pattern and tendency in agglomeration for China's high-tech industry from additional perspectives. These specific measures chosen to assess "comprehensive production capacity" include output-oriented measurements, i.e., the gross output value of high-tech products ("Output value"), the number of patents that has been issued ("Patent"), and input-oriented measurements, i.e., the capital stock of high-tech industry ("Physical Capital"), the man–years statistics for skilled labour in research and development (R&D) activities ("RDLT") and the R&D expenditure ("RDE"). The usage and effectiveness of these measures to distinguish industrial agglomeration from the perspective of material production, innovation generation and human capital accumulation have already been discussed and recognized in a vast literature (Sun 2003; Cruz and Teixeira 2009; Cutrini 2010). For parsimony, details explaining the construct of these measures are skipped here; interested readers are referred to Yu and Nijkamp (2009) for methods of deflating monetary measurements and calculating the capital stock in the high-tech industry.

Since it might seem banal to restate the temporal dynamics of China's high-tech industry as a phenomenon of rapid growth in comprehensive production capacity from 1995 to 2007, we pay greater attention to examining the capacity's spatial distribution and evolution. First of all, our preliminary impression on the spatial stability in high-tech industry agglomerations after 1995 is confirmed: no matter the output indicators or input indicators, a coastal high-tech "band" stands alone stretching from Liaoning to Guangdong, while it is accompanied by two neighbouring interior provinces – Sichuang and Shaanxi. Second, as the static spatial distribution of comprehensive production capacity remains, an underlying process of self-reinforcing development has progressed – in 1995, non-high-tech regions that are next to the coastal "band" made modestly minor contributions to the national high-tech industry input and output, shown from their relatively short stack bars in Fig. 3; in 2007, however, the industrial map became increasingly flat beyond the coastal areas in Fig. 4, where even the contributions from Sichuan and Shaanxi shrank drastically. This implies that the coastal regions' dominant position in high-tech production capability has been significantly strengthened, while the gaps between them and those interior regions have been enlarged during 1995–2007.

In view of this, China's high-tech industry resembles its counterparts in developed countries very much with respect to its character of agglomeration. On the one hand, associated industrial activities are highly concentrated in a relatively handful regions, where dense population and a great extent of urbanization quite often exist. On the other hand, once the industrial centers are formed, the strength of these places then becomes locked-in, self-reinforcing movements themselves, while marginalizing competing locations and effectively crowding them out of the field. This observation, therefore, further backs up the argument of Scott and Storper (2007) on the similarities between developing countries and developed countries in agglomeration patterns in knowledge-intensive industries.

2.4 *Specialization*

Another important discovery from the literature regarding the spatial pattern of high-tech industries of developed economies is that high technology varies significantly from place to place (Cortright and Mayer 2001; Maggioni 2002). This more delicate division of labour in high-tech industry has often been interpreted as the result of the increased interdependencies among firms in response to the more differentiated and constantly changing consumer demands on high-tech products (Scott 2006). Taking the US as an example, at the metropolitan statistical area (MSA) level, regions tend to specialize in certain technologies and have major concentrations of firms, employment and patenting in relatively few product categories. A few places, like Silicon Valley, may excel in many areas, while most of the high technology centers usually concentrate in relatively few products or technologies; for instance, Phoenix concentrates in hardware employment with strong innovation performance in electronics and Washington DC exceeds in software and biotechnology innovation. In this section, accordingly, we turn to explore the spatial dynamics of specialization in China's high-tech industry.

First, the provincial gross output value as a percentage of the national total in each sub-industry has been calculated and mapped in Fig. 5 for the year 1995 and in Fig. 6 for the year 2007. Following the categories of high-tech industry defined in the Appendix (Table 1), it can be observed that within the coastal high-tech "band", there exists very little variation in the way that each province contributes to every sub-industry's national output in 1995. Other than Guangdong province, which apparently leads in the production of Electronic and Telecommunications Equipment (EleTel) and the manufacturing of Computers and Office Equipment (ComOff), the rest of the "band" peers appear to tie in their outputs in these two segments. With regard to the Manufacture of Medical and Pharmaceutical Products (MedPha) and the Manufacture of Medical Equipment and Meters (MedMeter), all "band" provinces seem to split their output rather equally. In comparison, the two interior high-tech centers, viz. Sichuan and Shaanxi, account for a minor share in the EleTel

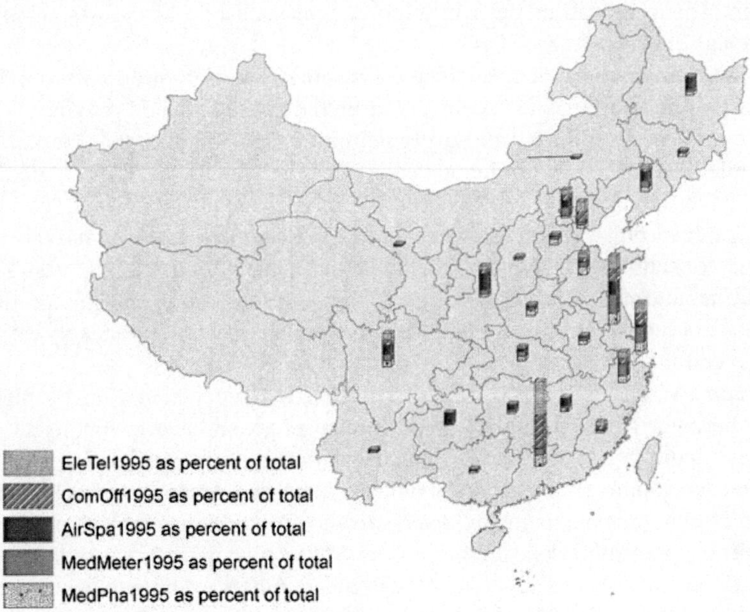

Fig. 5 The composition of high-tech industry output, 1995

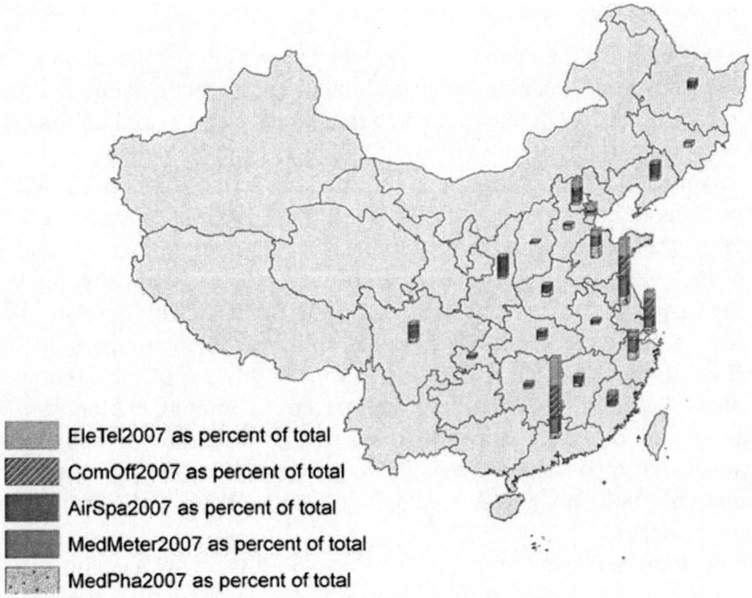

Fig. 6 The composition of high-tech industry output, 2007

and ComOff sub-industries but, yet a more significant role in the Manufacture of Aircraft and Spacecraft (AirSpa).

The preceding pattern of specialization is somewhat more pronounced in Fig. 8. First, the two interior high-tech centers outrun the coastal "band" provinces significantly in the AirSpa industry and overwhelmingly specialize in that market niche. Meanwhile, their previous production capacities in ComOff and EleTel, albeit very limited, have decreased to a trivial level. Second, inside the coastal "band", the formerly spread-out ComOff industry and EleTel industry have remarkably relocated and concentrated in Jiangsu, Shanghai and Guangdong, viz. the vast Yangze River Delta and the Pearl River Delta. However, in contrast, the production capacities in MedMeter industry and MedPha industry remain noticeably dispersed in a balanced manner among the band provinces.

In summary, our perception of the specialization process in China's high-tech industry between 1995 and 2007 is multi-faceted in general and is contingent on the specific sub-industry under concern. Compared to the experience of developed countries and other developing countries, for example, India and Mexico, China's high-tech industry exhibits a more complex spatial feature in specialization which therefore challenges the prevalent "delicate specialization" assertion in high-tech industry development. Concomitantly, it also suggests the existence of impacts from other unidentified or insufficiently recognized factors on the progress of high-tech industry specialization, aside from intensified firm linkages and demand diversification.

3 A Conceptual Framework

In order to account for the empirics in agglomeration and specialization which we previously discovered for China's high-tech industry, a conceptual model will now be constructed in this section to form an explanatory framework and then applied and assessed for the Chinese scenario. As shown from preceding evidence, the spatial dynamics of China's high-tech industry has exhibited both resemblances and distinctions in comparison with the ones of the developed countries – high-tech industry agglomerates in China – like it does in other countries – and is also subject to the rule of path dependency; however, while certain regions become specialized in sub-industries like AirSpa, ComOff and EleTel, most of the high-tech provinces still preserve a nearly equal ability in manufacturing MedMeter and MedPha products, which remarkably contrasts in view of the experiences of other countries. Acknowledging these, we attempt to build an exploratory framework that would not only draw on the existing literature in explaining the resemblances, but also augment or modify prevailing theories by incorporating those underlying unique Chinese elements that account for distinctions from the general experience.

Agglomeration and specialization have long been taken as accompanying one another and are widely believed to be driven by factors from two opposite ends – exogenous factors such as the location advantage, the endowment in scarce but necessary resources and the benefits from historical incidents, and endogenous factors, for example, the dynamic backward and forward linkage of firms in industrial systems, the formation of a rich local labour pool and the emergence of a local milieu conducive to spillover and knowledge transfer (Marshall 1890; Scott and Storper 2007; Cutrini 2010). Notwithstanding the numerous efforts of employing these two sets of factors in analyzing industrial spatial dynamics, a generic structure and operational procedures to associate these factors effectively with the constant variations in the industrial landscape are rarely seen. This is less problematic in the context of developed countries where the free market runs smoothly for most of the time with relatively less exogenous interference; hence the research focus of relevant studies could substantially concentrate on endogenous factors while controlling exogenous effects (Capello 1999). In contrast, when exogenous interferences become usual, for instance, as a result of frequent government intervention, the interactions between exogenous factors and endogenous factors and their direct and indirect effects on agglomeration and specialization would quickly make the analysis unmanageable without a clear conceptual structure and associated operation procedures. An explanatory analysis on the spatial dynamics of China's high-tech industry, therefore, would necessarily have be framed according to such a cognitive map, because of the significant role that government intervention has played in stimulating the formation of industry clusters by integrating resources and providing incentives and encouragement (Pearson 2007; Young 2000; Chen and Shi 2005).

The key features of the explanatory framework we proposed are outlined in Fig. 7. The driving forces of the formation and change in industrial spatial patterns

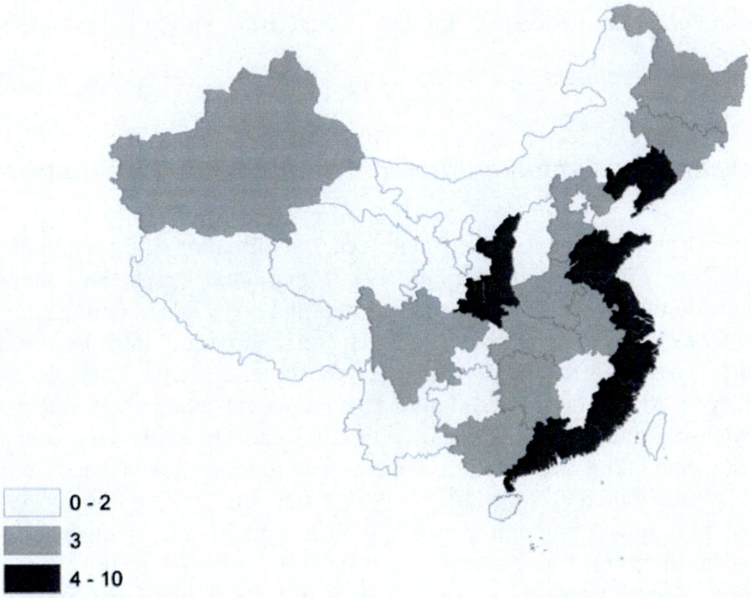

Fig. 7 Total numbers of HTIPs and EDZs, 1993

are generalized to comprise the exogenous factors and the endogenous factors, as is suggested by the vast amount of existing literature. Nevertheless, the occurrence of exogenous factors, specifically in the form of policies and regulations, is speculated to assume a recursive relationship with the endogenous factors in China, which derives from persistent government supervision and intervention (Young 2000). In particular, this speculation accentuates the reality that new industrial policies there are often initiated to confront the undesirable outcomes resulting from the interactions between the market and earlier interventions. Theoretically, this speculation can be also applied to free market economies, whereas government interventions are strictly restrained and thereby exogenous factors are found to include external shocks or structural changes such as post-war reconstruction, industrial revolution and economic downturn.

The interference of exogenous factors, in many occasions, can exert an enormous and immediate influence on resource allocation and may generate a significant "level effect" on the extent of agglomeration and specialization in certain industries and places. However, the level effect will not last to sustain the status of agglomeration and specialization, unless endogenous factors could create a "growth effect" on the degree of agglomeration and specialization in order to reinforce the "first-mover" advantage (Scott and Storper 2007; Capello 1999). As mentioned, the discussion on the level effect of exogenous factors and its impact on the initial status of agglomeration and specialization are often understated in the literature rooted in the experience of developed countries. Regarding the Chinese high-tech industry, in contrast, its spatial dynamics in terms of agglomeration and specialization is understood here to be composed by radical status changes ("Status") and incremental evolution ("Evolution"), and reflects the compound (level and growth) effects from policy interventions and market forces.

3.1 Explaining Agglomeration in China's High-Tech Industry

In view of our proposed conceptual framework, the formation of the agglomeration pattern of China's high-tech industry in 1995, from a static perspective, should first be attributed to the level effect of the most important government policy relating to technology development, viz. the "High-tech R&D Program" (also known as the "863 Plan" – named after the date of its establishment, March 1986) which was endorsed by the State Council and signified a government-directed effort to develop China's high-tech industry (Sutherland 2005; Chen and Shi 2005; Yu and Nijkamp 2008). The interaction between the government supports and China's fledging high-tech product market soon after revealed that the lack of institutions that would support new- and high-technology firm formation and mechanisms of technological diffusion had severely jeopardized the prospect of the program. In response, the "Torch Program" was put in place in 1988 with the main objective to deliver high-tech products, establish technology-oriented enterprises, and pave the way for the commercialization of innovations that would come out of major national science and technology programs. A major ingredient of the Torch Program was the establishment of High-tech Industry Parks (HTIPs), where most of the new- and high-technology commercialization efforts were expected to take place and where such efforts were to receive various forms of government subsidies. In March 1991, the State Council approved the establishment of 27 national level HTIPs, followed by yet another 25 in the following year and only 1 additional after 1993. These HTIPs together with the national level Economic Development Zones (EDZs), which are all deliberately designed to be equipped with superior infrastructure, R&D funding priority, highly preferential tax rates and streamlined bureaucracy – an array of exogenous factors that opens the "window of locational opportunity" in developing of high-tech industries (Scott and Storper 2007) – immediately became the engine of high-tech industry development across China. As shown in Fig. 8, the spatial distribution of all the HTIPs and EDZs in 1993 clearly resembles the agglomeration pattern of China's high-tech industry development revealed from Figs. 2–4. Therefore, government policies and their level effects appear to have exerted a structural impact on the geography of high-tech industry clusters.

Meanwhile, endogenous factors represented by the market force also matter to a great deal and especially in terms of the self-reinforcing process among agglomerations. First, since the very beginning, the design of the aforementioned government high-tech programs and particularly their location choices for HTIPs and EDZs are strictly subject to estimations of the candidate regions' indigenous capability to provide supporting industrial linkages, skilled labor forces and synergies between research institutions and industries (Hu 2007; Sutherland 2005; Yu et al. 2009). This in part demonstrates how market force is restraining the range and scale of government intervention in the development of China's high-tech industry. Second, as is illustrated in Sect. 2, the concentration of the comprehensive production capacity toward the coastal "band" regions has prominently characterized the agglomeration dynamics in the high-tech industry of China between 1995 and 2007. This process can be largely explained by the geographically focused growth effects that come into being in those regions – the

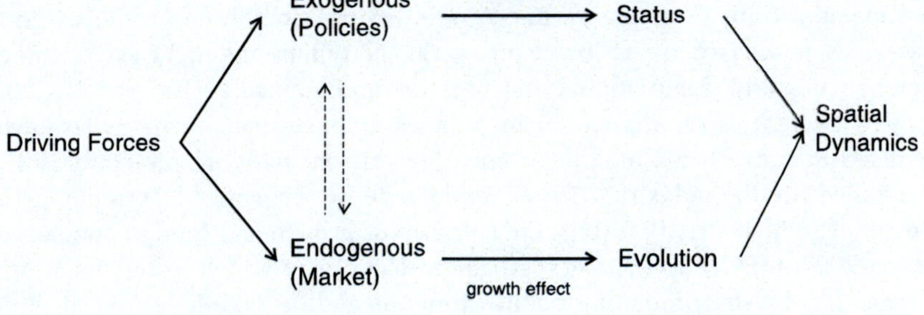

Fig. 8 Factors and processes shaping spatial dynamics

initially dispersed firms and production capacities are increasingly attracted to and absorbed by relatively few clusters that are equipped with conductive indigenous factors to achieve increasing returns effects. And these places begin to move ahead as their self-reinforcing concentrations of capital and labour make them progressively more efficient and competitive.

The most intriguing part of the agglomeration dynamics in China's high-tech industry is that it is not only incrementally reinforced by the growth effects of market forces, but also radically upgraded by new policy intervention's level effects – the rapid economic growth in China is well-known for its reliance on export (Gilboy 2004; Liu 2002); however, while traditional manufacturing sectors gradually began to experience declining terms of trade, the high-technology sectors were increasingly seen as the most dynamic areas of global demand. As a result the Chinese government was pressured to undertake a structural update of its export by moving away from labor-intensive low value-added manufactures. To gear its export trade towards the high-tech sector, therefore, HTIPs were intervened to accommodate high-tech products' export processing rather than R&D and to depend on FDI whose production capability is already in place rather than those domestic startups (Ning 2007; Story 2005). In early 2000, the Ministry of Science and Technology and Ministry of Foreign Trade approved 16 of the 53 national level HTIPs as a trial group of high-tech export bases. As a consequence of this revised national emphasis on increasing high-tech products export by engaging foreign producers, the expansion of comprehensive production capability in the "band" cluster has been remarkably accelerated because of its location advantage in attracting FDI. Next, success breeds success – the promotion of high-tech products export has exemplified the location advantage of coastal provinces in terms of attracting implanted foreign production capacities; and the level effects of this policy would be locked-in and create more favorable endogenous factors for both foreign and domestic firms to concentrate more densely in the "band".

3.2 *Explaining Specialization in China's High-Tech Industry*

Specialization is believed to happen only if the relevant regional and national governments refrain from policy intervention in the uninhibited trade of goods and services, in the freedom to locate production activities, and in the establishment of inter-regional or international risk-sharing arrangements (Bai et al. 2008). However, two types of protectionism policies against specialization have been recognized to prevail in China ever since the reform: national protectionism on the so-called lifeline industries, for example, defense, energy, telecommunication and banking, which strictly forbids the entrance of private and foreign competitors (Pearson 2007; OECD 2002), and regional protectionism which fosters local firms and industries by discriminating competitors outside the region (Bai et al. 2008; Holz 2009; Young 2000). Accordingly, the firm victims of protectionism can be distinguished by their ownership – domestic non-state-owned enterprises and foreign enterprises are normally discouraged by national protectionism policy, while only domestic non-state-owned enterprises suffer from regional protectionism policy, since foreign and state-owned enterprises are independently regulated by state agencies (OECD 2002).

Next, we will attempt to explain the specialization dynamics of China's high-tech industry by incorporating protectionism policies as exogenous factors in the explanatory framework. First, in Fig. 5, the interior centers constituted by Sichuan and Shaanxi are found to be specialized in AirSpa. As a sub-industry with high relevancy to national security, AirSpa has been completely operated by state-owned enterprises which historically sit in the hinterland. Due to the political geographical concern on safety, new investment in this sub-industry will still be largely directed to Sichuan and Shaanxi as long as it is state-owned. Meanwhile, the comparative advantages in endogenous factors, as we discussed previously, can to a large extent explain the specialization of the "band" regions who capitalized on their HTIPs and EDZs to attract emerging foreign and non-state-owned enterprises in civil sub-industries. Therefore, the separation line of specialization between the interior centers and the "band" regions in 1995 seems to be primarily drawn by the national protectionism policy and the market force.

In Fig. 6, the spatial dynamics of specialization from 1995 to 2007 is seen to comprise two notable changes (1) the interior centers and the "band" regions are increasingly specialized in AirSpa and the remaining civil sub-industries, respectively; (2) within the "band" regions, places like Jiangsu, Shanghai and Guangdong are evidently specialized in ComOff and EleTel industries, whereas a few provinces in fact concentrate in MedMeter and MedPha industries. The first change is fairly easy to understand when placed in our explanatory framework considering that market forces will reinforce the specialization status formed in 1995 and thus lead to a more explicit division of labour. In addition, the combined facts that foreign invested firms have dominated the production in China's high-tech industry while they are still excluded from the AirSpa business also suggest that they can only specialize in other civil sub-industries while locating the superior production capacities in their favorable coastal regions. In view of this, the increased segregation between the interior centers and the coastal "band" can be most effectively accounted for by market forces, continued national protection on AirSpa, and the soaring FDI in the high-tech industry after China has redirected the industry toward exports since the late 1990s.

The interpretation of the second change, however, is complicated by the recursive relationship between market forces and multiple policy interventions besides national and regional protectionism. First, in open and competitive industries like ComOff and EleTel, the market force determines the dynamics of specialization – local governments are motivated to match other provinces' preferential policies to attract foreign producers, who dominate the Chinese market because of their advantages in technology development and thus can substantially boost the local economy by creating jobs and tax revenues (Fung et al. 2004; Ran et al. 2007). As a consequence, foreign companies are enabled to be rather footloose in terms of acquiring a favorable policy environment across the country. This eventually allows their location choices to follow the economic rationale to a larger extent and facilitates the agglomeration of the ComOff and EleTel industry in the most competitive provinces (Wei et al. 2009; Yang 2009). In contrast, the nominally open market in the MedPha and MedMeter industries have been mostly turned into a battlefield for domestic firms, both state-owned and non-state-owned, because foreign enterprises in these sub-industries have exceedingly high concerns on intellectual property protection and are very cautious in market entry. As highly profitable sub-industries, local governments in China were motivated to develop and protect their own MedPha and MedMeter enterprises and thereby to generate more fiscal revenues by taxing the industry. The prevalence of local protectionism, therefore, continues to create fragmented industries by hampering an industry consolidation and conducting overlapping investment (OECD 2002; Young 2000; Yu and Nijkamp 2008).

In an attempt to further demonstrate the impact of the aforementioned local protectionism on the spatial dynamics of China's high-tech industries, we apply a shift-share analysis on 28 municipalities and provinces (Tibet, Qinghai and Ningxia are excluded due to severe missing value problems in their official statistics) for the period 1995–2002 and 2002–2008.[3] The standard formula was used (Polese and Shearmur 2006), which is considered adequate for the needs of this study. MedPha and EleTel are subjected to the shift-share analysis which represent the sub-industries affected by regional protectionism and vice versa.

Unlike a location quotient analysis which compares static distributions of production capacity, shift-share analysis looks at a dynamic change. In particular, we take its regional component as follows: as a measure of the relative GDP growth (or decline) of a given industry in a geographic analytical unit, compared to the national performance. As shown in Tables 2 and 3 in the Appendix, variations in the regional component across different municipalities and provinces are increasing both in the MedPha industry and in the EleTel industry, as is indicated by a comparison of their standard deviation between 1995–2002 and 2002–2008. However, when compared to the MedPha industry, the EleTel industry's variation in regional component is consistently and significantly larger, which highlights the divergence among regions in developing the EleTel industry – provinces like Jiangsu, Shandong and Guangdong have outstandingly high regional shares in the EleTel industry development, while others are mostly losing their regional components. The MedPha industry witnesses the same tendency of divergence, nevertheless in a much more moderate manner, which could be attributed to local governments' strict restrictions on cross-region business activities.

[3]The year 2002 has been chosen to separate these two periods, because China became a member of the World Trade Organization (WTO) since 2001, after which the magnitude of FDI in China has experienced a remarkable surge.

4 Conclusion

The spatial pattern of China's high-tech industry and its dynamics during the last two decades are a less studied theme compared to the industry's well known story of an economic miracle based on rapid growth. By constructing and mapping relevant numerical measures based on recently released official statistics, the general spatial trends and unique locational characters of China's high-tech industry are for the first time identified and highlighted in this research. They are particularly observed from the perspective of agglomeration and specialization. We then seek for explanations of these spatial dynamics by developing a generic conceptual model that attempts to take account of the variable impact of policy intervention and its interaction with market factors. This conceptual framework is later validated by tapping into a pool of historical, industrial and economic development policies which, in combination with indigenous market forces, have advanced our understanding of the multi-faceted spatial dynamics of China high-tech industry. For policy makers, the results of our analysis suggest that adjustments in China's economic growth strategy, which used to depend highly on export, the removal of regional market barriers and the abandonment of industry protection, may significantly affect the spatial pattern of China high-tech industry in the future.

Appendix

Table 1 The codes for high-tech industries in China's "industrial classification for national economic activities" (ICNEA)

	ICNEA GB/T4754-2002	ISIC revision 3
Manufacture of medical and pharmaceutical products	27	2423
Chemical pharmaceutical products	2710&2720	
Processing of traditional Chinese medicine	2730&2740	
Biology products	2750	
Manufacture of aircraft and spacecraft	376	353
Manufacture and repair of aircraft	3761	
Manufacture of spacecraft	3762	
Electronic and telecommunications equipment	40	32
Telecommunication equipment	401	
Telecommunication transmission unit	4011	
Telecommunication exchange unit	4012	
Telecommunication terminal unit	4013	
Radar and peripheral equipment	4020	
Broadcast and television equipment	403	
Electronic apparatus	405	
Electronic vacuum apparatus	4051	
Semiconductor separated parts	4052	
Integrated circuits	4053	
Electronic components	406	
Household audiovisual equipment	407	

(*continued*)

Table 1 (continued)

	ICNEA GB/T4754-2002	ISIC revision 3
Other electronic equipment	4090	
Manufacture of computers and office equipment	404&415	30
Computers	4041	
Peripheral equipment of computer	4043	
Office equipment	415	
Manufacture of medical equipment and meters	368&411	33
Medical equipment and instruments	368	
Instruments and meters	411	

Note: The terminology for each respective industry in accordance to the ISIC code above is: pharmaceuticals-2423; aircraft and spacecraft-353; radio, TV and communication equipment-32; office and computing machinery-30; medical, precision and optical equipment-33
Sources: OECD (2004, 2006)

Table 2 Shift-share analysis results for the MedPha industry

	1995–2002 MedPha			2002–2008 MedPha		
Province	National component	Industrial component	Regional component	National component	Industrial component	Regional component
Beijing	62.72	(27.97)	36.66	185.62	(38.67)	(31.65)
Tianjin	81.27	(36.24)	16.49	178.80	(37.25)	(54.08)
Hebei	165.09	(73.61)	27.02	350.33	(72.99)	(156.51)
Shanxi	37.44	(16.69)	1.40	70.04	(14.59)	(28.27)
Inner Mongolia	18.64	(8.31)	1.01	35.49	(7.40)	32.88
Liaoning	130.27	(58.09)	(49.93)	134.69	(28.06)	45.81
Jilin	96.47	(43.01)	20.53	214.15	(44.62)	3.54
Heilongjiang	87.62	(39.07)	13.79	184.84	(38.51)	(106.20)
Shanghai	232.49	(103.67)	(49.44)	319.30	(66.53)	(167.76)
Jiangsu	293.21	(130.74)	(7.16)	512.48	(106.77)	29.33
Zhejiang	168.09	(74.95)	79.31	459.77	(95.79)	(26.03)
Anhui	93.43	(41.66)	(49.55)	69.25	(14.43)	23.04
Fujian	50.98	(22.73)	(10.92)	69.85	(14.55)	(1.39)
Jiangxi	82.51	(36.79)	(15.08)	118.24	(24.63)	83.97
Shandong	206.83	(92.22)	(10.25)	351.16	(73.16)	484.52
Henan	158.48	(70.66)	(60.25)	164.82	(34.34)	191.21
Hubei	117.94	(52.59)	25.89	263.39	(54.88)	(102.91)
Hunan	49.50	(22.07)	(3.30)	82.35	(17.16)	84.65
Guangdong	293.89	(131.04)	(33.50)	461.30	(96.11)	(167.20)
Guangxi	63.16	(28.16)	(9.69)	94.19	(19.63)	(25.08)
Hainan	18.76	(8.37)	11.40	56.38	(11.75)	(39.47)
Chongqing	51.82	(23.10)	(0.77)	91.55	(19.07)	(1.81)
Sichuan	113.59	(50.65)	5.88	215.76	(44.95)	79.28
Guizhou	23.28	(10.38)	25.19	91.94	(19.16)	(17.14)
Yunnan	25.52	(11.38)	17.34	80.34	(16.74)	(21.25)
Shaanxi	77.59	(34.59)	19.78	178.74	(37.24)	(88.40)
Gansu	17.11	(7.63)	3.91	38.52	(8.03)	(16.48)
Xinjiang	11.56	(5.16)	(5.79)	9.26	(1.93)	(6.58)
Standard deviation	79.87	35.61	29.95	139.49	29.06	124.03

Note: Negative values are in parentheses
Source: Authors' calculation

Table 3 Shift-share analysis results for the EleTel industry

	1995–2002 EleTel			2002–2008 EleTel		
Province	National component	Industrial component	Regional component	National component	Industrial component	Regional component
Beijing	311.35	(4.65)	223.22	1,270.39	(119.08)	(87.74)
Tianjin	603.85	(9.01)	12.33	1,627.08	(152.52)	(1,010.71)
Hebei	49.44	(0.74)	(14.39)	102.59	(9.62)	(43.76)
Shanxi	7.49	(0.11)	0.29	20.45	(1.92)	52.30
Inner Mongolia	18.29	(0.27)	8.93	66.31	(6.22)	(13.19)
Liaoning	154.20	(2.30)	28.52	465.96	(43.68)	(225.18)
Jilin	20.30	(0.30)	(3.20)	47.51	(4.45)	(32.74)
Heilongjiang	20.68	(0.31)	(21.77)	11.56	(1.08)	(8.56)
Shanghai	669.90	(10.00)	12.77	1,803.24	(169.03)	(657.60)
Jiangsu	773.15	(11.54)	(32.47)	1,987.24	(186.28)	2,295.02
Zhejiang	354.69	(5.29)	(4.50)	932.33	(87.40)	(336.06)
Anhui	45.73	(0.68)	(7.43)	106.57	(9.99)	(32.14)
Fujian	229.49	(3.42)	61.35	731.10	(68.53)	(360.75)
Jiangxi	48.20	(0.72)	(41.67)	45.01	(4.22)	47.85
Shandong	254.74	(3.80)	56.06	787.58	(73.83)	763.19
Henan	91.01	(1.36)	(56.01)	130.10	(12.19)	(60.18)
Hubei	66.49	(0.99)	21.04	218.34	(20.47)	(137.46)
Hunan	33.04	(0.49)	18.45	124.40	(11.66)	(76.49)
Guangdong	2,011.30	(30.01)	153.45	5,643.03	(528.97)	343.28
Guangxi	33.22	(0.50)	(31.72)	25.05	(2.35)	35.33
Hainan	2.71	(0.04)	(1.95)	3.33	(0.31)	1.57
Chongqing	13.48	(0.20)	(6.33)	23.18	(2.17)	26.68
Sichuan	343.21	(5.12)	(178.63)	555.45	(52.07)	(206.90)
Guizhou	23.87	(0.36)	(10.79)	41.88	(3.93)	(30.44)
Yunnan	9.85	(0.15)	(11.44)	3.39	(0.32)	(0.19)
Shaanxi	225.50	(3.36)	(165.62)	268.94	(25.21)	(219.48)
Gansu	17.82	(0.27)	(6.96)	33.43	(3.13)	(35.27)
Xinjiang	1.21	(0.02)	(1.54)	0.15	(0.01)	9.61
Standard deviation	409.83	6.16	74.45	1148.75	107.68	539.40

Note: Negative values are in parentheses
Source: Authors' calculation

References

Audretsch DB, Feldman M (1996) R&D spillovers and the geography of innovation and production. Am Econ Rev 86:630–640

Bai C, Tao Z, Tong Y (2008) Bureaucratic integration and regional specialization in China. China Econ Rev 19:308–319

Capello R (1999) Spatial transfer of knowledge in high technology Milieux: learning versus collective learning processes. Reg Stud 33:353–365

Capello R (2009) Spatial spillovers and regional growth: a cognitive approach. Eur Plann Stud 17:639–658

Chen J, Shi H (2005) High-tech industries in China. Edward Elgar, Cheltenham, UK

Cheng S (2008) How can Western China attract FDI? A case of Japanese investment. Ann Reg Sci 42:357–374

Cortright J, Mayer H (2001) High tech specialization: a comparison of high technology centers. The Brookings Institution Survey Papers, pp 1–18
Cruz S, Teixeira A (2009) The evolution of the cluster literature: shedding light on the regional science-regional studies debate. Regional Studies, iFirst, pp 1–26
Cutrini E (2010) Specialization and concentration from a twofold geographical perspective: evidence from Europe. Reg Stud 44:315–336
Demurger S, Sachs J, Woo W, Bao S, Chang G, Mellinger A (2002) Geography, economic policy, and regional development in China. Asian Econ Paper 1:146–197
DeVol R, Wong P, Catapano J, Robitshek G, Institute M (1999) America's high-tech economy: growth, development, and risks for metropolitan areas. Milken Institute Santa Monica, CA
Feldman M, Audretsch D (1999) Innovation in cities: science-based diversity, specialization and localized competition. Eur Econ Rev 43:409–429
Florida RL (2005) Cities and the creative class. Routledge, New York
Fung K, Iizaka H, Tong S (2004) Foreign direct investment in China: policy, recent trend and impact. Global Econ Rev 33:99–130
Gilboy G (2004) The Myth behind China's miracle. Foreign Aff 83:33–48
Goetz S, Rupasingha A (2002) High-tech firm clustering: implications for rural areas. Am J Agric Econ 84:1229–1236
Holz C (2009) No Razor's edge: reexamining Alwyn Young's evidence for increasing interprovincial trade barriers in China. Rev Econ Stat 91:599–616
Hu G (2007) Technology parks and regional economic growth performance in China. Res Policy 36:76–87
Jefferson G (2005) R&D and innovation in China: has China begun its S&T takeoff? Harv China Rev 5:44–50
Keeble D, Wilkinson F (1999) Collective learning and knowledge development in the evolution of regional clusters of high technology SMEs in Europe. Reg Stud 33:295–303
Keeble D, Wilkinson F (2000) High-technology clusters, networking and collective learning in Europe. Ashgate, Aldershot
Liu Z (2002) Foreign direct investment and technology spillover: evidence from China. J Comp Econ 30:579–602
Liu X, Buck T (2007) Innovation performance and channels for international technology spillovers: evidence from Chinese high-tech industries. Res Policy 36:355–366
Lu J (2010) Agglomeration of economic activities in China: evidence from establishment censuses. Reg Stud 44:281–297
Lyons D (1995) Agglomeration economies among high technology firms in advanced production areas: the case of Denver/Boulder. Reg Stud 29:265–278
Maggioni M (2002) Clustering dynamics and the location of high-tech firms. Physica, New York
Marshall A (1890) Principles of economics. Macmillan, London
Ning L (2007) Economic liberalization for high-tech industry development? Lessons from China's response in developing the ICT manufacturing sector compared with the strategies of Korea and Taiwan. J Dev Stud 43:562–587
Norton R (2000) The geography of the new economy. The web book of regional science. Regional Research Institute, West Virginia University, Morgantown
OECD (2002) China in the world economy the domestic policy challenges. OECD, Paris, pp 125–193
OECD (2004) Chinese standard of industrial classification and its relationship to ISIC. Research of methodological issues in national accounts – China. OECD, Paris
OECD (2006) OECD science, technology and industry outlook 2006. OECD, Paris, pp 9–199
OECD (2007) Science, technology and innovation indicators in a changing world: responding to policy needs. OECD, Paris, pp 3–305
OECD (2008) OECD science, technology and industry outlook 2008. OECD, Paris, pp 168–169
Pearson MM (2007) Governing the Chinese economy: regulatory reform in the service of the state. Public Adm Rev 67:718–730

Polese M, Shearmur R (2006) Growth and location of economic activity: the spatial dynamics of industries in Canada 1971–2001. Growth Change 37:362–395
Ran Z, Voon J, Li G (2007) How does FDI affect China? Evidence from industries and provinces. J Comp Econ 35:774–799
Scott AJ (2006) Creative cities: conceptual issues and policy questions. J Urban Aff 28:1–17
Scott A, Storper M (2007) Regions, globalization, development. Reg Stud 41:191–205
Sigurdson J (2004) Technological superpower China. R&D Manag 34:345–347
Simon DF (1989) China's drive to close the technological gap: S&T reform and the imperative to catch up. China Q 119:598–630
Story J (2005) China: workshop of the world? J Chin Econ Bus Stud 3:95–109
Sun Y (2003) Geographic patterns of industrial innovation in China during the 1990s. Tijdschrift voor economische en sociale geografie 94:376–389
Sutherland D (2005) China's science parks: production bases or a tool for institutional reform? Asia Pac Bus Rev 11:83–104
Sutherland D (2007) China's 'national team' of enterprise groups: how has it performed? China Policy Institute's Discussion Paper
Wei Y, Lu Y, Chen W (2009) Globalizing regional development in Sunan, China: does Suzhou industrial park fit a Neo-Marshallian District model? Reg Stud 43:409–427
Yang C (2009) Strategic coupling of regional development in global production networks: redistribution of Taiwanese personal computer investment from the Pearl River Delta to the Yangtze River Delta, China. Reg Stud 43:385–407
Young A (2000) The Razor's edge: distortions and incremental reform in the people's republic of China. Q J Econ 115:1091–1135
Yu J, Nijkamp P (2008) China's prospects as an innovative country: an industrial economics perspective. Int J Appl Econ Econometrics 16:160–187
Yu J, Nijkamp P (2009) Ownership, R&D and productivity change: assessing the catch-up in China's high-tech industries. Stud Reg Sci 39:149–173
Yu J, Stough RR, Nijkamp P (2009) Governing technological entrepreneurship in China and the west. Public Adm Rev 69:s95–s100
Zhou Y, Sun Y (2006) Introduction: mediating state and market in China's competitive high technology industry. China Rev 6:1–7

Regional Psychological Capital and Its Impact on Regional Entrepreneurship in Urban Areas of the US

Ryan C. Sutter and Roger R. Stough

1 Introduction

Investigation of the determinants of entrepreneurship has been extensively pursued by academics in many disciplines. Two dominant but alternative approaches to understanding these determinants have been used. The first systematically studies individual entrepreneurs in an effort to identify characteristics or tendencies common to successful entrepreneurs. The second examines the role of various hypothesized structural attributes and conditions across regions in an effort to explain variation in levels of entrepreneurial activities. Both approaches reveal considerable insight into how and why entrepreneurial activities emerge, yet much remains unanswered as no study exists to confirm or refute the evidence pertaining to the manifestation of the psychological characteristics found at the individual-level at the meso or regional level.

The literature focusing on individuals tends to emphasize characteristics that are common to successful and/or nascent entrepreneurs. Many studies have found that characteristics such as: social networks, (Saxenian 1999; Sorenson 2003; Johannisson 1988; Larson 1991) work experience, educational attainment (Evans and Leighton 1990) and family history (Dunn and Holtz-Eakin 2000; Blanchflower and Oswald 1990) have an important influence on which and how many individuals engage in entrepreneurship. As well, many studies of entrepreneurs suggest that individual differences in human psychology play a large role in an individual's tendency to engage in entrepreneurship as discussed below.

Risk aversion is one psychological characteristic that has been widely studied. This body of literature consistently finds that entrepreneurs are risk prone (Kilstrom and Laffont 1979; Brockhaus 1980; Van Pragg and Cramer 2001). The basic supporting conceptual or theoretical argument is that risk prone individuals are more willing to bear the burden of "Knighterian" uncertainty, a fundamental characteristic of those who engage in entrepreneurship. Other individual-level

R.C. Sutter (✉) and R.R. Stough
School of Public Policy, George Mason University, Fairfax, VA 22030-4444, USA
e-mail: sutter.ryan@dol.gov; rstough@gmu.edu

K. Kourtit et al. (eds.), *Drivers of Innovation, Entrepreneurship and Regional Dynamics*, Advances in Spatial Science, DOI 10.1007/978-3-642-17940-2_12,

studies suggest that self-efficacy, or the belief in ones ability, plays a large role in an individual's willingness to engage in entrepreneurial activities (Markman et al. 2002). Still, other authors suggest that an individual's need for achievement and tolerance for ambiguity are characteristics of entrepreneurial individuals (McClelland 1961; Schere 1982).

The regional or meso level approach to entrepreneurship research necessarily leaves aside individual or micro level characteristics, focusing instead on regional-level structural factors that drive entrepreneurial activities. This body of literature argues that opportunities are not homogeneously distributed across space and, as a result, structural differences, and not individual differences, are driving regional variation in rates of entrepreneurship. This approach has focused primarily on factors such as: transport costs, human capital concentrations, employment characteristics, industrial structures, research and development activities (new knowledge creation), diversity and financial capital availability (see, e.g., Bartik 1989; Reynolds et al. 1994; Dunn and Holtz-Eakin 2000; Acs et al. 2009). As well, others have argued that structural factors like population (or employment) and income growth are important determinants of entrepreneurship (Acs and Armington 2002).

More recently, regional-level studies have begun to tilt toward an emphasis on pseudo-psychological factors,[1] which essentially underlie regional stocks of social capital (trust), creativity and tolerance. Coleman (1988, 1990) and Putnam (1993), among others, have argued that regional stocks of social capital make possible trust and cooperation among regional agents, which the individual-level studies have found to be important ingredients for entrepreneurship (see, e.g., Brusco 1982; Piore and Sable 1986; Saxenian 1999). Other psychological factors, such as, creativity and tolerance have crept into regional analyses of entrepreneurial activities via occupation-based indicators. The authors of these studies have suggested that agglomerations of creative individuals along with a tolerant and open regional culture exert significant positive impacts on levels of entrepreneurial activities (Lee et al. 2004; Florida 2002; Mellander and Florida 2006).

The basic arguments underlying the importance of these factors are that higher levels of social capital make entrepreneurship relevant social networks denser and thus strengthen the links between increasingly larger numbers of nodes, while creative and tolerant regional cultures facilitate entrepreneurship by advancing lower barriers to entry, thus making these regions more open to new or radical ideas and innovations. To a lesser extent, this latter body of literature has argued that "open" and "creative" environments contain larger stocks of individuals which are more apt to both demand and adopt new and innovative types of products with the implication being that such regional attributes attract these sorts of people. The probability for entrepreneurial activity, then, is higher in these types of places

[1]This is not to say that this is research into the regional expression of individual psychological attributes but to indicate that this line of research implies that there may be an underlying interrelated complex of psychological variables at work. If so, then one might argue that there is a sort of "psychological capital" at work and that this may underlie variables like regional social capital, creativity and trust.

because there is a larger willingness to produce and a larger ability to consume radical types of products. Essentially, there exist better markets for the types of products and processes often introduced by entrepreneurs in these places.

It is likely that positive reinforcement is an important attribute of these high social capital environments. The role theory model of learning is well validated and shows that it is powerfully associated with high performing group (classroom) learning. High social capital environments (regions) with large numbers of creative and talented people provide a context where there are more entrepreneurial role models and that these likely enable higher levels of learning about the entrepreneurial role among new comers and residents of such regions. This is important because it offers a theory about how high level presence of certain psychological types influences the behavior of others and positively reinforces increasing entrepreneurial behavior.

While the individual and regional level studies approach entrepreneurship from different perspectives, both seek to explain variation in entrepreneurial activities. The individual-level approach has found considerable evidence suggesting that psychological characteristics are important determinants of entrepreneurial behavior. The regional-level approach has yet to fully grasp the issue of psychological characteristics and their possible relevance. At the same time it has provided evidence that some social psychological-like characteristics, such as: social capital, creativity and tolerance, have important influences on entrepreneurial activities.

What these two approaches seem, then, to have in common is that certain psychological characteristics may underlie the regional level or intensity of entrepreneurial activity. This is especially interesting because an emerging field of psychological research, denoted "positive psychology" in the psychology literature, has begun to clamor that positive aspects of human psychology need to be better understood and integrated into the field's scholarly pursuits. This is relevant to scholars of entrepreneurship because the recent evidence suggests that pseudo-psychological characteristics are important ingredients of the process of entrepreneurship and that these characteristics are positively underlain by various but specific psychological characteristics. This brings about a new and interesting question relevant to entrepreneurship research. Does variation in positive psychological environments exist at the regional level, and if so, does this variation influence regional levels of entrepreneurship?

From here on in the paper we will refer to positive psychological environments simply as psychological environments with the understanding that the "positive" is implied. The reason for this is to ensure we do not confound the notion of psychology aspiring to more relevance, i.e., facilitating human and organization adaptation and change with the notion that such environments are necessarily positive. It is not difficult to imagine that a region may have a psychological profile or character that is negative or at least noticeably "less" positive, e.g., in declining regional economic environments that have hollowed out due to cyclical or technological change patterns.

In light of these arguments and findings, this paper seeks to add to the existing entrepreneurship literature in four important ways. First, this paper appears to be one of, if not the first, attempts at incorporating a measure of the regional psychological environment into an analysis of the determinants of entrepreneurship. Second, it formally investigates the interesting question of whether or not variation in psychological environments exists across US metropolitan areas. Third, it explores whether or not variation in psychological environments explain some portion of the variance in regional-levels of entrepreneurial activities, regardless of whether or not entrepreneurial opportunities are homogeneously or heterogeneously distributed. Lastly, this research explicitly incorporates the considerable model uncertainty inherent in these untried waters into the analysis. Considerable model uncertainty is likely associated with the specification of any empirical model explaining variation in entrepreneurial activities, but it is especially relevant to this paper, as the importance of psychological environments have no theoretical basis, to date, other than the supposition that role model learning may be driving a positive reinforcement effect that creates/maintains a regional psychological character or culture.[2]

The developments of Bayesian Model Averaging methodologies provide a quite suitable if not a breakthrough approach for advising development of theory in this area. In fact, much of the regional level entrepreneurship research should use this method more frequently, especially when considerable uncertainty exists with regard to some or all of the candidate explanatory variables.

The arrangement of this paper is as follows. Section 2 will discuss the dataset and the definitions of the variables used in this analysis. Section 3 lays out the methodological approaches, including discussions of the statistical issues surrounding spatial dependence and model uncertainty. Section 4 provides the estimation results while Section 5 contains a discussion and the conclusions.

2 Data

The effect of regional differences in psychological environments on entrepreneurship across metropolitan regions in the USA is investigated using data from a sample of Metropolitan Statistical Areas (MSAs). The dataset covers 173 MSAs, which is about one half of all MSAs. The reason that not all MSAs were used is due in part, to a lack of reliable data for each explanatory variable covering the entire set of MSAs in the USA. Specifically, the data availability for the psychological variables is limited as only a small number of individual responses exist for many MSAs. As a result, the ability of these data to adequately represent the entire MSA population is questionable. Thus, about half of the MSAs were dropped from the analysis, resulting in a final dataset containing 173 metropolitan regions. Although this loss would on the surface appear to be significant, the remaining 173 metros

[2]This last issue deserves attention in its own right as it has been largely ignored in all previous research on entrepreneurship.

contain nearly 85% of the US urban population, as many of the omitted MSAs were small. Further, given that the focus of the research is on entrepreneurship in technology intensive establishments and that over 90% of all technology enterprises (Stough and Kulkarni 2001) and entrepreneurship occurs in large metropolitan regions, then the narrowed list of 173 metropolitan areas would seem to be defensible.

2.1 *Entrepreneurship*

The dependent variable utilized in this analysis is high technology single establishment formations, a variable frequently used in the entrepreneurship literature to proxy Schumpeter's notion of "creative destruction" brought about by entrepreneurs. These data were requested and purchased from the US Bureau of the Census and were broken out by county and by five-digit North American Industrial Classification System (NAICS) codes. Year 2003 data was utilized, as it corresponded to the most recent available data, and was aggregated to the appropriate metropolitan definition using aggregations of relevant counties.[3]

The Census defines a single establishment as a single physical location where business is conducted or where services or operations are carried out. A single-establishment birth is defined to be a single establishment having no payroll in the first quarter of an initial year with a positive payroll in the first quarter of a subsequent year. In this paper, single establishment firm formations constitute entirely new agents of firm-level economic organization or activity.

High technology single-establishment births were isolated from the complete set for a number of important motives. First, it is well known that the technology intense sectors of the US economy are highly dynamic, urban centric (Stough and Kulkarni 2001) and transitory. These sectors embody the outcomes of the process of new knowledge commercialization, which incumbent firms were unwilling or unable to commercialize (Acs and Plummer 2005; Acs et al. 2009). Furthermore, these sectors are responsible for a large amount of the growth in aggregate US output; rendering them crucial players in the growth processes of the evolving "knowledge economy". Second, the high tech sectors epitomize the "Schumpeterian" sense of entrepreneurship, where new economic knowledge is being introduced in a highly competitive and dynamic environment. Thus, the "Schumpeterian" process of "creative destruction" is most certainly at work in these rapidly evolving sectors of the US economy. Third, these sectors exclude certain types of non-basic new establishments that are known to be approximately proportional to the population growth, such as: coffee shops, dry cleaners. Including these types of firms in the definition of entrepreneurship clouds the investigation of the determinants of new firms engaging in "creative destruction" as these types of firms are, in large

[3]MSA definitions correspond to the 2005 Office of Management and Budget (OMB) definitions.

part, simply serving demand created by local growth and include little or no new knowledge. While it is certainly the case that many non-high tech sectors of the US economy, such as business services, are engaging in creative destruction, the inclusion or exclusion of these sectors should have little impact on the fundamental results. This is because the process of creative destruction is certainly captured by the high tech sectors and so this measure serves the purposes of this paper well on its own.

High technology firms were isolated using Varga's (1998) definition of high tech industries, which was based in part on earlier work by Glasmeier et al (1983) and Stough et al. (1995). Varga's (1998) high tech industry criteria involved three elements (1) an above average research and development to industry sales ratio; (2) an above average percentage of mathematicians, scientists, engineers and engineering technicians compared to total industry occupations; and, (3) the total number of innovations per 1,000 employees. This work resulted in a list of SIC codes that met the criteria for being classified as high technology industries. The resulting list of three-digit SIC codes were then mapped to a list of five-digit NAISC codes. For the purposes of this research, high tech single establishment firm formations were those new firms in the set of unique high tech industry codes found in Appendix (Table 4). For standardization purposes, the number of high tech firm formations in each MSA was divided by the Census's 2000 population figures for that metropolitan region in order to standardize the measure with respect to size.

2.2 *Psychological Characteristics*

A growing body of literature has emerged regarding positive aspects of human psychology as discussed above. This approach has particularly interesting implications for entrepreneurship research efforts that, to date, have been unexplored.

Studies of individual entrepreneurs, such as those mentioned in the introduction, have provided arguments and evidence supporting the importance of psychological characteristics to entrepreneurial action. As well, many economists have argued that social capital characteristics, such as: trust and cooperation have important implications for collective action (Coleman 1988, 1990; Putnam 1993). Still, economists and scholars of urban environments have professed the importance of regional variation in human creativity, tolerance and ingenuity to regional economic outcomes (Florida and Gates 2001; Florida 2002; Lee et al. 2004).

While originating in different fields of research and for considerably different purposes, all of these efforts have a common theme or thesis; that human psychological disposition varies across regions and that this variation has important influences on general economic situations and circumstances. It is only a small step to consider the hypothesis that regional differences in human psychological orientation have important influences on entrepreneurial behavior and in turn entrepreneurship systems at the regional level. Much it seems may be added to understanding the determinants of regional technology intensive entrepreneurship by integrating the psychological study of the entrepreneur.

To do just that, data on psychological characteristics were obtained from the Positive Psychology Center in Philadelphia, Pennsylvania. These data cover 24 strengths of character contained in the Values in Action Inventory of Strengths (VIA-IS).[4] The strengths of Character are defined as positive traits reflected in thoughts, feelings, and behaviors existing in various degrees and can be measured as individual differences (Park et al. 2004). All of these measures were placed online at www.authentichappiness.org and www.positivepsychology.org/strengths (Park et al. 2004) along with associated geographic information represented by a three-digit zip code location.

Uniform tools exist to assess each of the positive traits in the classification, one of which is a 240 item self-reported questionnaire that asks individuals to report the degree to which statements reflecting each of these 24 strengths apply to themselves using a 5-point Likert scale (Park et al. 2004). Investigations have demonstrated acceptable reliability and validity for each of these 24 character strengths (see, e.g., Peterson and Seligman 2004). Peterson and Seligman (2004) conducted a validity study using the nomination known-groups procedure where individuals were asked to identify individuals they believed to possess a given strength to a notable degree. The nominated individuals were then asked to complete the questionnaire without being told why. People nominated as a paragon of a particular type of character strength tended to score higher on that strength than non-nominated individuals (Peterson and Seligman 2004).

Table 1 contains the 24 character strengths and their associated synonyms and definitions that are used to proxy personality characteristics in this paper. Park et al. (2004; 604) state that, "the identification of each strength with a list of synonyms was a deliberate strategy that attempted to capture the family of resemblance of each strength while acknowledging that the synonyms are not exact replicas of each other". The point was to provide descriptions of the 24 measures in a manner that would distinguish and describe exactly what the given attributes were attempting to measure.

A sample of 203,003 individual respondents was contained in the data concerning psychological characteristics. All respondents associated with the same three-digit zip code were averaged to create an average of each personality category for every three-digit zip code. The three-digit zip codes were then assigned to metropolitan statistical regions using a simple visual basic script run in ArcView 9.1. The script served to calculate the centroid coordinates associated with each three-digit zip code. These coordinates were then overlaid on an ArcView shapefile containing the geographic boundaries associated with each metropolitan region contained in the sample. All centriods falling into any given MSA boundary were

[4]The VIA-IS is an instrument bearing similarity to such methodologies as the California Personality Inventory and the Holland Self Directed Search as described above in footnote number 2.

Table 1 Definitions of personality variables

Variable	Synonym(s)	Description
Beauty	Awe, wonder, elevation	Noticing and appreciating beauty, excellence, and/or skilled performance in all domains of life, from nature to art to mathematics to science to everyday experience
Bravery	Valor	Not shrinking from threat, challenge, difficulty, or pain; speaking up for what is right even if there is opposition; acting on convictions even if unpopular; includes physical bravery but is not limited to it
Teamwork	Social responsibility, loyalty, citizenship	Working well as a member of a group or team; being loyal to the group; doing ones share
Creativity	Originality, ingenuity	Thinking of novel and productive ways to do things; includes artistic achievement but in not limited to it
Curiosity	Interest, novelty-seeking, openness to experience	Taking an interest in all of ongoing experience; finding all subjects and topics fascinating; exploring and discovering
Fairness	None given	Treating all people the same according to notions of fairness and justice; not letting personal feelings bias decisions about others; giving everyone a fair chance
Forgiveness	Mercy	Forgiving those who have done wrong; giving people a second chance; not being vengeful
Gratitude	None given	Being aware of and thankful for the good things that happen; taking time to express thanks
Hope	Optimism, future-mindedness, future orientation	Expecting the best in the future and working to achieve it; believing that a good future is something that can be brought about
Humor	Playfulness	Liking to laugh and tease; bringing smiles to other people; seeing the light side; making (not necessarily telling) jokes
Honesty	Authenticity, integrity	Speaking the truth but more broadly presenting oneself in a genuine way; being without pretense; taking responsibility for ones feelings and actions
Judgment	Open-mindedness, critical thinking	Thinking things through and examining them from all sides; not jumping to conclusions; being able to change ones mind in light of evidence; weighing all evidence fairly
Kindness	Generosity, nurturance, care, compassion	Doing favors and good deeds for others; helping them; taking care of them
Leadership	None given	Encouraging a group of which one is a member to get things done and at the same time maintaining good relations within the group; organizing group activities and seeing that they happen
Love	None given	Valuing close relations with others, in particular those in which sharing and caring are reciprocated; being close to people
Learn	None given	Mastering new skills, topics, and bodies of knowledge, whether on ones own or formally; obviously related to the strength of curiosity but goes beyond it to describe the tendency to add systematically to what one knows

(*continued*)

Table 1 (continued)

Variable	Synonym(s)	Description
Modesty	Humility	Letting ones accomplishments speak for themselves; not seeking the spotlight; not regarding oneself as more special than one is
Perseverance	Persistence, industriousness	Finishing what one starts; persisting in a course of action in spite of obstacles; "getting in the door"; taking pleasure in completing tasks
Perspective	Wisdom	Being able to provide wise counsel to others; having ways of looking at the world that makes sense to oneself and to other people
Prudence	None given	Being careful about ones choices; not taking undue risks; not saying or doing things that might later be regretted
Self-regulation	Self-control	Regulating what one feels and does; being disciplined; controlling ones appetite and emotions
Social intelligence	Emotional intelligence, personal intelligence	Being aware of the motives and feelings of other people and oneself; knowing what to do to fit in to different social situations; knowing what makes other people tick
Religiousness	Spirituality, faith, purpose	Having coherent beliefs about the higher purpose and meaning of the universe; knowing where one fits within the larger scheme; having beliefs about the meaning of life that shape conduct and provide comfort
Zest	Vitality, enthusiasm, vigor, energy	Approaching life with excitement and energy; not doing things halfway or halfheartedly; living life as an adventure; feeling alive and activated

used to create an average of each personality characteristic associated with the given metropolitan region.

The city-level averages represent the 24 strengths of character underlying the specific regional positive psychological environment index (PPE) used in this paper. The PPE index measures latent unobservable variation in a generalized positive psychological environment over a sample of US MSAs and is interpreted here as a form of psychological capital. It was created by extracting a single principle component from the responses on these 24 variables. A single component was used in the analysis because an examination of the associated eigenvalues revealed one dominant component. Every individual loading greater than 0.50 on the orthogonal rotation was included in the index.

The varimax component rotated factor loadings were then used as weights in the calculation of this index.[5] Table 2 contains the varimax rotated factor loadings for each of the 24 psychological attributes. Larger values of this index can be viewed as

[5]The varimax rotation searches for the linear combination of the original factors that maximizes the variance of the loadings.

Table 2 Rotated factor loadings

Characteristic	Loading
Beauty	0.398
Bravery	0.750
Love	0.672
Prudence	0.547
Teamwork	0.599
Creativity	0.529
Curiosity	0.407
Fairness	0.666
Forgiveness	0.493
Gratitude	0.795
Honesty	0.793
Hope	0.789
Humor	0.563
Perseverance	0.682
Judgment	0.530
Kindness	0.731
Leadership	0.825
Learning	0.207
Modesty	0.426
Perspective	0.741
Self control	0.620
Social intelligence	0.695
Spirituality	0.479
Zest	0.738

indicative of a more "positive" psychological environment and, hence, indicate larger stocks of psychological capital. One would expect the coefficient estimate associated with this variable to be positively related to entrepreneurial activities for the reasons discussed above.

A result such as this would be supportive of the existing research pertaining to social capital, the literature on creativity and tolerance as well as the discussed individual level entrepreneurship literature.

Lastly, it should be noted that every metropolitan area associated with less than 100 respondents was omitted to mitigate problems with bias that may result when too few respondents form the basis of the personality measures in the given metropolitan area. Furthermore, the distributions of the individual psychological characteristics tended to approach a bell-shape when a minimum of 100 individual respondents was defined as the minimum acceptable value.[6]

Figure 1 depicts the geographic distribution of the PPE index. Values for this index are highest in the southeastern US metropolitan areas. As well, there exists a cluster of cities in the south-central Midwestern region of the US that exhibit high

[6]Sensitivity analysis with respect to this minimum specification was carried out at several alternative thresholds (20, 50, 100, 150 and 200) and revealed that no notable differences in the mean parameter estimates or in the posterior distributions of those estimates existed.

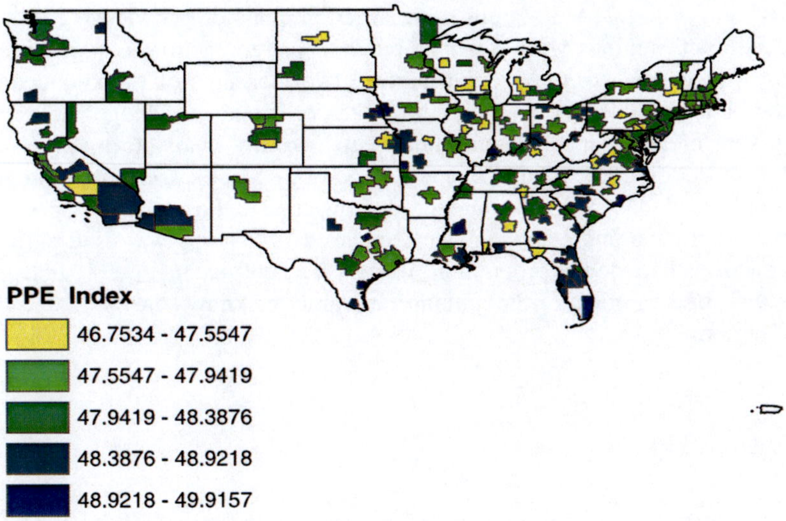

Fig. 1 The geographic distribution of the PPE index

values for the PPE. A handful of Southwest and Pacific coast metropolitan regions also exhibit higher PPE values.

2.3 Human Capital

It is well known that variation in human capital has important implications for both economic growth and for the underlying entrepreneurship ushering in that growth. To robustly account for this variation, two measures of human capital are included in the analysis. These measures correspond to the traditional measures of human capital, which are based on educational attainment. The first measure corresponds to the percentage of the 25+ 2000 population in each metropolitan region having obtained at least a bachelors degree. The second educational attainment measure is the percentage of the 25+ 2000 population having obtained a graduate or professional degree.

2.4 Knowledge

Potential entrepreneurs require an opportunity to exploit in the form of new firm formations (Shane and Venkataraman 2000). Therefore, some indicator of entrepreneurial opportunity should be included into any study of the determinants of entrepreneurship. The difficulty, however, is that opportunities cannot be directly observed until after they have been exploited. While this may be the case, it seems logical to assume that the regions with larger amounts of knowledge ought to be associated with regions with larger amounts of entrepreneurial opportunities. To sufficiently account for these opportunities, two measures of the knowledge are included into the analysis; one measuring the per capita knowledge and the other the growth of knowledge. Patents represent a readily available measure of the amount of codified knowledge existing in various places; therefore, patents are commonly used to measure the amount of knowledge in those places.[7] Specifically, the growth of patenting activities over the period 1975–1999 was used to measure the growth of knowledge production, while year 2000 patents per 10,000 individuals was used to measure the amount per capita of knowledge available in any given region.

2.5 Diversity

Many authors have suggested that diverse regions, or more specifically diverse cities, act as magnets of both talent and ideas (Jacobs 1961; Lucas 1988; Lee et al. 2004). The more diverse is the region, the more ideas and talent to draw from, thereby, facilitating more entrepreneurship in the form of new high technology firms. To adequately incorporate diversity into this analysis, two measures were used, both of which measure a particular variant of ethnic diversity. The first measure represents the percentage of the total 2000 population that is foreign born. The second measure corresponds to the percentage of the total 2000 population that is non-white. These measures are included to explain variation in entrepreneurship due to the impact of diversity.

2.6 Structural Control Variables

The last sets of variables included in this analysis correspond to structural variables that reflect a region's economic dynamic. These variables are: the average annual growth in the share of regional economic output coming from high tech industries when compared to the national average over the period 1990–2000, the percentage change in per capita income over the period 1990–2000, and the percentage change in total private employment over the period 1990–2000. These variables were introduced to control for systematic variation in entrepreneurial activities that result from: differences in industrial structures (or industrial legacies), differences in income growth and differences in the growth of the labor force. Including these structural variables captures systematic regional variation in high tech

[7]The author is well aware of the critiques associated with the use of patenting activity as a measure of knowledge, however, the alternatives are few and far in between. Therefore, patents are used in this paper in spite of the known weaknesses associated with doing so.

entrepreneurial activities due to economic differences. These control variables work to impede the positive psychological environment index from capturing variation in entrepreneurship that is really due to structural economic factors.

2.7 *Data Transformations*

A log transformation was used on the dependent variable vector to produce a distribution that looked to be more normally distributed. Second, all of the explanatory variables were studentized to accommodate the use of Zellner's g-prior (Zellner 1986), for reasons described in Sect. 3.

3 Methodology

There are two pertinent issues pertaining to regression modeling of high technology entrepreneurship that this paper explores. These issues are spatial dependence and model uncertainty. These issues are important because the existence of spatial dependence has been shown to cause bias in the resulting parameter estimates (LeSage and Pace 2004) while model uncertainty creates a situation yielding suboptimal results for a number of reasons (Raftery et al. 1997). For one, considerable uncertainty over which specific variables to include in the regression exists and estimates based on saturated regression models containing many possible variables does not address this issue in any consistently reliable theoretical framework.

Secondly, simply introducing all of the explanatory variables into the regression will result in the possibility of including irrelevant variables, which will tend to increase the dispersion of the estimated coefficients. This will make it difficult to identify the important variables influencing entrepreneurship. On the other hand, a strategy that relies on subsets of candidate explanatory variables will likely suffer from omitted variables bias if important variables are excluded. As well, the considerable uncertainty regarding which explanatory variables are truly relevant in explaining variation in high tech entrepreneurship can lead to explanatory variable matrices suffering from collinearity, further reducing the precision of the coefficient estimates (Belsley et al. 1980). Collinearity also yields instability with regard to the parameter estimates because the inclusion or exclusion of any single explanatory variable can dramatically alter the coefficient estimates associated with any given explanatory variable.

To handle these issues, Bayesian Model Averaging is employed in the context of a spatial autoregressive regression framework. This modeling approach is especially advantageous for this study because it adequately addresses the considerable model uncertainty inherent in this analysis of the determinants of entrepreneurship, while at the same time it accounts for suspected spatial autocorrelation of entrepreneurial activities.

3.1 Spatial Dependence

Numerous regional-level studies of entrepreneurship have noted that entrepreneurial activities are clustered across space, resulting in outcomes that are spatially correlated (see, e.g., Acs and Plummer 2005; Acs et al. 2009). The spatial clustering of these activities in previous empirical studies results in the expectation that spatial dependence may exist in the sample of data utilized in this paper, as entrepreneurial activities form the dependent variable. However, it is unclear whether or not the use of non-contiguous metropolitan data, further exacerbated by the exclusion of certain metropolitan regions, as was discussed above, has resulted in a sample of data that is unaffected by the existence of spatial dependence. To handle this situation, the Bayesian Model Averaging strategy discussed in Sect 3.2 will be carried out in the context of a spatial autoregressive regression framework. It is important to note here, that this strategy will produce parameter estimates statistically equivalent to the least-squares type approach should spatial dependence not be an important issue in this particular dataset.

Before introducing the Bayesian Model Averaging approach used in this paper, a brief discussion of the specification of the spatial weight matrix (W) is in order as this estimation strategy requires this. The standard specification is one based on first order contiguity (also known as the Queens criteria). However, the current set of sample data is not contiguous, inhibiting this type of specification. An alternative strategy is to specify this matrix such that it extracts the m nearest neighbors to any y_i. Under this specification, the individual elements of W, denoted w_{ij}, correspond to a value > 0 if y_j is contained in the set of nearest neighboring observations and to a value of 0 if y_j is not contained in this set. All $w_{i=j}$ are set equal to zero to prevent an observation from exhibiting dependence on itself. The matrix is then row standardized yielding a row stochastic matrix (since W is non negative). The purpose of obtaining a row stochastic weights matrix is that this type of weight matrix has nice numerical and interpretive properties (see LeSage and Pace 2004 for specific details). Once, W is specified in this manner, it will be used to represent the spatial relationships inherent in this analysis.[8]

3.2 Bayesian Model Averaging

Considerable uncertainty exists regarding how relevant each of the candidate explanatory variables are in explaining variation in high tech entrepreneurship over the sample of metropolitan regions used in this paper. While many empirical

[8]A spatial weight matrix extracting the five nearest neighboring observations was used throughout this paper. This specification was used because evidence suggests that knowledge absorbed by high tech firms tends to be bound to an area approximately 50–75 miles from its source. The specification of W, here, is based on this empirical evidence (see Anselin et al. 2002).

studies suggest that the economic and human capital measures utilized here, such as job growth or educational attainment are important predictors of high tech entrepreneurship, relatively little information exists to offer insights into which specific measures of educational attainment and diversity are appropriate. Furthermore, it is unclear whether or not the positive psychological environment index is truly relevant in explaining variation in the dependent variable. As a result, a considerable amount of model uncertainty exists with regard to the analysis carried out in this paper.

Fortunately a literature exists to address these issues in a consistent framework that has been empirically verified. The literature is called Bayesian Model Averaging and involves comparing alternative sets of explanatory variables. This methodology overcomes any problems associated with misspecification, collinearity and/or including irrelevant personality variables that this work might otherwise encounter. The basic theory was provided by Zellner (1971) and involves cases where there are a small number of alternative models to compare.

The basic process begins with the specification of prior probabilities for each of the *m* models as well as prior distributions for each of the model parameters. The priors for the models and parameters are then combined with the likelihood function, conditional on the parameters and models in order to produce a posterior distribution for each of the *m* alternative models under consideration. The posterior distributions are used to calculate posterior model probabilities for each of the *m* models, which are then used to compare the alternative model specifications.

While this procedure works well for cases with where *m* is small, its computational demands inhibit its application to cases where *m* is considerably larger, as is the case here. However, Madigan and York (1995) introduced a technique known as Markov Chain Monte Carlo Model Composition (MC^3) that enables the analysis to be carried out in cases where *m* is large, through a systematic sampling of the large model space. Work by Fernandez et al. (2001a, b) and Raftery et al. (1997) then extend this work to applications of econometric regression modeling.

The MC^3 procedure starts with an initial randomly selected set of explanatory variables then deriving a proposal model to compare to the initial model through the use of three steps, where the use of each step is equally probable (i.e., each step has a probability of 0.33 of being used). These three steps are a birth step, a death step or a move step. The birth step adds an explanatory variable to the model, the death step removes an explanatory variable from the model and the move step randomly switches an included variable with an excluded variable. The initial model is then compared to the proposed model through the use of a procedure known as the Metropolis–Hastings step. The Metropolis–Hastings step is used to compare the two alternative models where either the initial model or the proposed model is accepted. If the proposed model is accepted, it becomes the initial model and the process is repeated. If the initial model is accepted, it remains the initial model and the process is repeated. Madigan and York (1995) show that one can systematically walk though the large model space by repeating this procedure many times, essentially solving the problems associated with Bayesian Model Averaging in cases where m is large.

The key step in this process is the comparison of the initial and proposed models in the Metropolis–Hastings step. This comparison involves the calculation of the odds ratio, shown in relation (1).

$$\min\left[1, \frac{p(M_p|\, y)}{p(M_i|\, y)}\right] \tag{1}$$

In relation (1), M_p represents the proposed model and M_i represents the initial model, both of which are based on the inclusion of different sets of explanatory variables. The terms in this ratio can be obtained by first combining the priors ($\pi(M)$ and $\pi(M,\beta,\sigma/M)$) with the likelihood function ($p(y/\beta,\sigma,M)$), to arrive at the joint probability for the models and parameters, shown in relation (2),

$$p(M, \beta, \sigma|\, y) = \pi(M)\pi(M, \beta, \sigma|\, M)p(y|\, \beta, \sigma, M) \tag{2}$$

then by obtaining the joint posterior for the models and parameters shown in relation (3)

$$p(M|\, y) = \iint p(M, \beta, \sigma|\, y)d\beta, d\sigma \tag{3}$$

and analytically integrating β and σ out of the expression to arrive at a scalar expression for the numerator and denominator which are used in relation (1).

To arrive at the posterior model probabilities over the set of all unique models, it is necessary to save the log marginal density vectors for each unique model found by the sampling scheme. The models with posterior model probabilities that are greater than 0.001 were saved for use in constructing model averaged coefficient estimates, which account for the model uncertainty inherent in this application. To create these estimates each of the saved models are estimated (including an intercept term). The coefficient estimates associated with these models are then multiplied by their specific posterior model probabilities and summed to create a weighted average across all models with posterior model probabilities that are greater than 0.001.

It should be noted here that work by LeSage and Parent (2007) demonstrates that the least-squares model comparison inferences will be adversely affected by spatial autocorrelation. As a result, the basic model averaging strategy outlined above will have to be augmented to account for spatial autocorrelation. This can be done by relying on the strategy laid out by LeSage and Parent (2007) in the context of a spatial autoregressive model. This strategy is conceptually the same strategy as was laid out above, with the addition of the parameter ρ to expressions (2) and (3) with one computational difference. The difference is that the parameter ρ cannot be analytically integrated out of expression (3) as were β and σ. To handle this problem, LeSage and Parent (2007) suggest storing the vectors of the log marginal values for both the current and proposed models over a grid of values for the parameter ρ. These vectors can then be scaled and integrated with respect to this parameter to produce the odds ratio shown in (1).

The last two issues involved in the implementation of the model averaging strategies, outlined above, are the specification of the priors for the model parameters and diagnosis of convergence in the MC^3 sampling scheme. In accordance with the standard convention, Zellner's g-prior (Zellner 1986) is utilized for the parameter β (Fernandez et al. 2001a, b). A gamma prior is used for the parameter σ and following LeSage and Parent (2007), a beta prior is used for the spatial dependence parameter ρ. Fifty thousand draws were initially utilized and convergence in the sampling scheme was insured by carrying out the procedure twice and inspecting the results to see if the same results were obtained. Since the same results were, indeed, obtained the 50,000 draws were enough to ensure that the models with the most posterior support were among those sampled. Model averaged estimates can be obtained with confidence after ensuring that convergence in the sampling scheme has been attained.

4 Results

Several important findings were obtained via the particular methodological approach relied on in this paper. For one, the resulting posterior model probabilities lent empirical support for the preconception that a considerable amount of model uncertainty exists with regard to modeling the determinants of entrepreneurship. This was reflected by the fact that the most probable model was associated with a posterior model probability of only 0.25. This finding is interpreted as meaning that the most probable model only has a 25% chance of being the "true" data generating model. Furthermore, just under 1,000 unique models were found by the sampling scheme, yet only two of these were associated with posterior model probabilities greater than 0.1. The results suggest that no particular model dominated and that posterior support was spread rather evenly across the alternative model specifications. This suggests that substantial uncertainty exists regarding which set of the candidate explanatory variables are truly relevant in explaining variation in high tech entrepreneurship and reliance on any individual model, alone, may provide incorrect inferences. In general, this result supports our reliance on model averaged estimation as model uncertainty is particularly large in this application.

Table 3 contains the set of model averaged estimates. Column 2 contains the mean coefficient estimates, while columns 3 and 4 contain the upper and lower bounds of a 95% confidence interval computed around the means. The 95% confidence interval was computed to provide inferences regarding the statistical significance of the respective coefficient estimates, where intervals that do not contain 0 indicate such significance. The estimates were obtained by estimating the set of unique models that were associated with posterior model probabilities greater than 0.001 or 1/10th of 1%. The posterior probabilities correspond to the weights underlying the averaged estimates. Each individual model was estimated via a Bayesian heteroscedastic variant of the spatial autoregressive regression model, initially introduced by LeSage (1997), that is robust to the influence of

Table 3 Model averaged estimates

Variables	Coefficients	Lower 0.05	Upper 0.95
PPE index	0.0017	0.0002	0.0032
Tech. growth	0.0034	0.0012	0.0056
Foreign born	0.1058	0.0908	0.1209
Non-white	−0.0046	−0.0069	−0.0024
Pc. inc. growth	−0.0026	−0.0046	−0.0005
Job growth	0.1342	0.1164	0.1526
Pat. growth	−0.0358	−0.0482	−0.0232
Ba 25+	0.4567	0.414	0.4994
Pat. 10k pop.	0.0137	0.0082	0.0195
Gp deg.	−0.1953	−0.2358	−0.1543
Rho	0.0845	0.0401	0.1271

outliers and heteroscedasticity. Each model was estimated with an intercept term included and with 8,000 draws (4,000 were omitted for burn in purposes).

The set of model averaged estimates reveal several important findings. Beginning with spatial dependence, one can see that the coefficient estimate on the parameter, rho, is positive and statistically significant. This confirms the suspicion that entrepreneurial activities are correlated across space.

Regarding the structural economic variables, these results indicate that growth in high technology as a share of regional output is positively related to high technology entrepreneurship, as is total private employment growth. Per capita income growth, on the other hand, appears to be negatively related to the prevalence of high tech entrepreneurship. The two variables of ethnic diversity are both statistically significant as well, with foreign born being associated with a positive coefficient estimate while non-whites are associated with a negative estimate. With regard to the knowledge variables, the results suggest that the growth of patent activities over the period 1975–1999 is negatively associated with high tech entrepreneurship while the amount of knowledge per capita is positively related to it. Regarding human capital, the measure of the share of the 25+ 2000 population holding bachelor's degrees is associated with a positive coefficient estimate while this share holding graduate and professional degrees is associated with a negative estimate. Lastly, the PPE index is positively related to high tech entrepreneurial activities.

5 Discussion and Concluding Remarks

Several important conclusions are provided in this paper. This final section will discuss each of the individual results in considerable detail. The discussion will focus on how the coefficient estimates ought to be interpreted as well as what these coefficient estimates imply with regard to the determinants of high tech entrepreneurship.

Considerable model uncertainty was shown to exist between the alternative model specifications. This finding suggests that many empirical studies purporting to explain the determinants of entrepreneurship may be relying on models that have small probabilities of being correctly specified. This is particularly important because scholars of entrepreneurship have completely ignored this issue in all previous research on the determinants of entrepreneurship. Yet, these studies are increasingly being used to formulate entrepreneurship policy.

The results provide additional evidence that high technology entrepreneurship is a phenomenon clustered in space. This means that there exist latent unobservable sources of variation in entrepreneurial activities that are region specific. However, the impact of this variation is considerably small with regard to this particular dataset. The range of the possible values for this parameter is 0–1, where a 0 represents no spatial correlation and a 1 represents complete correlation.[9] Therefore, the coefficient estimate of 0.08 indicates that the spatial correlation that exists in this dataset is considerably small. However, this finding should be interpreted with caution due to a reliance on non-contiguity based data, which was further impacted, no doubt, by the exclusion of a large number of cities for reasons of data availability.

In the rest of this discussion, attention is focused on the explicit explanatory variables that underlie the current modeling framework, beginning with the structural economic variables. Three structural economic variables deemed important in the entrepreneurship literature were investigated in this paper. These variables were: growth in high technology output as a share of total regional output, per capita income growth and total private employment growth. All of these variables were measured in terms of their growth rates. This was deliberate as it facilitates an "apples to apples" comparison of the magnitudes of their coefficient estimates.

Of these structural economic variables, employment growth has, by far, the strongest relationship to high technology entrepreneurship. The coefficient estimate on employment growth was positive, indicating that cities with growing numbers of employees are associated with growth in high technology entrepreneurship. It is not surprising to find that the coefficient estimate on the growth of the share of high technology output is positively related to high technology entrepreneurship. This finding suggests that a history of an expanding share of output in technology intensive sectors has a positive relationship to high technology entrepreneurship. This implies two things; one, that the commercialization of new high technology knowledge (which is fundamentally what new high technology entrepreneurship is) is considerably path dependent and two, that the returns to the expanding share of output in high technology that occurs in any given city may in fact be recycled in that city, spurring the formation of new high tech entrepreneurial firms. The surprising result is that income growth is negatively associated with high tech entrepreneurship. This evidence seems to suggest the following; that high tech entrepreneurial activity tends to occur in less expensive places, holding constant all

[9] In actuality the range of possible values is -1 to 1, however, the range -1 to 0 represents that case of negative spatial correlation. Negative correlation was ignored in this discussion because such correlation would indicate that the presence of entrepreneurial activities in one region would discourage its presence in neighboring regions. This seemed to be particularly absurd when considering the literature on entrepreneurship and so this case was ignored here.

of the other factors. In short, technology entrepreneurs are, in final analysis, cost sensitive.

Two commonly used measures of diversity were included to control for variation in the dependent variable that results from concentrations of diverse populations. The results indicate that high concentrations of foreign born populations in a given city has a tighter relationship to entrepreneurial activities than do concentrations of non-white persons. There is considerable literature on the high entrepreneurship levels among immigrant populations that helps to explain this. For example, new immigrants have constrained access to participation in the local labor force and thus to compensate have a higher propensity to initiate new enterprises. Further, because they come from other cultural environments they often perceive opportunities that locals miss. This is evidenced by the fact that the mean coefficient estimate on foreign born populations is approximately ten standard deviations above zero, while the mean coefficient estimate on non-whites is only approximately two standard deviations above zero. Furthermore, concentrations of non-whites are associated with a negative impact on entrepreneurship, while the opposite is true with respect to concentrations of foreign born persons. These suggest an interesting possibility; that it's not racial diversity that is important to entrepreneurship (in fact the coefficient on racial diversity is negative), but rather that it's the diversity of the immigrant population that is important.

Two alternative measures of knowledge were included in this paper to proxy entrepreneurial opportunity. The differences between these two variables are that the per capita level of patenting activities measures the amount of recent codified knowledge (3 years ago in this case), as a percentage of the population, whereas the other measures the growth of codified knowledge over the last 25 years. The fact that the coefficient estimate associated with the former is positive while the latter is negative suggests that higher percentages of recent codified knowledge in a given population is positively related to high tech entrepreneurial activities while growth in the amount of codified knowledge over the past quarter century is not. This further suggests that it is the level of the availability of new knowledge in a city that relates to high tech entrepreneurship and not the growth of knowledge in past years.

As well, two alternative measures of human capital were included to account for variation in entrepreneurship resulting from differences in levels of talent. These were the shares of the 25+ 2000 population holding bachelors and graduate and professional degrees. The results indicate that 4-year degrees are positively associated with high tech entrepreneurship whereas advanced degrees have a negative association. This finding suggests two possibilities. One, that people who have obtained a graduate degree are less willing to leave their jobs to start new tech companies while those with bachelor's degrees are more willing to do so. Perhaps the incomes that persons with advanced degrees are earning make them less willing to bear the "Knighterian" uncertainty associated with entrepreneurship. Second, this could mean that small cities with large universities or research laboratories create observations where large concentrations of persons holding graduate and professional degrees exist, yet these places do not have enough access to the other necessary preconditions for emergence of high tech entrepreneurship. Therefore, the global impact shows up as negative.

Finally, focus will now be placed on the primary variable of interest in this paper; the PPE index. The results demonstrate that variation in psychological environments does exist across US cities; implying the existence of and variation in what we have called psychological capital. Furthermore, the results indicate that this variation has important influences on high technology entrepreneurship, holding all other explanatory variables constant.[10]

Many of the individual factors, which have been shown to be related to the latent unobservable variable (Psychological Capital), measured by the PPE (see Table 2), have precedence in the literature. The factors: honesty, leadership, teamwork, kindness and love all relate well to the contributors to social capital (see Table 1). Therefore, the importance of these factors in the PPE index provides evidence supporting conclusions drawn in the existing literature on social capital. The results also provide evidence in support of the importance of creativity and tolerance to high tech entrepreneurship. The individual factors: creativity, judgment and fairness match up appreciably well to creativity and tolerance (see Table 1) as defined in the literature. For instance, the factor creativity obviously relates well to the definition of creativity as laid out in the regional science literature. In this paper, the factor, fairness, is designed to measure the following: treating all people the same, not letting personal feelings bias decisions about others, giving everyone a fair chance (see Table 1). As a result, the factor, fairness, appears to be almost synonymous with the definition of tolerance espoused in the regional science literature.

However, while social capital, creativity and tolerance are important determinants of high tech entrepreneurship, this paper suggests that there is more to the story than just these elements. Other factors are also found to be important measures of latent psychological capital; some of which have precedence only in the individual-level approach to entrepreneurship research. Individual-level studies of entrepreneurs have found considerable evidence suggesting that risk taking, self efficacy and tolerance for ambiguity are important characteristics of entrepreneurial individuals. However, to date, these variables have not been investigated with regard to regional units.

The results contained in this paper have shown that the factors: bravery, hope and perseverance capture variation in the latent regional PPE (see Table 2). These particular findings, then, are especially interesting because the definition of the bravery factor corresponds well to the individual-level definitions regarding risk aversion and tolerance for ambiguity, while factor definitions pertaining to hope and perseverance correspond well to the definition of self efficacy. In light of this, it is evident that this research supports the cited individual-level studies at the regional-level and provides evidence that the proclivity of these characteristics in a particular population have a positive relationship with the amount of high technology entrepreneurship occurring in it.

[10]The relative size of this coefficient is rather small. However, it should be noted that the size of this coefficient is not comparable with the size of other coefficients. This is because the interpretation of the partial is complicated. The partial with respect to this variable reflects a change in a weighted average of Likert scales, thus, the small magnitude of this coefficient does not necessarily imply a small impact with regard to this variable.

Lastly, several of the remaining factors that this research has shown to contribute to the PPE have been unexplored in all previous research on entrepreneurship. These factors are: zest for life, gratitude, humor, perspective, self control and social intelligence. These factors capture variation in characteristics, such as: enjoying other people and one's life, an ability to control one's emotions, being gracious and exhibiting knowledge regarding what makes other people "tick". This set of psychological characteristics capture an unexplored area of psychological disposition that tends to reflect an ability to relate to other people along with a general "excitement" for life. These factors are related to the concept of social capital but go beyond the existing definition to include a more dynamic concept which we describe as "human energy".

The overarching goal of this paper was to examine several important issues pertaining to entrepreneurship that have not been adequately examined in the existing entrepreneurship literature. This paper has argued that psychological determinants of entrepreneurship are important factors. At the same time they may be inputs or underlying causal factors or vice versa (i.e., a reflection of these other factors) to other factors heretofore identified, i.e., social capital with considerable variation across MSAs in the US. This research has shown that social capital, creativity and tolerance are subcomponents of a more general socio-environmental determinant of entrepreneurship, denoted psychological capital. Additionally, this research provides empirical evidence that several factors found to be important at the individual-level are at work at the regional-level as well. This work has also demonstrated that regional variations in psychological environments and subsequently psychological capital exist and that they have important influences on entrepreneurial activities. While the evidence suggests that psychological environments have important influences on high technology entrepreneurship, in all likelihood, these environments are not, themselves, explicit producers of entrepreneurship, rather, they probably serve to enhance a regions propensity to commercialize new knowledge, i.e., they help produce a positive reinforcing environment for entrepreneurship and commercialization. The psychological environments or capital, then, are more likely "conductors" of entrepreneurial activity, increasing the probability of the commercialization of new knowledge.

Appendix

Table 4 High technology NAICS codes

Industry classification	Five-digit NAICS codes
Agricultural chemicals	32531, 32532
Aircraft and parts	33641, 54171
Audio and video equipment	33431, 33461, 51222
Communication services not elsewhere classified	48531, 51332, 51334, 51339
Communications equipment	33421, 33422, 33429, 33441

(*continued*)

Table 4 (continued)

Industry classification	Five-digit NAICS codes
Computer and data processing services	33461, 44312, 51121, 51419, 51421, 51811, 51821, 53242, 54151, 81121
Computer and office equipment	33331, 33411, 33441, 33451, 33461, 33994
Construction and related machinery	33243, 33299, 33312, 33313, 33392, 33651
Electric lighting and wiring	33221, 33511, 33512, 33593, 33632
Electrical industrial apparatus	33531, 33599
Electronic components and accessories	33422, 33431, 33441
Electronic distribution equipment	33531
Engines and turbines	33361, 33639
General industrial machinery	31499, 33299, 33341, 33361, 33391, 33399
Guided missiles and space	33641, 54171
Household appliances	33329, 33341, 33521, 33522, 33999
Industrial inorganic chemicals	21111, 32512, 32513, 32518, 32599, 33131
Industrial organic chemicals	32511, 32513, 32519
Measuring and controlling devices	33331, 33451, 33911
Medical instruments and supplies	32229, 33299, 33451, 33911
Medicinals and botanicals	32541
Metal working machinery	33221, 33299, 33351, 33399, 33531
Miscellaneous chemical products	31194, 32518, 32519, 32551, 32552, 32591, 32592, 32599
Miscellaneous electrical equipment and supplies	33319, 33361, 33399, 33461, 33512, 33591, 33599, 33632
Ophthalmic goods	33911
Ordnance and accessories not elsewhere classified	33299
Paints	32551
Petroleum refining	32411
Photographic equipment and supplies	32599, 33331
Plastic materials and synthetics	32521, 32522
Railroads	33391, 33651
Research and development and testing services	54138, 54171, 54172, 54191, 54194
Search and navigation equipment	33451, 33911
Soap	32561, 32562

References

Acs Z, Armington C (2002) The determinants of regional variation in new firm formation. Reg Stud 36:33–45

Acs Z, Plummer L (2005) Penetrating the Knowledge Filter in Regional Economies. Annals of Regional Science 39(3): 439–456

Acs Z, Plummer L, Sutter R (2009) Penetrating the Knowledge Filter in Rust Belt Economies, Annals of Regional Science 42: 238–264

Anselin L, Acs Z, Varga A (2002) Local geographic spillovers between university research and high-technology innovation. J Urban Econ 42:422–448

Bartik T (1989) Small business start-ups in the United States: estimates of the effects of characteristics of states. South Econ J 55:1004–1018

Belsley D, Kuh E, Welsch R (1980) Regression diagnostics. Wiley, New York, NY

Blanchflower D, Oswald A (1990) What makes an entrepreneur? J Labor Econ 16:26–60

Brockhaus R (1980) Risk taking propensity of entrepreneurs. Acad Manage J 23(3):509–520
Brusco S (1982) The Emilian model: productive decentralization and social integration. Camb J Econ 6(2):167–184
Coleman J (1988) Social capital in the creation of human capital. Am J Sociol 94:S95–S120
Coleman J (1990) Foundations of social theory. Harvard University Press, Cambridge, MA
Dunn T, Holtz-Eakin D (2000) Financial capital, human capital, and the transition to self-employment: evidence from intergenerational links. J Labor Econ 18:282–305
Evans D, Leighton S (1990) Small business formation by unemployed and employed workers. Small Bus Econ 2:319–330
Fernandez C, Ley E, Steel M (2001a) Model uncertainty in cross-country growth regressions. J Appl Econometrics 16:563–576
Fernandez C, Ley E, Steel M (2001b) Benchmark priors for Bayesian model averaging. J Econometrics 100:381–427
Florida R (2002) The rise of the creative class. Basic Books, New York
Florida R, Gates G (2001) Technology and tolerance: the importance of diversity to high-technology growth. The Brookings Institution, Washington DC
Glasmeier A, Hall P, Markusen AR (1983) Defining high technology industries, Working paper no. 407, Institute of Urban and Regional Development, University of California at Berkeley
Jacobs J (1961) The death and life of great American cities. Random House, New York
Johannisson B (1988) Business formation: a network approach. Scand J Manag 4(3–4):83–99
Kilstrom ER, Laffont J (1979) A general equilibrium entrepreneurial model of firm formation based on risk aversion. J Polit Econ 87:719–748
Larson A (1991) Partner networks: leveraging external ties to improve entrepreneurial performance. J Bus Venturing 6(3):173–188
Lee S, Florida R, Acs Z (2004) Creativity and entrepreneurship: a regional analysis of new firm formation. Reg Stud 38:879–891
LeSage JP (1997) Bayesian estimation of spatial autoregressive models. Int Reg Sci Rev 20:113–129
LeSage JP, Pace KR (2004) Introduction to advances in econometrics. In: Fomby TB, Hill RC (eds) Spatial and spatiotemporal econometrics, vol 18. Elsevier, Oxford
LeSage JP, Parent O (2007) Bayesian model averaging for spatial econometric models. Geogr Anal 39(3):241–267
Lucas R (1988) On the mechanisms of economic development. J Monetary Econ 22:3–42
Madigan D, York J (1995) Bayesian graphical models for discrete data. Int Stat Rev 63:215–232
Markman GD, Balkin DB, Baron RA (2002) Inventors and new venture formation: the effects of general self-efficacy and regretful thinking. Entrepren Theor Pract 27(2):149–165
McClelland D (1961) The achieving society. Collier-Macmillan, New York, NY
Mellander C, Florida R (2006) Human capital or the creative class: explaining regional development in Sweden, KTH/CESIS Working Paper Series in Economics and Institutions of Innovation
Park N, Peterson C, Seligman M (2004) Strengths of character and well-being. J Soc Clin Psychol 23(5):603–619
Peterson C, Seligman M (2004) Character strengths and virtues: a classification and handbook. Oxford University Press/American Psychological Association, Oxford/Washington, DC
Piore MJ, Sable CF (1986) The second industrial divide. Basic Books, New York
Putnam R (1993) Making democracy work: civic traditions in modern Italy. Princeton University Press, Princeton, NJ
Raftery A, Madigan D, Hoeting J (1997) Bayesian model averaging for linear regression models. J Am Stat Assoc 92:179–191
Reynolds P, Storey D, Westhead P (1994) Cross-national comparison of the variation in new firm formation rates. Reg Stud 28:443–456
Saxenian A (1999) Silicon valley's new immigrant entrepreneurs. Public Policy Institute of California, San Francisco, CA

Schere J (1982) Tolerance of ambiguity as a discriminating variable between entrepreneurs and managers. In: Proceedings of the academy of management, New York
Shane S, Venkataraman S (2000) The promise of entrepreneurship as a field of research. Acad Manag Rev 25(1):217–226
Sorenson O (2003) Social networks and industrial geography. J Evol Econ 13(5):513–527
Stough R, Popino J, Campbell HS, Fuller S, Haynes KE (1995) Technology in the Greater Washington Region, Report to the Greater Washington Board of Trade and the Greater Washington Initiative, Washington D.C
Stough RR, Kulkarni R (2001) Planning issues and the new generation technology economy: comparative regional analysis and the case of the US national capital region. In: Williams JF, Stimson RJ (eds) International planning settings: lessons of success. Elsevier Science Series, Amsterdam, pp 395–430
Van Pragg C, Cramer J (2001) The roots of entrepreneurship and labor demand: individual ability and low risk aversion. Economica 68(269):45–62
Varga A (1998) University research and regional innovation. Kluwer, Boston
Zellner A (1986) On assessing prior distributions and Bayesian regression analysis with G-prior distributions. In: Goel PK, Zellner A (eds) Bayesian inference and decision techniques; essays in honor of Bruno de Finetti. North-Holland/Elsevier, Amsterdam
Zellner A (1971) An Introduction to Bayesian Inference in Econometrics, J. Wiley and Sons: New York

Incubators in Rural Environments: A Preliminary Analysis

Peter Schaeffer, Shaoming Cheng, and Mark Middleton

1 Introduction

State policy makers and local leaders have long placed a high priority on local economic development (Isserman 1993; Pulver 1989; Ekstrom and Leistritz 1988), but the changing structure of traditional industries and the impact of those changes on local communities have challenged the efficacy of established policies and strategies. Many of the forces responsible for past economic changes continue to have an impact. One of these changes was the emergence of computer-based technology in production, administration and information, which has reduced the role of economies of scale in many sectors. Studies by Loveman and Sengenberger (1991) and Acs and Audretsch (1993), for example, have shown a shift in industry structure toward decentralization and an increased role for small firms. This was mainly due to changes in production technology, consumer demand, labor supply, and the pursuit of flexibility and efficiency. These factors led to the restructuring and downsizing of large enterprises and the entry of new firms. Brock and Evans (1989) provide extensive documentation of the changing role of small businesses in the US economy, which are likely the result of responses to structural adjustments.

In addition, new patterns of consumer expenditures and demand resulting from rising living standards contributed to the emergence of fragmented consumer markets that also favor small consumer-oriented firms over high volume, production-oriented firms. Thus, new business opportunities in small and medium size enterprises resulted as large firms downsized in response to a changing environment. The emerging view among policy makers is that small business and

P. Schaeffer (✉) and M. Middleton
Assistant Professor, Division of Resource Management and Regional Research Institute West Virginia University, P.O. Box 6108, Morgantown, WV 26506-6108, USA
e-mail: peter.schaeffer@mail.wvu.edu

S. Cheng
Professor, Department of Public Administration, PCA 350A, Florida International University, 11200 SW 8th Street, Miami, FL 33199, USA
e-mail: scheng@fiu.edu

K. Kourtit et al. (eds.), *Drivers of Innovation, Entrepreneurship and Regional Dynamics*, Advances in Spatial Science, DOI 10.1007/978-3-642-17940-2_13,

entrepreneurship are key elements for generating economic development. This paradigm shift has brought about a revival in small businesses promotion and entrepreneurial initiatives at local, national and international levels.

The hypothesis of this research is that a dynamic economic environment poses a special challenge to rural regions to adapt to changing conditions because of their distance from centers of innovation and venture capital. The small size of rural markets and their distance from major consumer and supplier markets add to the challenge for rural entrepreneurs trying to exploit emerging opportunities. The cost advantage of rural over urban locations does not necessarily compensate for such disadvantages, particularly when offshore locations offer the promise of even lower production costs. To counter such disadvantages and help rural regions adjust with due speed, programs supportive of entrepreneurs, entrepreneurship, and new businesses may be needed. Public support for such programs may be justified since it is likely that some rural disadvantages are the result of market failure (Scorsone and Weiler 2004; Tödtling and Wanzenbock 2003; Weiler 2000).

The support program investigated in this article is the business incubator. The purpose of business incubators is to improve the probability of creating new firms, as well as their likelihood of survival and success relative to new firm formation without such assistance. If the hypothesis that in a dynamic economic environment rural areas are at a disadvantage relative to urban areas is true, then it follows that rural incubators are faced with a more difficult task than their urban counterparts (Cheng and Schaeffer 2009). Therefore, first, is it possible that incubators are relatively less common in rural than in urban regions? Second, if incubators are relatively less common in rural areas, what alternatives exist to providing services and support to entrepreneurs and new businesses in their place? These are the two main questions this article will try to answer.

The rest of the article is organized as follows. Section 2 provides a brief introduction to business incubators. This is followed by an analysis of the geographical distribution of incubators by rural versus urban counties. Section 4 presents possible alternatives to business incubators, and Sect. 5 provides a summary and conclusions.

2 Introduction to Business Incubators

Business incubators provide a nurturing environment, through an array of business support resources and services, where entrepreneurs, start-ups, and small businesses can commercially validate and transform their ideas and concepts into viable and tangible products and services (Acs and Armington 2006; Storey 2003; Yu et al. 2006). Business incubation, consequently, has been increasingly recognized as a viable approach for promoting new business formation and accelerating new business growth.

The first incubator in the United States appeared in Batavia, NY, in 1959 (Adkins 2001; Lewis 2002) and the number grew to only 12 by 1980 (ASME

1996–2008). In the mid-1980s the US Small Business Administration (SBA) supported several business incubator initiatives, as well as the founding of the National Business Incubator Association (NBIA). It was estimated that more than half of the currently existing incubation projects were established since the late 1980s, in part because business incubators were treated as a means to utilize idle manufacturing facilities and create jobs in response to economic recession (Adkins 1996). Hackett and Dilts (2004), however, maintained that the fundamental reason for the flourish of business incubators in the late 1980s and early 1990s was the passage of the Bayh-Dole Act that expanded and amplified commercialization of federally funded research and hence produced significant profit opportunities. Lalkaka (2000) added that incubators in the 1980s essentially offered affordable space and shared facilities, while incubators in the 1990s provided a wide range of professional counseling, skill enhancement, and networking services to seed capital, suppliers, and potential buyers. The number of business incubators in the EU-15 has also increased to currently over 900 (Costa-David et al. 2002). Our own research in the United States found 726 incubators. This number should be viewed with caution, however, because the definition of what constitutes an incubator is not rigorous. Thus, some claim some 1,000 incubators in the United States (ASME 1996–2008), but based on our research, we suspect that not all of them are offering the services of business incubators but may be little more than landlords to start-up businesses.[1]

Incubators are often funded by public resources. About 90% of incubators in the United States have non-profit status; in the EU-15 77% are organized as non-profit organizations (ASME 1996–2008; Costa-David et al. 2002). Universities and colleges sponsor 25% of all US incubators, economic development corporations 15%, and governments 16%. Nineteen percent have no sponsor; multiple sponsorships exist in 5% of all US incubators (ASME 1996–2008). The rationale for publicly funded business incubators lies ultimately in addressing market failures, i.e., gaps and deficiencies in the support structure available to new and small firms. These market failures stem from the relatively high costs and risks of providing services to new and small companies compared with larger firms and the unwillingness of the private sector to assume these costs and risks given the often modest returns and/or the fact that private incubators need "deep pockets" to survive for often considerable periods of time before returns sufficient to even recover costs can be achieved (for relationship between high fixed costs and market provision of services, see Waldfogel 2007). Most existing business incubators are publicly funded, despite visions that incubators should become financially self-sufficient and profit oriented (Bearse 1998; Quittner 1999). Large numbers of for-profit

[1]We established our database with information on incubators from lists such as the kept by the NBIA and by state organizations. We cross-checked the lists and then attempted to establish contact with the listed incubators, to find out if they were still active. We had to drop quite a large number of incubators from our database at that point. After this paper was completed in summer 2009, we conducted a survey of the incubators in our data base; the survey was completed in spring 2010. In summer 2010 we plan visits to a sample of incubators to achieve more insight and learn more about metro/urban and rural differences.

incubators were founded only during the dot.com boom of the 1990s and not many of them survived the eventual bust. Apparently, most current for-profit incubators are sponsored by large companies (ASME 1996–2008).

Services provided by incubators almost always include facilities, that is, space. Beyond this basic offering, incubator services span a range that includes financing, business support services, know-how and technical support, and management consulting (ASME 1996–2008). Table 1 lists incubator sponsors and the types of services that incubators provide.

While some incubators focus on a particular sector (e.g., manufacturing) or industry (e.g., "high tech"), others accept a mix of businesses. According to ASME (1996–2008), 47% of US incubators belong to the latter group, 37% specialize in technology industries, and 7% in manufacturing.

3 Geographical Distribution of Incubators in the United States

Previous research indicates that incubators in rural regions do not perform as well as those in urban regions. Compared to their urban peers, business incubators in economically challenged and rural regions typically operate in a much more challenging context, e.g., small and often insufficient budgets, and fewer qualified and promising new businesses. In a competitive market, this would imply that business incubators are relatively more common in urban regions. As we have learned in the previous section, however, incubator sponsors are not necessarily entirely market-driven and only approximately one-fifth of all US incubators are without sponsors. Therefore, we cannot assume that market forces alone decide the geographical distribution of incubators.

Our main interest in this research is the difference between urban and rural areas. Unfortunately, the concept of rurality is ambiguous and different definitions exist (Isserman 2005, 2007; US Census Bureau 2005). We suspect that specific results with respect to density are sensitive to the choice of within-region geographical unit of analysis. To check this suspicion, we look at the incubator data using different urban and rural definitions.

3.1 Geographic Distribution of US Business Incubators by Major Region

We first look at US Census Divisions as major regions (Fig. 1), utilizing the Office of Management and Budget (OMB) definitions for Metropolitan Statistical Area (core area containing a population nucleus of 50,000 or more) and Micropolitan Statistical Area (core area containing a population nucleus of between 10,000 and 50,000). In both definitions, an additional criterion is that adjacent communities

Table 1 Incubator sponsors and incubator services

Sponsors	Facilities and equipment	Access to capital	Business support services	Management consulting and training	Know-how and technology
Universities and colleges	Office space	Investment/strategic partner linkages	Secretarial services	Business training	Research
Economic development corporations	Manufacturing space	Loan guarantees	Bookkeeping and accounting	Business plan preparation	Access to researchers
Local government	Conference room	Grant writing assistance	Legal services	Management team development	Product design assistance
State government	Computer laboratory	Financial proposal preparation assistance	Tax advice and services	Product and product design evaluation	Access to research libraries and collections
Federal government	Research laboratory	Loan application assistance	Marketing	Mentoring programs/ shadow boards	Intellectual property rights assistance
Private investors (*For-profit and Not-for-profit*)	Kitchen		Management services		Technology commercialization
Multiple sponsorships are possible	Warehousing space		Regulatory compliance		
	Broadband access		Information systems management		
	Hardware		International trade assistance		
	Software				
	Business equipment				
	Video conferencing				
	Furniture				

Source: Based on ASME (1996–2008)

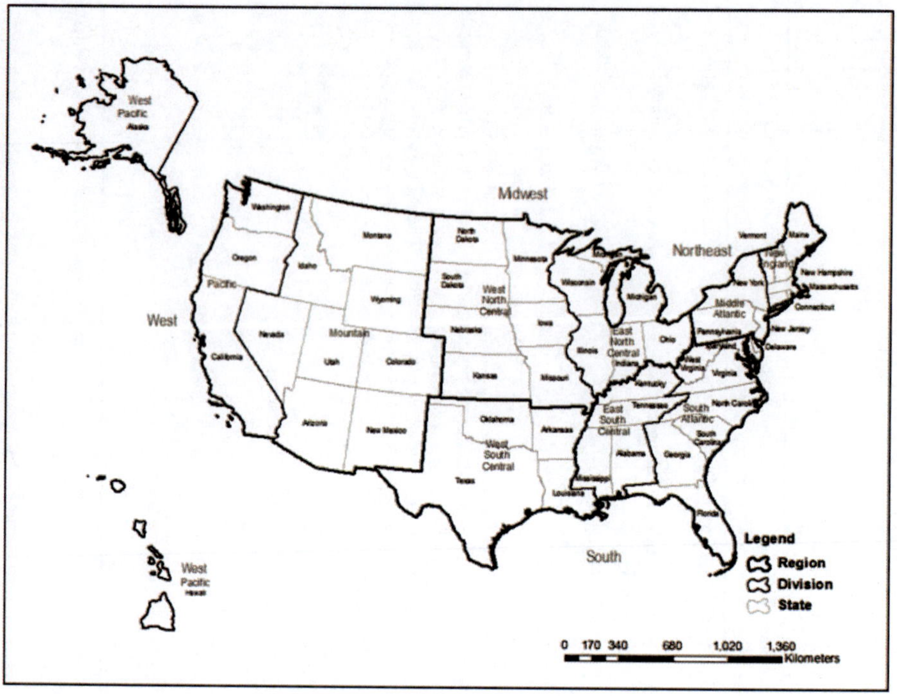

Fig. 1 Census regions and divisions

have a high degree of economic and social integration with the core. All other regions are referred to as Outside Core Based Statistical Areas (OCBSAs) (HAC 2008). Table 2 provides information about the geographical distribution of incubators, measured in number of incubators per 100,000 inhabitants.

Table 2a presents a mixed result. The density of incubators relative to population size differs as much between major regions as between OMB areas within a region. Overall, the density is lowest in the Pacific region and almost as low in the Mountain region. The similarity between these two regions ends there, however. Although the Pacific region has a very low incubator density in OMB Metro areas, the density in OMB Micro areas is the fifth-highest and the density in OCBSAs is the fourth-highest in the nine regions. By contrast, the density in the Mountain region ranks last in all three OMB areas. Thus, it appears that in the Mountain regions business incubators are comparatively rare, while in the Pacific region they are relatively rare only in the OMB Metro areas.

The East South Central region has the highest total business incubator density per population by a clear margin. However, the density in its OCBSAs is the second lowest in the United States, indicating that in this region, business incubators are largely a core urban phenomenon. In fact, the data suggest that incubators in the East South Central region are primarily located in OMB Metro areas.

Table 2 Distribution by major US region and by (a) OMB area (business incubators per 100,000 population); (b) HAC area (business incubators per 100,000 population)

(a)

Region	OMB metro area	OMB micro area	OMB OCBSA	Total
East North Central	0.2684	0.3923	0.3621	0.2901
East South Central	0.3171	0.1865	0.0373	0.5409
Middle Atlantic	0.2766	0.3690	0.1231	0.2798
Mountain	0.0864	0.0257	0.0151	0.1043
New England	0.1968	0.0926	0.7693	0.2155
Pacific	0.0926	0.2379	0.2504	0.1022
South Atlantic	0.2824	0.2579	0.3682	0.2859
West North Central	0.4534	0.3904	0.0562	0.3691
West South Central	0.2324	0.0801	0.0401	0.3526

(b)

Region	HAC urban	HAC rural	OMB OCBSA	Total	HAC rural rank	OMB OCBSA rank
East North Central	0.2745	0.3437	0.3621	0.2901	1	3
East South Central	0.3171	0.2238	0.0373	0.5409	5	8
Middle Atlantic	0.2781	0.2966	0.1231	0.2798	3	5
Mountain	0.0872	0.0378	0.0151	0.1043	9	9
New England	0.1994	0.3177	0.7693	0.2155	2	1
Pacific	0.0933	0.2162	0.2504	0.1022	6	4
South Atlantic	0.2844	0.2915	0.3682	0.2859	4	2
West North Central	0.4994	0.1870	0.0562	0.3691	7	6
West South Central	0.2324	0.1202	0.0401	0.3526	8	7

The West North Central region has the highest incubator density per population in OMB Metro areas and the second-highest density in OMB Micro areas, just barely behind the nation-leading East North Central region. Although it ranks fifth in the nation, the density in the OCBSAs is very low, almost as low as those of the regions ranked sixth through eighth. Only the last ranked Mountain region has a density in its OCBSAs that is much lower (more than three times lower) than that of the East North Central region.

In three regions incubators are relatively more common in non-core areas (OMB OCBSAs) than in core urban areas (OMB Metro and Micro areas). In the East Central region this difference is not very pronounced. In the other five regions, incubators are relatively more common in the core urban areas. The highest incubator density in non-core urban areas occurs in the New England region, with a density more than twice that of the second-ranked South Atlantic region.

Table 2b presents the same information as Table 2b, except that we use the urban and rural definitions of the Housing Assistance Council (HAC 2008). To check for sensitivity of results to the choice of definitions, we focus on the comparison of columns HAC Rural and OMB OCBSA in Table 2a, b.

The comparison reveals significantly different densities between HAC Rural and OMB OCBSA areas. For example, when HAC definitions are used, the ratio of urban-to-rural density is higher in the Mountain, Pacific, West North Central, and West South Central regions than in New England, which ranks first when the OMB definitions are used. However, New England still has the second-highest rural incubator density when the geographical unit of analysis is HAC areas, but it has the highest density by a factor of more than two when the unit of analysis is OMB. The highest rural incubator density occurs in the East North Central region (HAC definition); this region is third when using the OMB OCBSAs, just barely lagging the South Atlantic region.

Table 2b confirms the existence of substantial differences in business incubator density by major region. Some of the differences between using HAC Rural or OMB OCBSA are large, but qualitatively the results in the two table point in the same general direction. This is indicated by the value of the rank correlation between the two columns, which is $0.77\bar{7}$ and statistically significantly different from 0 with level of confidence better than 0.01 (the correlation between the raw densities is 0.679). The difference one minus the correlation coefficient reflects that the definitions in the two tables measure related but not identical urban and rural (or metro and non-metro)[2] areas. We also calculated the rank correlation between OCBSA and Total ranks in Table 2a and obtained a value of $-0.28\bar{3}$(the correlation between the raw densities is −0.290). The result indicates a certain trade off tendency in major regions in the United States between business incubator densities in Outside Core Based Statistical Areas and in core urban areas, respectively. The rank correlation between HAC Rural and Total ranks is 0.000 (the correlation between the raw densities is 0.172), which indicates at best a weak relationship between incubator densities in rural and non-rural areas between major US regions and no tradeoff, unlike the tradeoff between core urban areas and the rest of the country.[3]

3.2 *Distribution of US Business Incubators by State*

Because of great population and economic size differences between states, a meaningful comparison between them is one by densities. Figure 2 presents a map that reflects the regional differences in business incubator density that we already observed in Sect. 3.1. Specifically, incubators are comparatively less common in the Pacific region than elsewhere. In this region, California and Washington belong to the states with the lowest incubator densities.

The Mountain region is much more diverse in its use of incubators, measured by incubators per 100,000 inhabitants, than the earlier Sect. 3.1 suggests. While Arizona and Nevada are among the states with the lowest incubator densities, Idaho and New Mexico belong to the states with the highest densities. The Upper Midwest also has high densities, particularly North Dakota and Wisconsin. The two rural states Maine (New England) and West Virginia (South Atlantic) also belong to the eight states with the highest densities. By contrast, Arkansas and Texas in the

[2]Although rural and non-metro are sometimes used as synonyms, their definitions differ.

[3]A significance test has not been performed because the degree of freedom is too small to justify using an approximate t-distribution.

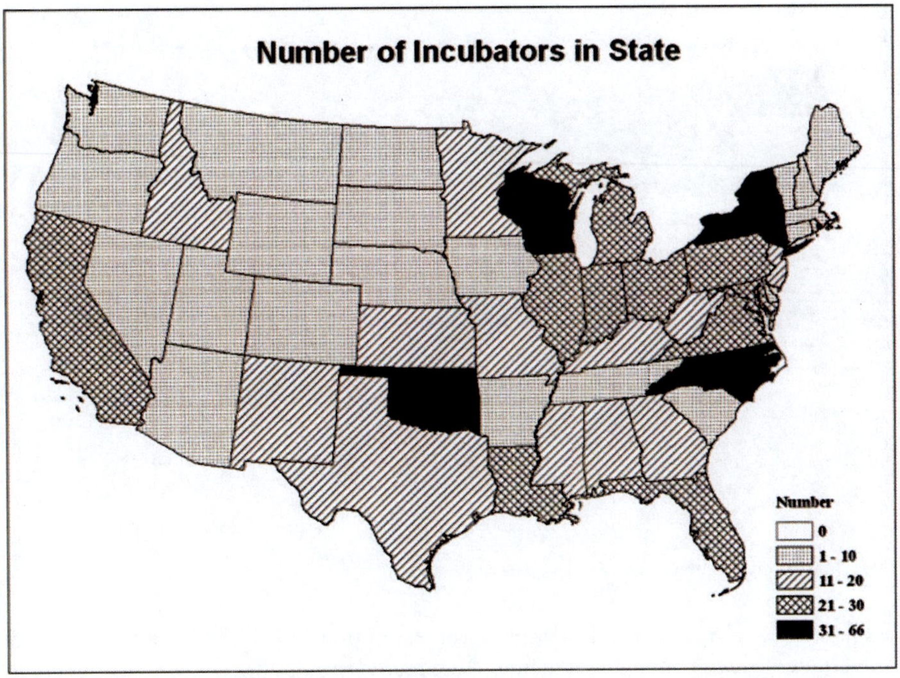

Fig. 2 Business incubators per 100,000 inhabitants by state

West South Central region, South Carolina in the South Atlantic region, and Massachusetts in the New England regions have among the lowest densities.

3.3 *Distribution of Incubators by Rural Versus Urban Areas*

The third look at the spatial distribution is by rural versus urban areas (Fig. 3). In 9 of 48 states, at least half of all incubators are located in rural areas (Oregon, Montana, Wyoming, Oklahoma, Mississippi, Kentucky, Vermont, New Hampshire, and Maine). In 13 states, we found no record of business incubators in rural areas (Oregon, Nevada, Arizona, Colorado, Texas, Arkansas, Tennessee, Florida, New Jersey, Connecticut, Rhode Island, Massachusetts, and Alaska – not shown on map).

Some of these findings are expected. For example, New Jersey is a primarily urban and Vermont is maybe the most rural state in the United States. Other results, however, are not as easily explained. For example, there is no obvious reason why there should not be any incubators in rural areas in Oregon, Texas, or Arkansas. There is also no easy explanation for the stark difference between the neighboring states Kentucky and Tennessee, both of which have counties located in rural

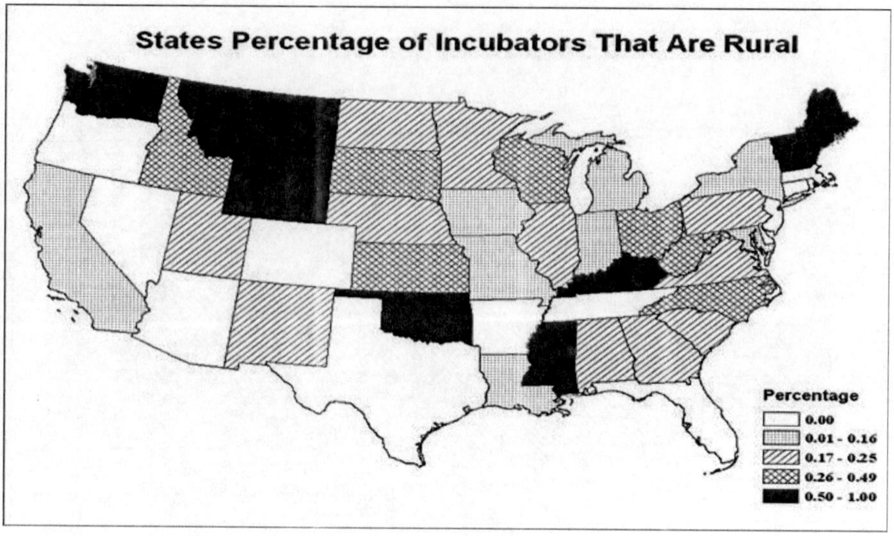

Fig. 3 Business incubators in rural versus urban areas

Appalachia, a region that is targeted for economic development by the federal government through the Appalachian Regional Commission.

The correlation coefficient between HAC Rural and HAC Total incubator densities at the state level is 0.803, and the correlation coefficient between OMB OCBSA and OMB Total is 0.680. This result differs from that obtained at the Census Division level and indicates that states that use incubators more frequently (higher incubator density) also use them more frequently in rural areas. The correlation coefficients are even higher at the county level and have the values of 0.981 (HAC definitions) and 0.907 (OMB definitions). These results indicate no significant differences in the use of incubators between core urban and metro areas on one side, and non-core statistical and rural areas on the other side. The result shows, however, how important the choice of the geographical unit of analysis can be.

3.4 *Distribution of Business Incubators by County*

Four-hundred and sixty-five counties, or 14.8% of the 3,141 counties in the United States, have at least one incubator. However, only 6.6% of the 2,213 rural counties (HAC definition) have at least one incubator, compared to 40% of 463 urban counties. This suggests that business incubators fit better in urban than in rural areas, but the evidence is inconclusive because it is possible that incubators serve a geographically larger area to compensate for thinner markets in rural regions. Of the 726 incubators that we were able to identify, 54.9% located in 138 counties with multiple facilities. All but 12 of these multi facility counties are classified as OMB-

Table 3 Number of counties with (a) at least one incubator; (b) incubator counties; (c) non-incubators counties

		OMB rural/urban		HAC rural/urban		
	Metro	Micro	OCBSA	Urban	Rural	Total
(a)						
Incubator	317	100	48	309	156	465
	29.11%	14.84%	3.48%	40.03%	6.59%	14.80%
Non-incubators	772	574	1330	463	2,213	2,676
	70.89%	85.16%	96.52%	59.97%	93.41%	85.20%
(b)						
Per capita income	20,794.00	16,700.50	15,434.50	20,883.00	16,342.00	19,219.00
Household median income	40,421.00	32,780.00	29,958.50	40,617.00	32,132.00	37,485.00
Family median income	50,111.00	41,002.00	36,821.50	50,196.00	39,719.00	46,452.00
Percentage of families below poverty level	8.06%	10.08%	11.16%	8.00%	10.33%	8.81%
Percent colleges	93.06%	86.00%	47.92%	93.53%	73.72%	86.88%
High school degree	83.40%	80.73%	77.80%	83.46%	79.38%	82.52%
BA degree	25.27%	17.23%	13.05%	25.39%	16.01%	22.47%
(c)						
Per capital income	18,653.50	16,782.50	15,817.50	19,516.00	16,246.00	16,680.50
Household median income	39,523.50	33,460.00	30,534.00	41,541.00	31,943.00	33,001.50
Family mid income	46,337.00	40,526.50	36,918.50	48,674.00	38,455.00	39,926.00
Percentage of families below poverty level	7.55%	9.57%	10.96%	6.93%	10.27%	9.69%
Percent colleges	46.63%	53.66%	15.11%	66.95%	25.26%	32.47%
High school degree	80.11%	78.45%	76.61%	81.62%	77.18%	78.54%
BA degree	16.17%	14.10%	12.68%	19.20%	13.02%	13.79%

metro counties. The 12 non-metro counties tend to be counties with large land masses and more than one large area of clustered development.

The economic profile (Table 3) of counties with incubators indicates that they have per capita, household, and family incomes above the national median. When separated into metro and non-metro counties, we find that the former account for the higher income level, whereas the latter tend to be below the national median. Using the OMB definition, metro areas with incubators have substantial higher economic indicators for per capita, household median and family median incomes than counties without incubators. The same does not hold for OMB micro and OCBSA incubator counties, which show income levels near or below the national and non-incubator counties median incomes. Incubator counties also tend to have a higher percentage of families living below the poverty level than counties without incubators. This is the case for urban and rural, and metro and non-metro counties and suggests that incubators do locate in response not only of economic opportunities, but also economic development need.

The population in counties with incubators tends to have higher levels of education and they are also far more likely to have a college offering at least a 2-year degree than non-incubator counties. This may partially account for the seeming paradox of higher education levels and higher poverty in rural incubator compared to rural non-incubator counties. In some counties with a college, students may account for a sizeable percentage of the population. Since they usually have very low incomes while in school, statistics show a higher degree of poverty than is actually the case.

Forty-seven percent of OMB OCBSA incubator counties and 73% of HAC rural incubator counties have a college, compared to 15% of OMB OCBSA and 25% of HAC rural counties without an incubator. At least in part, this reflects the significant role of colleges as sponsors of business incubators. The number of residents in incubator counties with a high school degree is 2% higher in rural and 3% higher in urban counties than in their non-incubator counterparts. Similarly, the percentage of the population with a college degree is also higher (Table 3).

3.5 *Evidence of Clustering*

We found evidence of clustering of incubators, and most clusters are found in the Northeastern United States and Great Lakes Region, the so called rust belt of the country (Fig. 4). This seems to support a view that incubators are created in response to industrial decline. The p-value that these clusters are chance events is very low. The observed mean distance is 0.34 km versus an expected mean distance of 0.91 km.

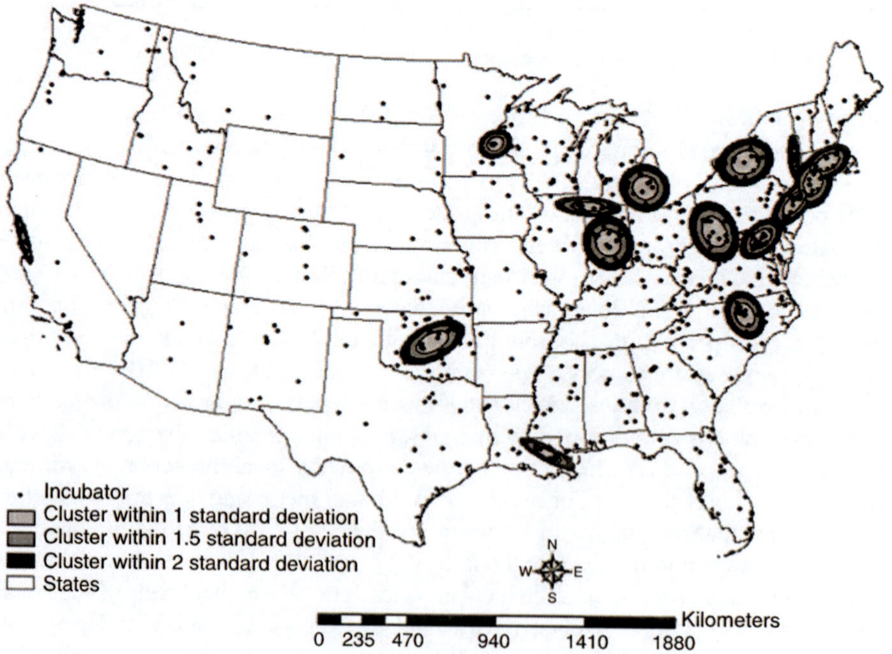

Fig. 4 Incubator clusters

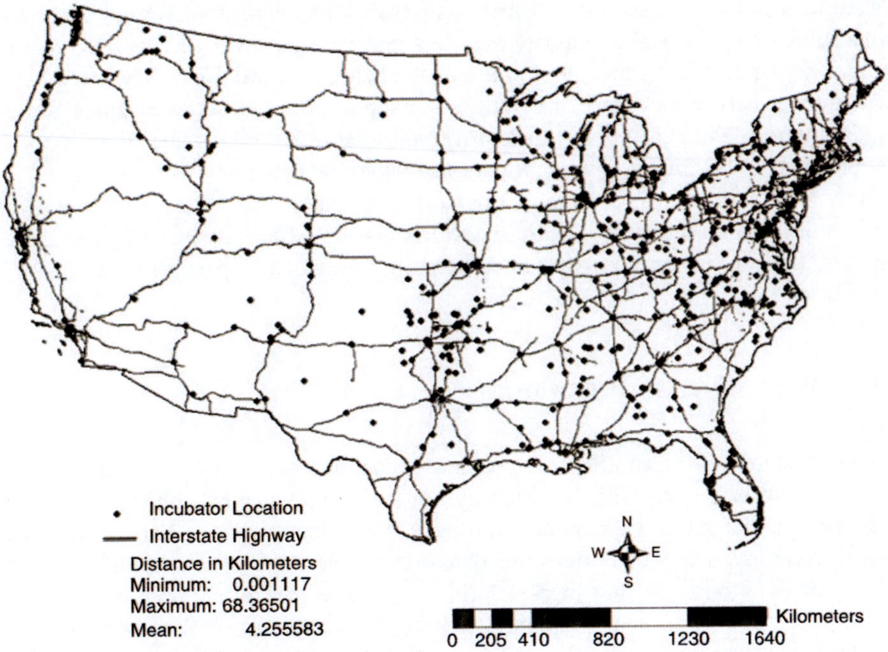

Fig. 5 Incubator clusters and interstate highways

Since accessibility is important to all businesses, we also checked for clustering of incubators along Interstate highways, and such clustering is indeed readily apparent (Fig. 5). The clusters along highways replicate some of the clusters shown in Fig. 5, particularly in the Northeast. The likelihood that the incubator location patterns relative to Interstate highways are a chance event is extremely low. Thus, Fig. 5 provides some plausible insight into why remote rural areas may be at a significant disadvantage, particularly if they are not served by an Interstate highway.

3.6 *Summary of Review of Geographical Distribution*

There are indications that incubators are used somewhat relatively less frequently in rural than in urban areas. Additional insights result from the use of the rural-urban continuum codes of the United States Department of Agriculture (USDA) (data not shown). The data based on these definitions show that in rural areas, business incubators are relatively more common (density is higher) in rural counties that are adjacent to a metro area than elsewhere, even a bit greater than in metro areas. This could indicate a greater need for business incubation in these than in metro counties and possibly better conditions for incubator success than in non-adjacent rural counties.

Since rural areas have fewer inhabitants than urban areas, we are not surprised that rural incubators are a majority in only a few states, but we are surprised that in about a quarter of all states we found no incubators in rural areas. The very large differences between Census Divisions, and even more those between states, are of great interest. The reasons for these differences are not obvious and deserve to be explored because they might be reflections of specific programs, agency initiatives, regional or state needs, or legal frameworks that favor incubators, in general, or urban over rural incubators. Thus, an exploration of differences might yield information relevant for an improved understanding of incubator performance.

4 Alternatives to Incubators

We considered the possibility that business incubators may not be as well suited to rural as to urban areas. This is generally not the case, however, although there are indications that the most remote rural areas, those that are not adjacent to metro areas, have lower incubator densities than other areas. As Weiler's (2000) research suggests, it is possible that market failure results in too few opportunities being recognized and exploited in sparse markets. This applies to many remote rural areas. It is also likely that some markets are just too sparse to support an incubator, even with sponsors picking up some of the costs. In such areas, where according to Weiler (2000) and Scorsone and Weiler (2004) informational market failures are relatively likely, alternatives to business incubators will be most valuable.

4.1 Results of Literature Search

Our literature search for alternatives to business incubators yielded few relevant results.[4] The most interesting idea, and one particularly applicable to sparse markets, is that of a virtual incubator (Zedtwitz 2003). Schwartz et al. (2008) describe the Palestine-Israel Virtual Incubator, which is to be focused on information and technology businesses. A major aspect of the proposal put forward by the authors is the establishment of a knowledge network, primarily on-line based. In addition to facilitating access to and the exchange of technical information, in general, a particular advantage of the proposed network is that it allows trading of tacit knowledge, such as how a process is best organized, that is not readily available otherwise without regular face-to-face contact. Thus, the network overcomes one of the major disadvantages of remote or otherwise isolated regions, which is their effective distance from technology, knowledge, and innovation centers. A somewhat related idea is currently being promoted by Microsoft in its Mobile Incubation Week program (Hoskins 2009) and

[4]We conducted searches on Google, Business Source Premier, and EconLit.

some incubators include virtual tenants, that is, start-up companies that are not co-located with the incubator but that receive services.

Cooper and Dunkelberg (1986) list and analyze the various paths someone might take to business ownership, such as start-up, purchase, or inheritance. Each path comes with different needs for entrepreneurial skills. A new owner of an existing business has some, but not all, of the same needs for information, knowledge, and support as a start-up entrepreneur, and therefore could also benefit from some of the services provided by an incubator (see Table 1). In the case of the purchase of a business, for example, the network providing access to sources of capital may be as valuable to the transfer of the management-ownership of an existing business, as it is in the building of a new business. In the case of an inheritance, the new owner may not be well prepared take over and run the business, and might benefit from management services. Thus, the study of entrepreneurship and its support by incubator organizations should include management ownership transfers. After all, it is well known that independently owned business are particularly vulnerable during their start-up, and again during the transfer-of-ownership phase. Mucalov and Mucalov (not dated) mention that 70% of family businesses in Canada do not survive the transition to the second and only 10% survive the transition to the third generation.

Cooper and Dunkelberg (1986) analyzed a sample of some 1,800 owner-managers. They found that 49% (890 firms) had started, 28% (504 firms) had inherited, and 15% (275 firms) had purchased the business. The remaining 8% had been promoted, brought in, or had acquired ownership through other means. What is noteworthy is that the parents of half of those who started a new business had also been business owners. In the case of purchases, this percentage was 43. Not surprisingly, with 83%, it was highest in the case of inheritance. These percentages suggest that entrepreneurial skills can be passed on from parents to children. If this is indeed the case, then maybe entrepreneurial programs in schools should consider involving the students' parents to increase the success of such programs.

The importance of the entrepreneurs' prior activities in the start-up of is one of the factors influencing success of knowledge intensive start-ups in the business service sector (Koch and Stahlecker 2004). The owner brings knowledge and connections from previous experiences and activities, and spatial proximity was found to be important to the exchange of knowledge. This does not come as a surprise, as it is well known that many high technology firms were started by former employees of other computer or software firms. Early in the twentieth century, the same phenomenon was observed in the automobile industry. Thus, firms/previous employers seem to fulfill some of the same functions as an incubator.

Related to the contributions of Cooper and Dunkelberg (1986) and Koch and Stahlecker (2004), Zander's (2007) research calls attention to the reasons of why new firms are being established. While we have learned much about why firms exist, the reason of why they are being started has not yet been well researched.

We found some interesting adaptations of the concept of an incubator. One of these is Bucketworks in Milwaukee, Wisconsin, which was started in May 2002. On its website, it refers to itself as a health club for the brain (Bucketworks 2007–2009). Its facility has 22,000 square feet that house a café, art gallery, theater, business combinator,[5] workshops and labs for a variety of crafts and arts, computer labs, print shop, and a library. The organization also has a garden. The geographical target area of Bucketworks is its community. The idea is that bringing together individuals with a wide range of skills and interests, and by offering training programs, the organization will create an environment that produces new ideas, encourages their implementation, and supports risk-taking.

Bucketworks claims 700 members who, in the last 5 years, "... have created 28 new companies, 65 new jobs, a high school, a student film festival, 7 theatre companies, 3 technology companies, and innumerable pieces of art, items for sale, performances, gatherings, shows, and events – there were over 863 events a Bucketworks in 2006" (Bucketworks 2007–2009). Members represent a wide range of professions, from lawyers and accountants to artists. Membership fees range from $20 a year for students to $200 for corporations and membership is open to anyone at least 14 years old. Educational and training programs range from business and entrepreneurship to dance and theater. Bucketworks also rents out space. The organization's name alludes to the water bucket that was passed from person to person to put out fires before more effective methods came about. It reflects Bucketworks' philosophy that economic and community revitalization require a holistic approach and the involvement of all members of the community.

Finally, we know of instances where local governments purchased an old building and turned it into space for rent to new and existing businesses. This type of cohabitation provides only the most basic of incubator services, space, but by bringing a number of such businesses together under one roof, they may be able to share services and facilities, and to the extent that they share challenges, it is easier to provide them with services to deal with them.

4.2 *Other Start-Up Support Organizations*

Our research alerted us to the importance of prior family entrepreneurial experience, and of the new entrepreneurs' experiences in previous jobs and activities. This suggests that parents and employers play a role as business start-up facilitators. We also find some of these roles played by established members and/or the management of cooperatives, when new members join. This applies particularly to agricultural cooperatives, because each individual producer has an interest in the competence, knowledge, and success of the other producer-members. This may

[5]Unfortunately, the term is not defined on Bucketworks' homepage, but it seems to be a business accelerator or something similar.

also be true for cooperatives that bring together craftsmen and/or artists for the purpose of joint marketing and selling.

Non-profit economic development organizations other than incubators can also provide services to facilitate business start-ups. For example, one of the authors was part of an effort that resulted in the establishment of a non-profit supported by a local government that provides funding to new and existing small businesses. Although the organization provides loans, the major objective is to help businesses to become "bankable" by assisting them with the preparation of the business plan and other for a sound start. The organization also seeks to facilitate the application for conventional loans, by performing much of the work of a loan officer, so that a bank would only have to meet due diligence requirements and not have to spend time and resources on other activities that normally make small loans expensive to provide. In the start-up phase, the non-profit organization may give the new firm a line of credit to finance operating expenses, but leave the financing of facilities and/or equipment to conventional lenders who are more inclined to give loans if they are backed by assets such as real estate or tradable equipment. This reduces the funds requirement for the non-profit and serves local and regional banks to make more business loans, develop new clients, as well as meet the requirements of the Community Reinvestment Act.

For businesses that receive start-up funding from venture capitalists, the venture fund may enforce sound business practices by placing representatives in the top management of the start-up firm, particularly in the position of chief financial officer (CFO). Thus, in addition to capital, some venture funds also provide managerial expertise to protect their investment.

Finally, there are a number of organizations that provide management and entrepreneurial training. Land grant universities are particularly well positioned to provide such services, and most of them do, through their Extension service. The advantage of this delivery organization is that Extension has a state-wide presence and can therefore bring expertise and knowledge to, or close to, where it is needed. Local Extension offices may also have facilities where interactive online courses can be delivered. Other colleges also provide entrepreneurial training and many also have incubators. Increasingly, business schools host new business competitions among students to encourage them to start their own business. Many high schools offer related programs to their students. Last but definitely not least, the United States Small Business Administration (SBA) provides many services to small businesses that are also relevant for start-up businesses. One of these is SBA's Service Corps of Retired Executives, which brings managerial and marketing experience to small businesses. SBA also maintains Small Business Development Centers (SBDC), most of them located on university campuses. The range of SBA services includes financing, from loans to grants. For small business owners and those interested in starting a business, SBA offers online training courses. Finally, the mission of SBA includes the facilitation of small businesses' access to government contracts (SBA not dated).

5 Summary and Conclusions

Although the focus of this research was incubators, we do not want to lose sight of the ultimate goal, which is the facilitation of business start-ups, business survival, and entrepreneurship, in general. A look at the services provided by incubators in Table 1 makes it clear that most of those services can also be delivered through other types of organizations, though maybe not in the same combination or not always as effectively. The reason we are interested in this issue is because we do not think that incubators are equally viable in all environments. Although we did not discover a systematically reduced presence of incubators in rural areas, we did find evidence that the most remote rural areas, that is, those not adjacent to a census metro area, have among the lowest business incubator densities. Since these regions already face many challenges to adapt to changing economic conditions, such as sparse or thin markets, distance from centers of innovation and often also from suppliers and customers, searching for substitutes for and alternatives to incubators serves an important policy purpose. In this article, we provide preliminary results and some first ideas of incubator alternatives and substitutes.

We also leave some questions unanswered, particularly the question of why different census regions and states use incubators to very different degrees. The answers might provide new information about conditions that are supportive or incubators in particular, and entrepreneurship in general. Factors responsible for different degrees of popularity of incubators may include tax provisions for start-ups, rules related to initial public offerings (IPO), including the tax treatment of the proceeds, degree of complexity of complying with regulations, access to venture capital, entrepreneurial training in schools and colleges, availability of appropriate facilities, existence of industrial clusters, or the characteristics of the available labor force. In addition, we should not forget need. Unfortunately, public policy, in particular, becomes active only after a problem has occurred. In the case of new start-ups, it is harder to create new businesses when the economy is down than when it is doing well. This is also an important suggestion for those responsible for policies supportive of entrepreneurship.

References

Acs Z, Armington C (2006) Entrepreneurship, geography, and American economic growth. Cambridge University Press, Cambridge, MA

Acs ZJ, Audretsch DB (1993) Introduction. In: Acs ZJ, Audretsch DB (eds) Small firms and entrepreneurship: an east-west perspective. Cambridge University Press, Cambridge

Adkins D (1996) A decade of success: NBIA 10th anniversary survey of business incubators, 1985–1995. National Business Incubation Association, Athens, OH

Adkins D (2001) A report for the Japan association of new business incubation organization: summary of the US incubation industry. National Business Incubation Association, Athens, OH

ASME (1996–2008) Business incubators. American Society of Mechanical Engineers, Professional Practice Curriculum online. http://www.professionalpractice.asme.org/SecuredModules/business_functions/incubators/index.htm

Bearse P (1998) A question of evaluation: NBIA's impact assessment of business incubators. Econ Dev Q 12(4):322–333

Brock WA, Evans DS (1989) Small business economics. Small Bus Econ 1(1):7–20

Bucketworks (2007–2009) The bucketworks story. http://bucketworks.org/content/the-bucketworks story

Cheng S, Schaeffer PV (2009) Evaluation without bias: a methodological perspective of performance measures for business incubators. Regional Research Institute, West Virginia University, Morgantown, WV

Cooper AC, Dunkelberg WC (1986) Entrepreneurship and paths to business ownership. Strat Manag J 7:53–68

Costa-David J, Malan J, Lalkaka R (2002) Improving business incubator performance through benchmarking and evaluation: lessons learned from Europe. Paper presented at the 16th international conference on business incubation. National Business Incubation Association, Toronto, Canada, 28 April–1 May 2002

Ekstrom B, Leistritz FL (1988) Rural community decline and revitalization: an annotated bibliography. Garland, New York

HAC (2008) What is "rural"? Working towards a better programmatic definition. Housing Assistance Council, Washington, DC

Hackett S, Dilts D (2004) A systematic review of business incubation research. J Technol Transf 29(1):55–82

Hoskins B (2009) Calling all start-ups! Announcing mobile incubation week. Microsoft Corporation. http://www.microsoftstartupzone.com/Blogs/brian_hoskins/Lists/Posts/Post.aspx?ID=7

Isserman AM (1993) State economic development policy and practice in the United States: a survey article. Int Reg Sci Rev 16(1–2):49–100

Isserman AM (2005) In the national interest: defining rural and urban correctly in research and public policy. Int Reg Sci Rev 28(4):465–499

Isserman AM (2007) Getting state rural policy right: definitions, growth, and program eligibility. J Reg Anal Pol 37(1):72–79

Koch A, Stahlecker T (2004) Firm foundations in the knowledge-intensive business service sector: results from a comparative empirical study of three German regions. Institut für Angewandte Wirtschaftsforschung, Tubingen, Germany, IAW discussion paper no. 17

Lalkaka R (2000) Assessing the performance and sustainability of technology business incubators. Paper presented to new economy and entrepreneurial business creation in mediterranean countries. International Centre for Science and High Technology, International Centre for Theoretical Physics, and Third World Academy of Sciences, Trieste, Italy

Lewis D (2002) Does technology incubation work? A critical review. National Business Incubation Association, Athens, OH

Loveman G, Sengenberger W (1991) The re-emergence of small-scale production: an international comparison. Small Bus Econ 3(1):1–37

Mucalov J, Mucalov G (not dated) Passing down a business need not be a recipe for its failure. Harper Grey LLP, Vancouver, Canada. http://www.harpergrey.com/articles-51.html

Pulver GC (1989) Developing a community perspective on rural economic development policy. J Community Dev Soc 20(2):1–4

Quittner J (1999) Can business incubators justify their existence? Business Week, October 25. http://www.businessweek.com/smallbiz/news/coladvice/trends/tr991025.htm

SBA (not dated) US small business administration webpage. www.sba.gov

Schwartz D, Bar-El R, Malul M (2008) A joint virtual advanced technology incubator: a new pattern of Israeli-Palestinian economic cooperation. Peace Econ Peace Sci Publ Pol 14(2), article 3

Scorsone E, Weiler S (2004) New markets as informational asymmetries. Econ Dev Q 18(3): 303–313

Storey D (2003) Entrepreneurship, small and medium sized enterprises and public policies. In: Acs Z, Audretsch D (eds) Handbooks of entrepreneurship research. Kluwer Academic, Boston, pp 473–511

Tödtling F, Wanzenbock H (2003) Regional differences in structural characteristic of start-ups. Entrepren Reg Dev 15(4):351–370

U.S. Census Bureau (2005) The urban and rural classifications. Chapter 12 of Geographic areas reference manual. www.census.gov/geo/www/garm.html

Waldfogel J (2007) The Tyranny of the market: why you can't always get what you want. Harvard University Press, Cambridge, MA

Weiler S (2000) Information and market failure in local economic development: a new role for universities? Econ Dev Q 14(2):194–203

Yu J, Cheng S, Stough R (2006) Does entrepreneurship matter in China's economic growth? Paper presented at the 53rd annual North American meetings of the regional science association international, Toronto, Canada

Zander I (2007) Do you see what I mean? An entrepreneurship perspective on the nature and boundaries of the firm. J Manag Stud 44(7):1141–1164

Zedtwitz MV (2003) Classification and management of incubators: aligning strategic objectives and competitive scope for new business facilitation. Int J Entrepren Innovat Manag 3(1,2):176–196

Creative, Intellectual and Entrepreneurial Resources for Regional Development Through the Lens of the Competing Values Framework: Four Australian Case Studies

John Martin

1 Introduction

Stories of regional development are resplendent with examples of businesses and associations failing because many of the basics of business were simply not followed. The entrepreneur who over extends themselves bringing a new product to market; the business that fails to develop local support networks; the established business that does not seek new and innovative products soon becomes irrelevant and fails; as does the business that fails to find new markets for its products are common stories. Our experience is that the same issues apply to associations running local community events. Why is it that community event committees become entrenched in a particular modus operandi and fail to see that there are other business processes which need attention to ensure long-term business sustainability?

Such behaviour applies equally to the creative entrepreneur who develops a new service or product but does not attend to proper business management practices as it does to the manager whose focus is attending to business relations within the firm at the expense of developing new services and products in an ever changing market. New and established businesses contributing to regional development are, at some time in their history, inappropriately oriented in their management and organisation values in terms of the needs of the firm and the market. So too with community organisations. We are interested in ways in which we can assist local event committees involved in regional development to obtain the strategic insight into their modus operandi – and the needs of their market – to ensure they continue contributing to sustainable regional economic development.

In this paper we first outline the Competing Values Framework before applying it to four case studies to show how businesses need to balance competing values if they are to be successful the long run contributing to sustainable regional

J. Martin
Law and Management, Centre for Sustainable Regional Communities, La Trobe University, PO Box 199, Bendigo, VIC 3552, Australia
e-mail: John.Martin@latrobe.edu.au

K. Kourtit et al. (eds.), *Drivers of Innovation, Entrepreneurship and Regional Dynamics*, Advances in Spatial Science, DOI 10.1007/978-3-642-17940-2_14,

development. We conclude with a discussion of these events in terms of the Competing Values Framework asking what can governments do to ensure regional events such as outlined and analysed remain viable over the long term.

2 The Competing Values Framework

Steven ten Have and his colleagues (2003) rank the Competing Values Framework as one of the 40 most important frameworks in the history of business. The originators of the concept, Bob Quinn and John Rohrbaugh (1983) did not set out to develop a conceptual framework based on competing values. In analysing earlier research by John Campbell and his colleagues who "created a list of thirty-nine indicators that he claimed presented a comprehensive set of all possible measures of organisational effectiveness" to "determine if patterns or clusters could be identified" (Cameron and Quinn 1999, p. 30) Quinn and Rohrbaugh identified two major dimensions that organised the indicators into four main clusters.

> One dimension differentiates effectiveness criteria that emphasise flexibility, discretion, and dynamism from criteria that emphasise stability, order, and control. ... The second dimension differentiates effectiveness criteria that emphasise an internal orientation, integration, and unity from criteria that emphasise an external orientation, differentiation, and rivalry. (Cameron and Quinn 1999, pp. 30–31)

These values "compete" because individual managers, and thus organisations, are typically oriented toward either one of these dimensions. Diagrammatically the Competing Values Framework is represented as set out in Fig. 1. Over the last two decades Cameron and Quinn and their colleagues at the University of Michigan have refined and applied the Competing Values Framework such that it is, as ten

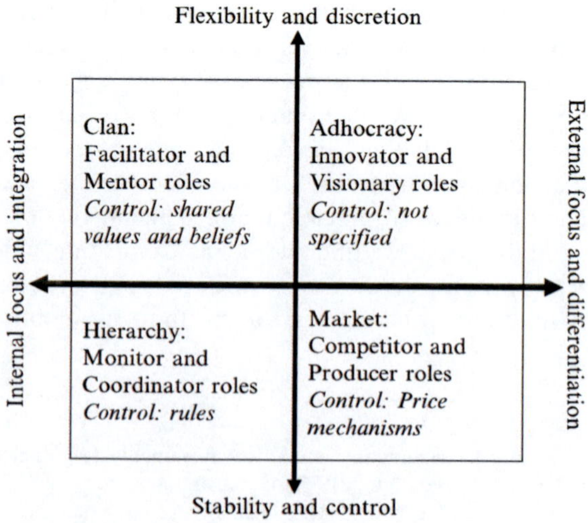

Fig. 1 The competing values framework (Cameron and Quinn 1999, p. 32) [additional control emphasis from Langfield-Smith (1995)]

Have and his colleagues (2003) have identified, an important framework for students of business and management. We have previously used this framework to assess the leadership styles or mayors and CEOs Australian local government (Martin and Simonds 2002).

As Fig. 1 shows Cameron and Quinn have identified four leadership styles characterising the four main clusters of the Competing Values Framework: clan, adhocracy, hierarchy, and market. They refer to these as "culture types" and organisations led by managers with these leadership styles create organisational cultures reflecting their style. Cameron and Quinn have summarised these four leadership styles as follows:

> Leadership roles in the clan culture seeks consensus seen as facilitator or mentor. The facilitator is people and process oriented. This person manages conflict and seeks consensus. His or her influence is based on getting people involved in decision-making and problem-solving process. Participation and openness hallmark actively pursued. On the other the mentor is caring and empathetic. This person is aware of others and cheers for the needs of individuals. His or her influence is based on mutual respect and trust. Morale and commitment are actively pursued. (Cameron and Quinn 1999, p. 114)

The stereotypic view of entrepreneurs and innovators does not fit this description of the clan leadership style. Yet we will argue in this paper that there are circumstances where this leadership style is entirely. More typically found in successful community organisations the clan leadership style is effective in appropriate settings.

> Leadership roles in the adhocracy culture are either the innovator or the visionary. The innovator is clever and creative. This person envisions change. In his or her influence is based on anticipation of a better future and generates hope in others. Innovation and adaptation are actively pursued. The visionary is future oriented in thinking. This person focuses on where the organisation is going and emphasises possibilities as well as probabilities. Strategic direction and continuous improvement of current activities is a hallmark of this style. (Cameron and Quinn 1999, p. 114)

This leadership style fits well with the stereotypic view of the entrepreneur and innovator. Being externally oriented to the possibilities in the world around them looking for opportunities to add value is a common view of the way in which entrepreneurs and innovators typically act.

> Leadership roles in the hierarchy culture layout the monitor and the coordinator. The monitor is technically expert and well-informed. This person keeps track of all details and contributes expertise. His or her influence is based on information control. Documentation and information management is actively pursued. The coordinator is dependable and reliable. This person maintains the structure and flow of the work. His or her influence is based on situational engineering, managing schedules, giving assignments, physical layout. Stability and control are actively pursued. (Cameron and Quinn 1999, p. 114)

Neither does this leadership style fit with our expectations of the entrepreneur and innovator. Their internal focus sees them preoccupied with ensuring the organisation is properly managed and functions according to a set of rules and regulations.

> Leadership roles in the market culture are the competitor and the producer. The competitor is aggressive and decisive. This person actively pursues goals and targets and is energised by competitive situations. Winning is a dominant objective, and the focus is on external competitors and marketplace position. The producer is task-oriented and work-focused. This person gets things done through hard work. He is or her influence is based on intensity and rational arguments around accomplishing things. Productivity is actively pursued. (Cameron and Quinn 1999, p. 114)

This leadership style also fits with our view of the typical entrepreneur and innovator. Once again the active external orientation is what confirms this view.

In this paper is we argue that each leadership style is appropriate and important for innovative and entrepreneurial regional development. In the four case studies presented below we see that striking the right balance between these competing tensions is required for success. It is a function of the leader's ability to work across all four styles and the competing values that underpin them that will determine the success of their regional development project.

Cameron and Quinn developed The Organisational Culture Assessment Instrument (OCAI) (1999, pp. 18–27) to assess important aspects of an organisation's underlying culture. The instrument has been used extensively and has been found to predict organisational performance. It consists of six categories covering:

1. The dominant characteristics of the organisation
2. Organisational leadership
3. Management of employees
4. Organisation glue
5. Strategic emphases
6. Criteria of success

Within each category the respondent has four choices reflecting the four cultures of the Competing Values Framework. They are asked to allocate 100 points across each choice as to how they believe the organisation responds now.

As part of the organisational analysis the categories are reviewed twice, first time to identify what they believe the organisation culture is like now, the second time what they would prefer the culture to be like in 5 years time. In our analysis of the four case studies below, based on our extensive evaluation of each event, we have OCAI assessment process to each case, as it currently exists, and as we believe it should be in 5 years time if each organisation is to continue run viable, and sustainable regional events.

The assessment of an organisation's culture is mapped on the diagram found in Fig. 2. This diagram shows that there is an inherent tension between opposing quadrants. For example between clan and market culture there is a collaborate versus cooperate tension. Between the hierarchy and adhocracy culture there is a control versus create tension. Cameron and his colleagues (2006) refer to the "entrepreneurial cycle" as the inevitable process organisations progress through to deal with these cultures and the inherent tensions between them.

We will use this schema to chart the current and the preferred future organisational culture of the four case studies presented in this paper.

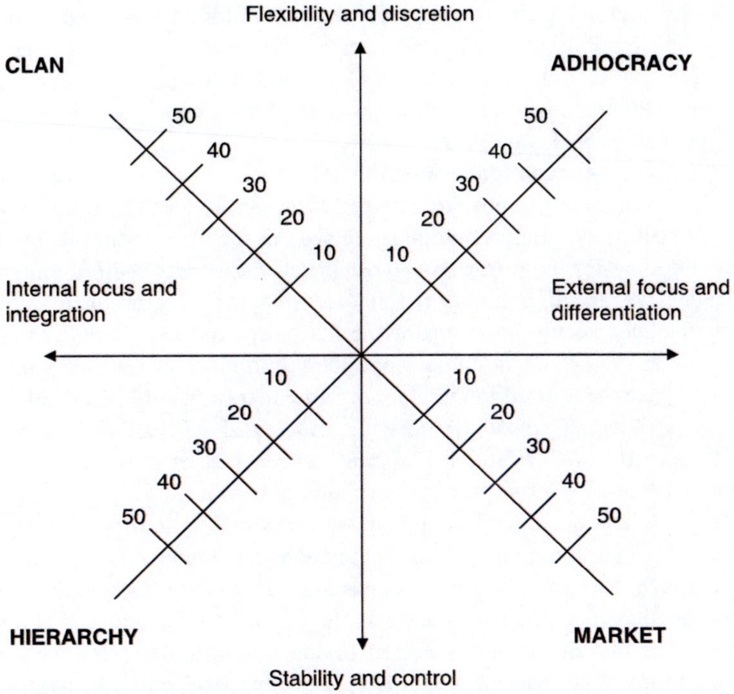

Fig. 2 The schema for identifying the outcomes of the organisational culture assessment instrument (adapted from Cameron et al. 2006)

3 Case Studies

These case studies, described and analysed in terms of the Competing Values Framework, were evaluated for their economic impact on the regional economy of central Victoria on behalf of the City of Greater Bendigo. In undertaking this research we gained a close appreciation of the organisation and management of the respective associations responsible for each case. This paper provides them with additional information for them to consider in the strategic management of their events in the future.

3.1 The Elmore Farm Machinery Field Days

The Elmore Farm Machinery Field Days (field days) held every year in October attracts around 40,000 visitors over 3 days. The Elmore field days are the largest of their type in the southern hemisphere displaying the latest range of products and services of interest to farmers and the agriculture industries. The field days have been in operation for over 40 years and are a major source of income for community groups in the region. Staffed largely by volunteers income is generated through admission and exhibitor fees. The field days committee is an incorporated body which employers several staff all year round and owns and operates the events centre, some 100 acres just outside the town of Elmore which now hosts a number of other types of events each year.

As a community-based organisation the field days committee is fortunate to have high levels of volunteering provided by members of the local Elmore community. This is reflected in the relatively high score seen in the clan culture which is also a preferred outcome for the future. As an organisation that has been running for over four decades we see this tradition reflected in strong hierarchical culture that underpins the successful operations of the field days and the events centre. While this operational culture needs to continue the committee recognises that a greater creative presence is required in the planning and operations of the annual field days if it is to renew its offering to visitors. To this end the field days committee has created the position of "Features Director" whose task it is to ensure a topical, contemporary feature is organised for each annual field days. Last year this director created a major new event, CO_2 and You, which brought attention to the important issue of effective carbon management in the context of understanding what farmers can do to reduce their greenhouse gas emissions. Given the importance of this topic the committee has installed this feature as the key theme for the next 5 years.

In Fig. 3 the current versus preferred organisational culture for the field days committee shows a shift in emphasis from a hierarchical culture towards greater creativity found in adhocracy. Given the success of the field days committee over several decades developing a well organised approach to the management and operations of the field days and the events centre for other functions the field days would benefit from a focus on creativity and innovation around new ways of engaging visitors to the event.

A similar shift from a focus on collaboration towards the competitive culture of a market orientation will also give new life to the field days. With an extensive support base in the local community as well as exhibitors who faithfully return year after year the field days event can rely on this support while the committee focuses on more effective ways to market the field days to a wider audience. With over 40,000 visitors over the 3 days of the field days the event is traditionally well supported by the farming community.

A shift in the entrepreneurial cycle toward the creativity of adhocracy and the speed and urgency of competition in a market orientation will add value to the efforts of the field days committee making this event for sustainable as it addresses challenging issues facing farmers in south-eastern Australia.

3.2 *The Heritage Food and Wine Festival*

Central Victoria is world-famous for the high-quality of the wines produced in the region. Finding ways to showcase this product is important to the regional wine

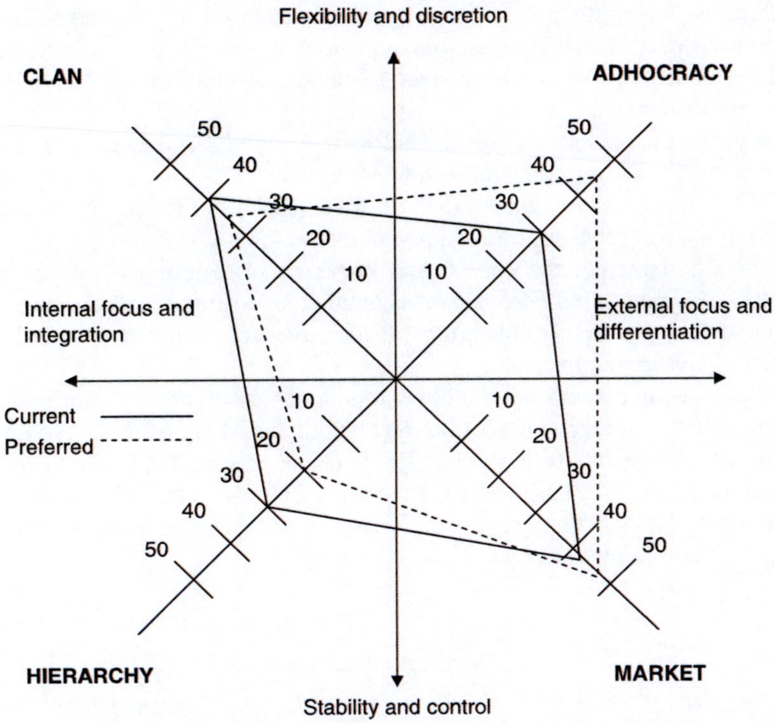

Fig. 3 The current and preferred organisational culture assessment of the organisation and management of the Elmore Field Days committee

industry and the Heritage Food and Wine Festival, known as "Heritage Uncorked" is an important event in this regard. Places are limited, primarily due to the amount of seating at restaurants and tickets to the event are sold out early. It is held over a weekend and participants have the opportunity to taste wine in a range of restaurants and heritage venues in the heart of Bendigo, a regional city of 100,000 people with a history built on the discovery of gold in the 1850s. Bendigo's defining architectural heritage from the Victorian era was built on the rich deposits of gold which lay under the city and are still mined today.

Heritage Uncorked is successfully embedded in the Bendigo events calendar. From the point of view of the patrons of the event, the weekend in Bendigo is very highly rated. The wines are the major draw card for what is essentially a mature and slightly female biased customer profile. And nobody complains about the price.

The event revels in the heritage buildings of the inner city. Patrons have high expectations of the event and those expectations are generally met. However the evaluation report (Nanere and Reimers 2006) reveals that the major group attending the festival are "enthusiastic locals" who return year after year to sample the local fare. These locals, in fact, know the local winemakers and believe they are supporting the local industry simply by their attendance. They are not great purchases of wine at this event.

While the large minority of visitors from outside the region staying in local accommodation and making generous purchases of wine from the festival, more could be done to market the festival to these outsiders through creative and innovative campaigns.

The visitors to the city for a festival also commented that they did not see much of the heritage of the city while they were here. This is clearly a marketing opportunity for the city council and the festival organisers to open up more of the heritage buildings for these events for visitors to enjoy.

In Fig. 4, depicting the current and preferred organisational culture for the Heritage Food and Wine Festival we recommended a shift in emphasis along the control and create tension toward more creative and innovative strategies for conducting the annual festival.

The organising committee could look to more innovative ways of offering wine tasting and local produce in other heritage buildings. Many of these buildings are owned and managed by the city council, are open on weekends (the Art Gallery, the Information Centre – a heritage listed property which was the original post office for the city). The Art Gallery already serves food so a wine tasting opportunity would require minimal changes.

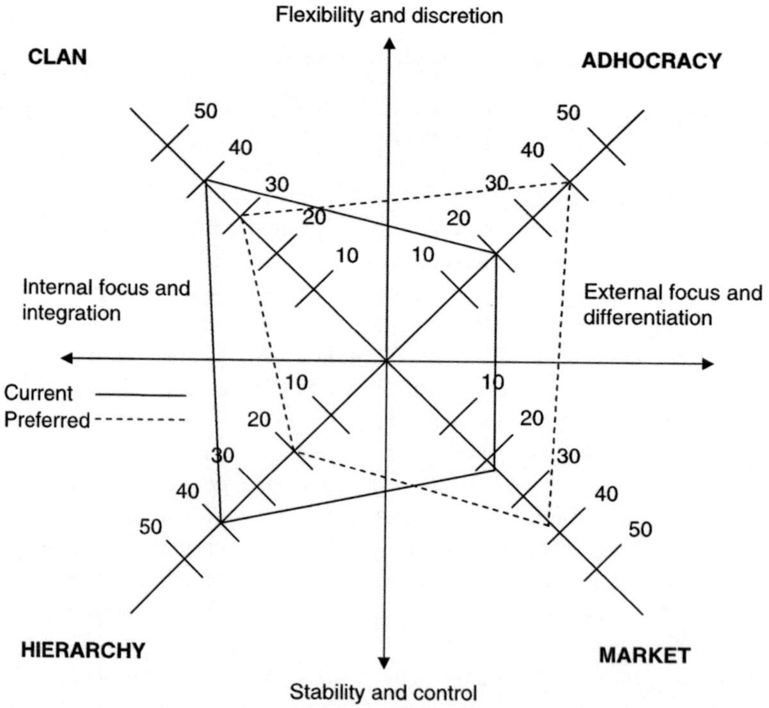

Fig. 4 The current and preferred organisational culture assessment of the organisation and management of the heritage food and wine festival

We also recommended a shift in emphasis along the collaborate and compete tension towards a greater focus on competing for external markets to encourage the more lucrative weekend visitors to the region. This could be done via conventional marketing programs, through wine distribution outlets, media advertising, special discounts for early birds, and so on. For many small businesses that make up the region's winegrowers association this may require a greater investment than they have typically made.

The strong support provided the festival by local wine enthusiasts is expected to continue under this strategy. With a greater emphasis on creative marketing potential visitors beyond the region will provide a greater opportunity for the festival to grow and develop into an even more successful event.

3.3 The Bendigo National Swap Meet

The Bendigo National Swap Meet (BNSM) draws, largely men in their 50s, from far and wide in search of rare motorcar and motorbike spare parts. While two thirds of attendees are male our research informs us that their female partners also take the opportunity to shop in retail stores while the men browse the avenues of motor vehicle paraphernalia at the showground. People attending the BNSM usually do so looking for specific items to assist them in the maintenance and re-creation of typically vintage vehicles. They are very single-minded about this task and in previous years have not responded well to ancillary events put on for their entertainment (Guo et al. 2006).

Most event patrons come from outside Bendigo with around one quarter being locals. As we have already noted 75% of patrons are male, as to be expected with an event revolving around swapping old motorcar parts. They are more likely to be above 30 years old and have high school and diploma education. Importantly the majority of patrons were repeat customers. This is a well-established event with a faithful clientele returning each year.

The most mentioned reasons for attending the event are "general interest", "to buy collectables" and "to buy car parts" and the majority of attendance depends on "word of mouth" and, to a lesser extent, on "newspapers/industrial magazine" and "motoring clubs". This event is clearly for those men who have a particular interest in restoring and maintaining motorcars. While they are not big spenders outside of the swap meet venue our research discovered that their wives and partners made up any shortfalls in spending in the city by attendees at the meet.

In Fig. 5 depicting the current and preferred organisational culture for the organisation and management of Bendigo National Swap Meet we recommended a shift in emphasis along the collaborate and compete tension towards a greater market orientation. This event suffers through its success in having a large group of loyal supporters who attend on a regular basis. It is a well-established event with a clientele who faithfully return each year in search of elusive spare parts for their favourite pastime. They do so with very clear intentions to comb the stalls at the

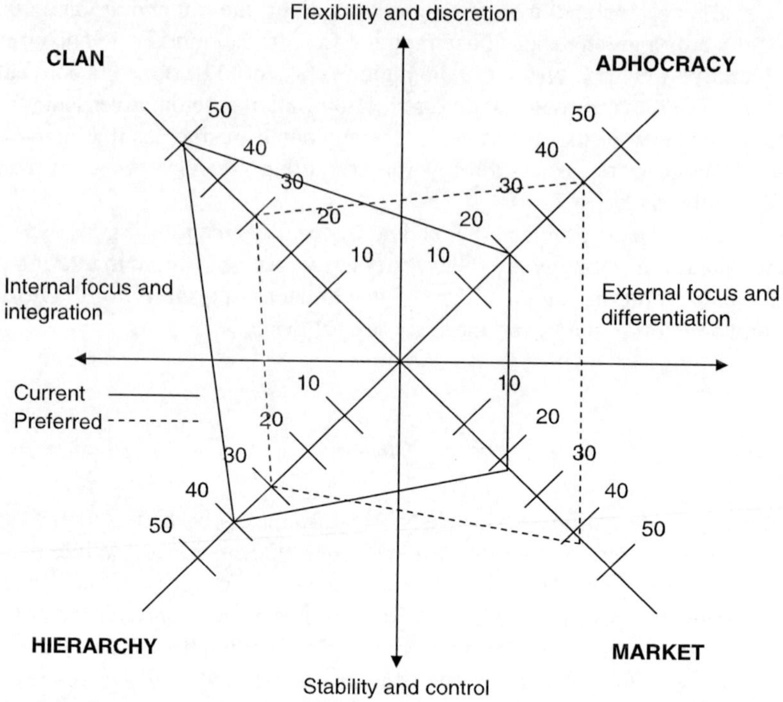

Fig. 5 The current and preferred organisational culture assessment of the organisation and management of the Bendigo National Swap Meet committee

showground looking for particular items of also being open to the possibility of finding other treasures. The emphasis should be on ways in which stallholders at the meet can market their wares more widely.

Our research also reveals a high proportion of attendees who simply enjoy wandering through the stalls curious to see what is on offer. The nature of the swap meet is such that the organisers cannot predict just who will return each year to sell their wares. This challenges the organising committee in their attempts to identify what we will be available at the swap meet each year. It requires a creative marketing strategy designed to recognise these needs. For example, is it possible to advertise prior to the meet the range of goods that will be available for purchase? This needs to be done in such a way as not to deter potential visitors to the city. It also needs to be done in such a way which recognises that a large part of the fun attending the swap meet is the serendipity which goes with finding unique and rare items amongst the bric-a-brac.

On the control and create tension we recommended a shift towards more creativity and innovation in operation of the actual swap meet. As the event currently functions the meet is simply row upon row of small store holders laid out on a first-come basis. While this adds to the excitement of serendipitous shopper it is also a source of frustration for the individual in search of a particular item.

In recent years it is not uncommon for people to wander through the stalls carrying a billboard advertising the items they are after. It is, we believe, this type of creativity and innovation which is to be encouraged by the organising committee such that people visiting the swap meet are more likely to be able to find the items they come in search of.

We also recommended a shift in emphasis along the control and create tension toward creative and innovative strategies especially for the partners of those attending the swap meet. They typically have little interest in old motorcar parts and are also looking to have an enjoyable weekend in the city. Apart from extended trading hours for specialty retailers other events associated with Bendigo's galleries and theatres could also make visiting the city for these people a more memorable occasion.

The challenge for the Swap Meet organising committee and the city officials is to add value around creativity in running the event and prior marketing without impacting on what is already regarded as a highly valued event by the loyal visitors who faithfully return each year.

3.4 The International Madison

The Bendigo International Madison is a premier event on the cycling calendar for central Victoria (Martin et al. 2006). Held in March each year this event draws around 20,000 people over the weekend culminating in the Madison cycling race on the Sunday evening. It is patronised by small groups of adults and while children do accompany adults to the event, these types of social groups represent a minority. Patrons tend to be within the ages of 20 and 50 and are marginally more likely to be male, and to have completed some or all of a secondary school education (Jorgensen 2007). Jorgensen's evaluation summary of the event provides the following picture of the Madison.

Patrons tend to be resident in the City of Greater Bendigo. Nonetheless, a sizeable proportion of patrons from outside the City did attend. These patrons tended to hail from other regional centres in Victoria and from Melbourne. However, there was little evidence to suggest that visitors to Bendigo stayed overnight in paid accommodation, or that they made use of the City's many tourism experiences. Rather, shopping and visiting friends were the most popular activities for non-residents.

The patrons of the International Madison demonstrate significant frequency and reliability of attendance in general. The majority are not new to the Madison experience, having visited on at least one occasion in the past. In fact, the bulk of frequent visitors to the Madison make regular attendances at the event. While past attendances did underpin the likelihood of future attendance, the extent of attendance across the weekend of activities was also important. Individuals who attended Madison events and activities on days other than Sunday spent more money and, in some cases, held higher expectations of the Madison, were more satisfied with the events they saw, and were more likely to recommend the Madison to others. There are significant benefits to be gained from promoting patrons involvement over the Madison weekend.

The main reason motivating patrons to attend the Madison concerned the visual spectacle of the event and its entertainment value. Other motives included catching up with friends and family and supporting friends and family who were participating in events. These family/friendship motivations proved to be an important justification for future attendance at the Madison, but patrons in this group tended to experience less satisfaction compared with patrons motivated toward other goals. For some patrons, friends and family relationships may invoke obligations to attend the Madison.

For the most part, patrons were very interested in the Madison cycle race. It is clearly a cornerstone event in the Madison programme and is symbolic of the atmosphere of the event as a whole. Although, there was substantial interest in events and activities other than the Madison cycle race, these were accompanied by a significant lack of awareness and uncertainty. As might expect, the greatest amount of uncertainty was associated with the corporate area, but there was also reasonable deficits in awareness about the criterium (or street cycle race), the community cycle, and the concert at Rosalind Park.

Patrons' interest in cycling reflected interest in other cycling events such as the Herald-Sun Tour and road racing in general. There was also significant interest shown in non-cycling pursuits such as concerts, the arts, and sports in general.

Facilities and services at the Madison were rated between "fair" and "good" on average. Overall, the Madison experience was highly rated on average, and only moderately influenced by patrons' evaluations of facilities and services. Key areas of concern were, however, the quality and accessibility of toilets, food choices and accessibility, the lack of shade, and uncomfortable seating. Security was also an issue for some patrons, particularly those who either perceived or experienced problems associated with youth drinking. And, as noted earlier, those patrons who sought to catch up with friends and family tended to be less satisfied with the International Madison compared with patrons holding other motivations.

Patrons' expectations of the Madison and their satisfaction with the events on offer were high on average. These positive evaluations tended to be based on patrons' interest in cycling, their past attendance, their extent of involvement in the Madison, and their appreciation of the fun and exciting atmosphere of the event. These general feelings of satisfaction had flow-on effects with respect to patron's intentions to recommend the Madison and attend again in the future. The only significant limitation upon intentions was negative experiences with other patrons (such as youth drinking).

In our analysis of the Madison Committee in terms of the OCAI we concluded that its strengths are a strong local network of supporters and sponsors and a systematic approach to staging the even each year. In Fig. 6, depicting the current and preferred organisational culture for the organisation and management of the Bendigo International Madison, we recommended the Committee make a shift in emphasis along the collaborate and compete tension towards a greater market orientation.

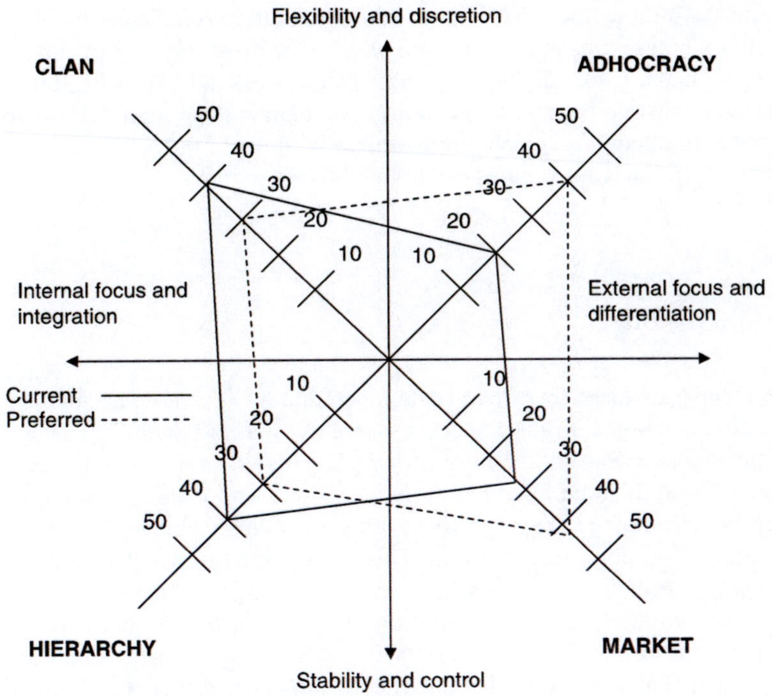

Fig. 6 The current and preferred organisational culture assessment of the organisation and management of the international Madison committee

The Madison weekend is seen as an event for cyclists and runners, the former being, by far, in the majority attending the event. Yet the circular track cycle racing under lights is an exciting event appreciated by non-cyclists. Recommendations from the evaluation research (Jorgensen 2007) also indicate the need to develop a communication strategy that integrate the interests of the Bendigo International Madison Committee and the City of Greater Bendigo in its desire to use the event as a wider marketing event for the city.

Part of this recommendation recognises the need to broaden and/or link existing communication sources so that more comprehensive information might be rendered accessible to potential visitors. Improving attendance and length-of-stay over the whole Madison weekend, given the benefits shown to be associated with these outcomes, should be a strategy of interest to the city council and the business community.

With the control and create tension we recommended a shift towards more creativity and innovation in the events across the weekend. The recently developed Tom Flood stadium, home of the Madison and other major regional track cycling events throughout the year now provides the opportunity to cater for the needs of others attending the event with friends and family less interested in cycling. Other options might be to enable patrons to sample particular products (local wine and food), to learn more about tourism activities and attain some experience of them, possibly a Chinese dragon performance prior to the fireworks, and to purchase tour tickets and make bookings for other parts of the weekend. Activities for children can be as simple as providing (or enabling) face painting opportunities, or providing the chance to obtain autographs from athletes and photo opportunities with them. More resource intensive options might include programming of racing events and competitions for children.

4 Conclusion

When comparing these four, long standing community events it is clear they have similar issues impacting their success, viability and sustainability. They are all characterised as events which have a strong supporter base and preoccupied with control and coordination functions at the expense of more creative and innovative approaches to running their respective events, including a greater focus on marketing to bring new people to each event. This is not surprising as all are long standing events having been in operation over several decades. Given that all four events are well patronised the organising committees might believe that this is sufficient in itself for them to maintain their current course and approach to organisation and management. This, of course, is their choice. However if their vision is to enrich and develop their particular event the Competing Values Framework provides a rationale for doing so. To engage the creative, intellectual and entrepreneurial talents of their supporter base in this process will create opportunities for each event to grow in value to all involved; the patrons, the organisers, and the regional community from which, in most cases, the majority of supporters come from.

For the city council, a major supporter of these events, the challenge is to engage in a dialogue with the respective management committees to show them that striking the right balance between the competing values of organisation and management will do much for the viability and sustainability of their event.

Each of the four case studies analysed in this paper are valued events in the central Victorian regional community. Each has a different heritage, offer different yet unique experiences for the people who attend and patronise these events. Together they provide a diverse patchwork of events contributing to the economic and social sustainability of the region. They are also at different stages of development, in terms of the entrepreneurial cycle, and each will benefit from considering the competing values inherent in alternative approaches to leadership and management of their event.

The analysis reported in this paper has been done by third party evaluators employed by the city council to assist them in their role as a major sponsor of these events. The council's sponsorship occurs by allowing access to council property as well as financial grants to fund the development of each event. It is, of course, different for each event. Assuming the key players working in the committees associated with each event was to make a similar assessment the question remains: how do they make the shift in the entrepreneurial cycle from a hierarchical culture towards greater creativity found in adhocracy, and from the collaborative clan culture towards the competitive culture of a market orientation? We believe there are several considerations the city council might consider in their negotiations with the respective event committees.

Should the change be incremental or systemic? In the absence of wholesale systemic change we believe each event committee should be looking for incremental steps towards the competing values we have outlined. This could be done through the committee membership or by recruiting people on to the committee who have experience and skills in innovation around events and marketing to wider audiences. In any case a strategic review of each committee and event strategic and business plan is essential.

Should change be driven internally or via an outside facilitator? Committee members currently have specific roles relating to the annual planning and operation of each event. To take on this broader strategic role requires a different view, additional energy, time and overall commitment if it is a member of the committee who assumes this role. Importantly the committee should be integral to strategic and business planning. If they are not so involved there is less likelihood that change will occur. This must be a key criteria for any facilitator engaged to lead this task.

Should the new work be done by volunteers or by paid staff engaged for specific purposes? The services of the latter could be obtained via a state government grant, either through Regional Development Victoria, or other government departments which support local community events. The theme of the changes would largely determine which state department to approach.

A key issue for all committees is their motivation to engage in a change process. We hope that the analysis completed by local university staff and the interpretation given in this paper using the Competing Values Framework is reason enough to review "the ways things are done around here". An attitude pervades Australian culture, one which we joke about, but can been seen in many community-based activities: the "she'll be right mate" attitude can limit the preparedness of community groups to make the changes we are suggesting in this paper. They key, we believe, is for them to understand and appreciate the findings and analysis presented in this paper.

Finally, and most importantly, *each event needs to develop a key set of indicators to assess the performance of their annual event.* This information should be used as the basis for the type of change they desire for their event and the way their committee functions in staging their annual event.

We believe the Competing Values Framework is a useful concept when considering entrepreneurship and innovation in regional development. It reveals both the inherent tensions in the leadership, organisation and management of community-based organisations contributing to sustainable regional development. It also reveals to those running these organisations and events they have choices, over time, in the way in which they orient their activities to achieve desired outcomes. This will always be reflected in the personal leadership styles of key individuals, however the success of their organisation is reflected in their capacity to consider how their collective leadership and management style influences organisational effectiveness and the ultimate outcomes they seek to achieve.

References

Cameron K, Quinn R (1999) Diagnosing and changing organizational culture: based on the competing values framework. Jossey-Bass, San-Francisco

Cameron K, Quinn R, DeGraff J, Thakor AV (2006) Competing values leadership: creating value in organizations. Edward Elgar, Northampton, MA

Guo W, Martin J, Jobling E, Nanere M (2006) The Bendigo National Swap Meet Survey Report, CSRC Report. LaTrobe University, Bendigo

Jorgensen B (2007) The Bendigo international madison: a survey of visitor perceptions, CSRC Report. LaTrobe University, Bendigo

Langfield-Smith K (1995) Organisational culture and control. In: Berry AJ, Broadbent J, Otley D (eds) Management control theories, issues and practices. Macmillan, London

Martin J, Simonds R (2002) Managing competing values: leadership styles of Mayors and CEOs. Aust J Publ Admin 61(3):65–75

Martin J, McCaig R, McLean J, Miles T (2006) Building on Bendigo's cycling bookends: the Madison and the Herald Sun Tour. Paper presented at the thinking on two wheels conference, Adelaide, January

Nanere M, Reimers D (2006) Heritage uncorked: an attendee survey for the Bendigo & District Winegrowers Association, CSRC Report. LaTrobe University, Bendigo

Quinn RE, Rohrbaugh J (1983) A spacial model of effectiveness criteria: towards a competing values approach to organizational analysis. Manage Sci 29:363–377

ten Have S, ten Have W, Stevens A, Vander Elst M, Pol-Coyne F (2003) Key management models: the management tools and practices that will improve your business. Prentice-Hall, London

Regional Growth in the United States: Correlates with Measures of Human and Creative Capital

William B. Beyers

1 Introduction

This paper presents a new classification scheme for looking at industrial structure in the United States, and explores patterns of growth related to this new classification scheme, with an eye towards examining the role of human and creative capital in the growth process. The paper begins with a background discussion that is intended to motivate the need for the development of a new industry classification scheme. This section is followed by a description of the process of developing this new industry classification scheme, a description of regional trends in the clusters developed in this paper, and then by a section that reports on correlates of regional growth and indicators of human and creative capital. The paper ends with some concluding comments.

This paper is motivated by dissatisfaction on the part of the author with current groupings of industries by the federal statistical establishment in the United States. The current NAICS system relies largely on the definitions of what types of goods or services are produced by business establishments, rather than on the way that these goods or services are produced (such as a grouping of establishments by their technical input requirements or production functions), or on the basis of their use of particular categories of labor. Scholars such as Bryson and Daniels have cautioned us about the changing nature of production processes, that are blurring the lines between goods and services industries (Daniels and Bryson (2002)). In this paper the author has used data on the occupational structure of industries to create new groupings of industries based on their distinctive occupational structure. As will become evident later in this paper, this classification scheme produces new ways of organizing economic activity, and it also allows us to define industries with strong concentrations of workers with high levels of human capital, or with high concentrations of people engaged in creative work.

W.B. Beyers
Department of Geography, University of Washington, 426 Smith Hall, Seattle, WA 98195-3550, USA
e-mail: beyers@uw.edu

K. Kourtit et al. (eds.), *Drivers of Innovation, Entrepreneurship and Regional Dynamics*, Advances in Spatial Science, DOI 10.1007/978-3-642-17940-2_15,

2 Background

2.1 Industry Classifications

Pioneering work by Clark (1957) and Fisher (1939) documented the transformation of national economies in the first part of the twentieth century from dominant employment in "primary", and "secondary" industries into new forms of production that became labeled as "tertiary", or as we say today "services." Beyers (2007a, b, 127) shows that over half of US employment has been in the service category since 1929, and that this share has expanded continuously to the present. Clark and Fisher's simple tripartite classification scheme was found to be too simple by many scholars, who documented the need to decompose the "service" category into more meaningful components. Greenfield (1966), Browning and Singelmann 1975, and Petit (1986) examined types of markets being served by various service industries, and proposed functional classifications differentiated on this basis In particular, this work differentiates between services with household markets as opposed to business markets. Greenfield and Petit focused on the growing importance of services with business markets, which became labeled "producer services."

In the United States the Standard Industrial Classification (SIC) system grouped services into six broad categories (transportation, communications and utilities; wholesale; retail; finance, insurance, and real estate; services; and government). Within services, the statistical establishment did not develop industry categories matching the functional categories suggested by Greenfield, Singelmann and Browning, or Petit. Scholars interpreted these SIC codes to be correspondent with conceptions such as producer services, but there was not unanimity as to how inclusive to make the producer service classification (Beyers 1989; Daniels 1985). With the change in the USA from the SIC classification scheme to the North American Industry Classification System (NAICS) in 1997 this problem of defining groupings of services became even more problematic. Not only was there not a perfect cross-walk between the SIC and NAICS classification scheme, making longitudinal analysis very difficult, there was also movement of some categories that had been part of manufacturing into new categories such as "Information", as well as movement of some categories that had been classified in services into manufacturing categories (such as software reproduction). There have been revisions to the NAICS classification scheme that have also made it difficult to undertake longitudinal analysis with the NAICS classification scheme. In 2002 there were major revisions to the wholesale trade and telecommunications categories, including the creation of new categories with no priors (such as internet service providers).

2.2 Occupations

At the same time as there has been this changing focus on the classification of industries, there has also been recognition of the shifting composition of occupations.

Greenfield (1966, 72–76) observed the high proportion of professional, technical, and scientific personnel in selected producer services, and also the relatively large number of clerical workers in producer services. Greenfield was writing before the development of modern computers, that have had a dramatic impact on the size of the clerical workforce, and have increased the share of professional, technical, and scientific personnel in the producer services. A recent paper by Wyatt and Hecker (2006) focused on occupational changes in the US economy over the long 1910–2000 time period. Figure 1 is based on data presented in this paper, and it clearly shows the rise of professional, technical, and kindred; service workers; clerical and kindred; and managers, officials, and proprietors occupations. This figure also shows the decrease in the relative importance of farm workers, private household workers, laborers, and operatives.

Wyatt and Hecker address the strong expansion of professional, technical, and kindred workers:

> Industrialization, technological development, and the growing size and complexity of organizations; rapid growth in healthcare education, and social services; and the expanded role of government all contributed to the increase." "Computer specialists did not exist in 1910, and there were few if any, in 1950, so they do not appear in the 1950 census or the IPUMS classification system. The first commercial electronic computer was delivered in

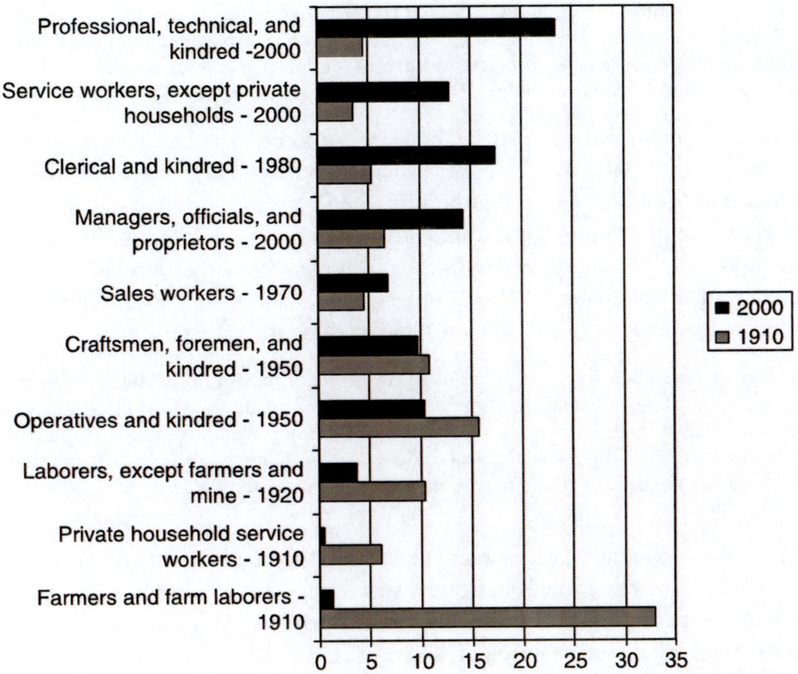

Fig. 1 Composition of occupational employment in the United States 1910–2000 (maximum year for percent of employment included with category label)

1951, and employment data on computer specialists were first collected in the 1960 Census (Wyatt and Hecker 2006, 38, 42).

Wyatt and Hecker discuss reasons for the expansion of a number of other lines of professional, technical, and kindred workers, but space does not permit detailed discussion of all their observations. They argue that accountants growth was due to a combination of increasing complexity of business and government operations, more sophisticated management techniques that require more accounting data, greater government regulation, and more complex tax laws (Wyatt and Hecker 2006, 42). They note that engineers grew due to rapid industrialization and growing technical sophistication, the growth in R&D laboratories, the growth of some lines of manufacturing and defense spending. They note slower growth in engineers in recent years:

> "Slower growth after 1970 reflects the slower growth of manufacturing, in which engineers are concentrated, and the use of computers in design work, which increased engineers' productivity. The 1990–2000 trend also reflects a drop in defense spending with the end of the Cold War." "Growth (in health care occupations) occurred as improved medical technology permitted many more medical problems to be treated, or to be treated more aggressively, greater wealth and the spread of health insurance made healthcare more affordable, and a more long-lived population increased the need for healthcare." "Lawyers and judges increased one-and-a-half times as a proportion of total employment between 1910 and 2000, with almost all growth coming since 1970." ". . ..demand for lawyers increased as many more laws were enacted, business activities became more complex, and society became more litigious. Civil rights legislation for minorities, women, and older and disabled persons; laws regarding the environment, employer–employee relations, product safety, and consumer protection; and higher crime and divorce rates all contributed to the growth of lawyers and judges." (Wyatt and Hecker 2006, 42–43)

Service workers outside private households have also had strong growth, as indicated in Fig. 1. Wyatt and Hecker provide interpretations of reasons for growth of selected sectors. They note that building and grounds and maintenance occupations grew rapidly "due to the rapid growth in office buildings, hotels, stores, healthcare facilities, apartment buildings, schools, and other structures requiring cleaning and maintenance" (Wyatt and Hecker 2006, 44). Food preparation and serving occupations have also grown rapidly; they argue that:

> Greater income made food prepared away from home more affordable; the advent of automobiles, improved roads, and greater urbanization made food and drink purveyors more accessible; and an increasing percentage of women working outside of the home intensified the need for prepared meals. More nursing home and assisted-living facility residents and an expansion of school lunch programs also stimulated growth. (Wyatt and Hecker 2006, 44)

Protective service workers also expanded strongly, and Wyatt and Hecker argue that this was primarily growth in police, sheriffs, guards, and marshals. "Increased urbanization, more motor vehicle traffic, higher crime and incarceration rates, more properties and other assets to protect, and more laws to enforce all contributed to the growth" (Wyatt and Hecker 2006, 44). Personal care and service occupations also grew strongly, primarily professional and personal services attendants, such as teachers aides and childcare workers.

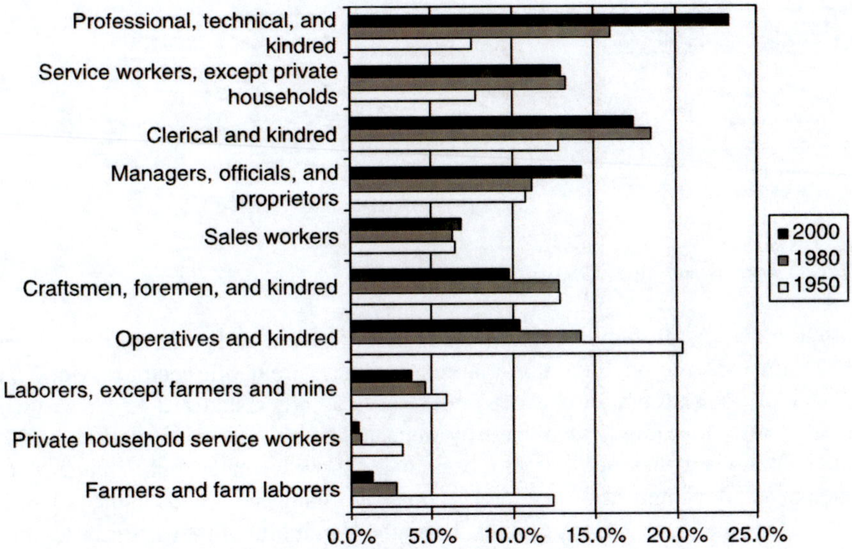

Fig. 2 Occupational structure United States 1950–2000

Another perspective on occupational structural change in the United States is presented in Fig. 2. This figure covers the more recent 1950–2000 time period. Clearly, the most dramatic structural change is the increase in the professional, technical, and kindred occupations. These are the categories just discussed using the work of Wyatt and Hecker, and as will be seen below they are strongly associated with the growth of several clusters of service industries. Figure 2 indicates a turnaround in the proportion of clerical workers, which had their peak share of total employment in 1980. The decline since 1980 reflects impacts of computers and information technologies on the need for clerical labor. Managerial employment can be seen to have grown in Fig. 2, while the share of non-household service workers and sales workers have remained relatively constant since 1980. Decreases are evident in the share of blue-collar workers in the 1950–2000 time period, mirroring the longer-run pattern reported in Fig. 1.

2.3 *Industry × Occupations*

The simultaneous consideration of industries and occupations has also been a theme in scholarly work, as well as a characteristic of data provided about employment. Greenfield (1966) discussed these dimensions in his pioneering work on producer services. The work of Gershuny and Miles (1983) also provided evidence on shifting patterns of employment and occupations, as summarized in Fig. 3. This figure, depicts in a stylized way changes in patterns of consumption, industrial

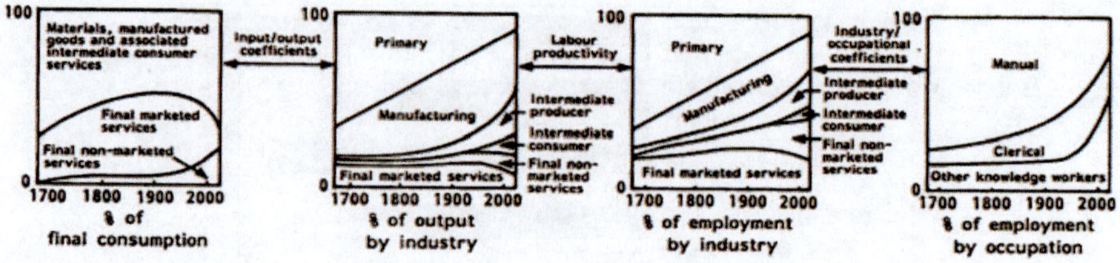

Fig. 3 Gershuny and Miles model
source: Gershuny and Miles 1983, p. 252

output, industrial employment, and occupation. Several categories in this diagram need some explanation. Final non-marketed services are public sector services, paid for out of taxes. Intermediate consumer services are visualized to be servicing investments and equipment owned by households or other final market, such as repair or advisory services. It has been over 25 years since publication of this schema, and it is interesting to reflect on possible changes in trends in the tail end of each of Gershuny and Miles graphics. From the standpoint of industry employment, the decline in manufacturing and primary may have been more dramatic than depicted by Gershuny and Miles, and final non-marketed services may not have grown as much as anticipated, while final marketed services may have grown more than expected (arts, culture, sports, entertainment, gambling), fueled by information technologies not present at the time Gershuny and Miles were writing (the Internet, PDAs, ISPs, DVDs, etc.). However, from an occupational standpoint there clearly has been a trend as envisaged by Gershuny and Miles. The structural links between "other knowledge workers" in Gershuny and Miles schema could be linked more strongly back to categories of industrial production, differentiating between knowledge workers in fields such as producer services from knowledge workers in consumer industries. This differentiation will become clearer in the analysis to follow.

While not developing an industry × occupation perspective, Illeris (1996) tried a decade ago to sketch out a vision of an advanced service society, in which the professional producer service sectors were providing coordination in production with other industries, as depicted in Fig. 4. Leisure personal services; trade, transport, and communication; education, welfare, and health; government; finance; and marketing and public relations are visualized to have direct links to households (Gershuny's final marketed services). In some ways, it will be seen that Illeris' schema is similar to that developed later in this paper.

2.4 *Problematic Labels*

We have used various words to describe groupings of industries and groupings of occupations, and in many cases these labels have been problematic. Let me touch

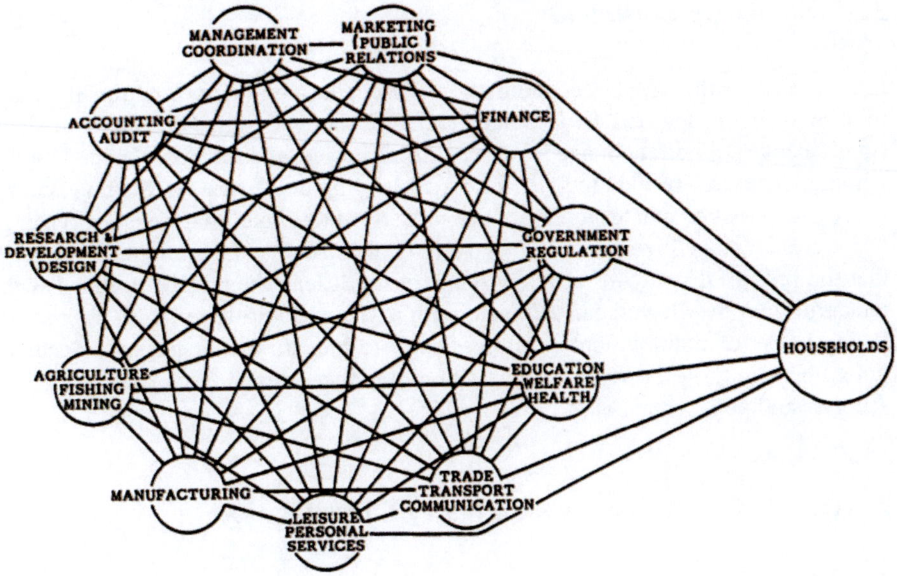

Fig. 4 Illeris' service society
source: Illeris 1996, p. 185

on a few of these problematic labels very briefly. Markusen and Schrock (2006) have made a strong case for using occupational data to identify those working in cities in artistic activity. They note that industry statistics fail to isolate people employed in artistic or creative work, and given the large number of self-employed people in some of these lines of work, employment statistics fail to count their effort. They used Census data on occupations to better identify the numbers of people working in artistic activities, and made a strong case for the relevance of these measures. Elsewhere, Markusen and her coauthors have wrestled with alternative definitions of the cultural economy, and they make it clear that there are difficulties associated with any definitional choice due to the way in which official statistics are collected (Markusen et al. 2008). Other scholars have wrestled with these classification issues as well, and have suggested phrases such as "the cognitive-cultural dimension" (Scott 2007), or the "new urban cultural economy" (Hutton 2007) as ways of defining cultural industries.

Florida's work is clearly rooted in the recognition of shifting occupational structures in advanced economies, and his conception of a "creative class" is directly tied to a set of occupations contained in Figs. 1 and 2 (Florida 2002). Florida's arguments have spawned a huge amount of research trying to either validate or extend his assertions and analyses, and there has also been a strong set of critical voices arguing that his analysis is not based on sound primary evidence (see, e.g., Scott 2007 or Markusen 2006. Florida does not utilize industry × occupation data in his analysis; he confines his definitions of the creative class to selected occupational categories.

2.5 *Summary Comments*

It is quite clear that the US economy is experiencing ongoing occupational structural change. As suggested by Gershuny and Miles, as well as by Greenfield, there is simultaneous structural change in the mix of industries, and the mix of occupations within industries. The existing schema for classifying industries are ripe for review, to provide possible updates to the service industry taxonomies that we are still using that were developed over 40 years ago by scholars such as Greenfield. It is likely that the industrial structure of many industries will lead them to be grouped with industries that they have been affiliated with in the past. However, with the rise of strong levels of employment in professional, technical, and managerial occupations, there may be newly emergent categories that bear closer examination by scholars and policy-makers.

3 Development of a New Industry Classification

To evaluate current patterns of occupations across industries, the 2006 industry × occupation matrix developed by the US Bureau of Labor Statistics was accessed. This matrix was accessed with four digit NAICS industry codes (295 industries), and six digit occupational codes. There are hundreds of occupational codes in this matrix, including 21 two-digit codes, and then within each two digit industry there are detailed codes for occupational subcategories. In many cases these detailed codes are related to very few industries, while in other cases they are found in almost every industry. The methodology used here estimated the equivalent of location quotients, for the concentration of occupations within an industry compared to the overall national composition of occupations across all industries.

An initial grouping of the 295 industries was undertaken with two-digit occupation codes, using Ward's algorithm for cluster analysis in SPSS. This test did not yield a very robust grouping of the 295 industries. On the other hand, the matrix with all the six digit occupations included many occupations with a small number of workers. Therefore, an aggregation of a number of the six digit codes was made, attempting to create aggregates that were relatively large in terms of number of employees. Frequently these were aggregates using the three-digit occupational groupings. Appendix (Table 10) reports the occupational groups created through this aggregation process, and the level of employment estimated by BLS in these occupational categories. Occupational quotients were calculated for each of these groups, and Ward's Hierarchical Algorithm was used to classify the industry occupational distributions using SPSS software. A dendrogram was created as a part of this cluster analysis, which provides a visual means of identifying the groupings of industries based on their occupational structure. Figure 5 presents an aggregate version of this dendrogram. The original has all 295 industries in it, and in this aggregate version the industries have been collapsed into 31 cluster groups, that are in turn part of 18 clusters that form the most aggregate level in the present analysis.

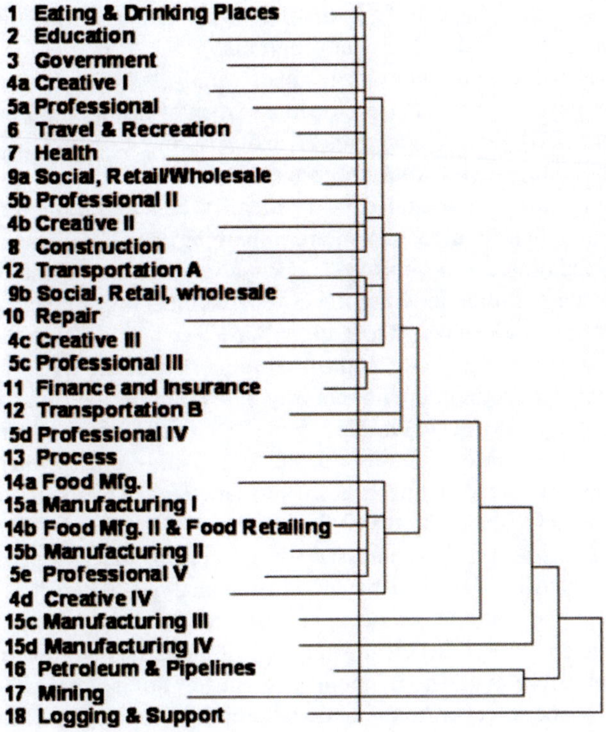

Fig. 5 Dendrogram from cluster analysis

There have been previous studies that utilized hierarchical cluster analysis to engage in classifications of industries and regional economies. A few examples will be cited here. Noyelle and Stanback (1983) used this type of cluster analysis to classify the industrial structure of cities. Beyers used the same technique as Noyelle and Stanback to classify the industrial structure of the BEA regions; the BEA regions are nodal regions encompassing all counties in the United States (Beyers 1989, 2007a, b). However, these analyses clustered industrial structure across regions, and did not make use of occupational data. Barbour and Markusen (2007) used industry × occupation data to evaluate the structure of California metropolitan areas, compared to the USA as a whole, but their analysis did not involve the use of cluster analysis to group either industries or occupations. Barbour and Markusen (2007) also used industry × occupation data to identify industries that were considered to be part of the "creative economy." Their analysis focused on the concentration of selected artistic and creative occupations to define core industries in the creative economy. The resulting classification is quite similar to one of the clusters reported later in this paper.

Some years ago Thompson and Thompson (1993, 1985, 1987) suggested methods for considering simultaneously industry and occupational characteristics, along with local economic characteristics, to define through a process they called "cross-hairs targeting" to identify industries in particular communities that were candidates for economic development programs.

Another application of hierarchical cluster analysis to occupational data was undertaken by Feser (2003). He developed a matrix of occupations and knowledge indicators, and used hierarchical cluster analysis to group the occupations and knowledge indicators into "knowledge-based clusters." This analysis was then applied to regional occupational data, to identify concentrations of employment in the knowledge clusters across 38 metropolitan regions. Feser does not group the knowledge based clusters to industry × occupation matrices at the regional scale, but this would be possible knowing the cluster membership of each occupation in selected metro area industry × occupation matrices.

The dendrogram in Fig. 5 is a faithful reproduction of the SPSS groupings, but the stems in this dendrogram are not entirely equivalent to the stems that appear in the SPSS dendrogram. For example, the SPSS dendrogram clusters groups 1 through 9a in Fig. 5 into a single cluster. Within this cluster there are in fact separate clusters, whose structure is more similar to other members of this cluster, than to other clusters created by the SPSS program. To tease apart these differences it was necessary to order the data used by the cluster analysis program into the sequence of industries classified by the cluster analysis. Then it is necessary to visually inspect the patterns of occupational concentrations by industry within clusters. This procedure yields clear groupings of industries within what outwardly appear to be undifferentiated components of a cluster. For example, the food cluster is dominated by about ten times the national employment average in food preparation and serving occupations. Another example is the logging and support cluster, that has occupational coefficients between 149 and 224 times the national average in farming, fishing, and forestry occupations. The names of the clusters in Fig. 5 were given by the author; they are not an output of the SPSS software.

Not all clusters have such distinctive degrees of concentration of employment as the two mentioned above. For example, group 4b, part of the cluster labeled as creative, has a strong concentration in people engaged in repair; arts, design, entertainment, sports, media; and computer-related occupations. This group also has selected concentrations in other professional and technical occupations. The repair-related occupational concentration is the defining basis for this group.

Many of the clusters are quite similar to "labels" used for decades: These include: health, construction, transportation, retail and wholesale, finance, manufacturing, food manufacturing, petroleum and pipelines, mining, and logging and support activities for primary activity. One of the most interesting results of the clustering scheme reported here is the classification of industries in various lines of services. With the flowering of occupations within professional, technical, and service workers except private households since 1980, it is very clear that the US economy has been on a different industrial and occupational path in the last 25 years. The cluster analysis has broken up lines of industries commonly classified in the past as producer and consumer services into what appear to be new groups. In the present analysis the two groups that seem to stand out are what have been labeled as "creative" and "professional." Closely related to these two is the group labeled as "travel and recreation." The "food" and "education" groups are also distinctive.

Summary information on the occupational concentrations associated with selected clusters will now be presented. While the cluster analysis groups activities with similar occupational structures, there are so many permutations and combinations of these structures, such that it is not always possible to generalize without leaving out some basis for specialization within a cluster.

The creative cluster's members are reported in Table 1. This cluster has three subgroups. Group 4a has strong employment concentration in Standard Occupational Classification (SOC) 37: arts, design, entertainment, sports, and media occupations. Group 4b has strong concentrations in computer support specialists, computer systems analysts, database and other computer systems administrators, sales representatives, and repair, installation, and maintenance services. Group 4c has strong concentrations of managers; arts, design, entertainment, sports, and

Table 1 The creative cluster

Subgroup	NAICS	Industry title
4a	519100	Other information services
4a	611600	Other schools and instruction
4a	711200	Spectator sports
4a	711300	Promoters of performing arts, sports, and similar events
4b	423400	Professional and commercial equipment and supplies merchant wholesalers
4b	443100	Electronics and appliance stores
4b	515200	Cable and other subscription programming
4b	517100	Wired telecommunications carriers
4b	517200	Wireless telecommunications carriers (except satellite)
4b	517300	Telecommunications resellers
4b	517400	Satellite telecommunications
4b	517500	Cable and other program distribution
4b	517900	Other telecommunications
4b	811300	Commercial and industrial machinery and equipment (except automotive and electronic) repair and maintenance
4c	323100	Printing and related support activities
4c	453100	Florists
4c	511100	Newspaper, periodical, book, and directory publishers
4c	511200	Software publishers
4c	512100	Motion picture and video industries
4c	512200	Sound recording industries
4c	515100	Radio and television broadcasting
4c	516100	Internet publishing and broadcasting
4c	518100	Internet service providers and web search portals
4c	518200	Data processing, hosting, and related services
4c	541400	Specialized design services
4c	541500	Computer systems design and related services
4c	541800	Advertising and related services
4c	711100	Performing arts companies
4c	711400	Agents and managers for artists, athletes, entertainers, and other public figures
4c	711500	Independent artists, writers, and performers

Table 2 Travel and recreation cluster

Cluster group	NAICS	Industry description
6	481100	Scheduled air transportation
6	487100	Scenic and sightseeing transportation, land
6	712100	Museums, historical sites, and similar institutions
6	713100	Amusement parks and arcades
6	713200	Gambling industries
6	713900	Other amusement and recreation industries
6	721100	Traveler accommodation
6	721200	RV (recreational vehicle) parks and recreational camps
6	721300	Rooming and boarding houses
6	813400	Civic and social organizations

media occupations; and service sales representatives. Cluster 4a has some concentrations in computer related occupations, while cluster 4b also has concentrations in arts, design, entertainment, sports, and media occupations. The word "creative" was chosen to characterize the overall character of this cluster. Note that it includes some sectors not typically thought of as part of cultural industries, such as florists, printing, other schools and instruction, electronics and appliance stores, and professional and commercial equipment and supplies merchant wholesalers. There are many similarities between the industries included in cluster 4 with the group of industries considered as cultural industries by Markusen et al. (2007).

The travel and recreation cluster is defined in Table 2. This cluster has strong concentrations of employment in SOC 39, personal care occupations, as well as personal care and building and grounds cleaning and maintenance. This includes occupations such as airline stewards and stewardesses, hotel employees, gambling table occupations, personal trainers, and museum staff.

The large professional cluster is defined in Table 3. This cluster has five subgroups. Group 5a is characterized by a variety of occupational concentrations, but they are largely in management, professional and technical, and computer occupations. However, there are individual concentrations beyond these common specializations. For example, land subdivision has a strong presence of architects; services to buildings has a strong complement of SOC37 (building cleaning and maintenance); and investigation and security services has a strong complement of protective service occupations. Several categories of industries are included in this group outside of traditionally considered producer services, including land subdivision, reproducing electronic and optical media, electronic shopping, and grant making and giving services. These industries are included given their occupational concentrations in the core activities in this subgroup of cluster 5. Cluster 5b also has a strong concentration of employment in management and professional/technical workers, but with stronger concentrations of architectural and engineering, and scientific employees than in cluster group 5a. Legal services has an extraordinary concentration of legal occupations (66 times the national average). Cluster group 5c is dominated by a concentration in tax-related occupations. As with all cluster groups in this cluster, there are also significant concentrations of executive,

Table 3 The professional cluster

Cluster subgroup	NAICS	Industry description
5a	237200	Land subdivision
5a	334600	Manufacturing and reproducing magnetic and optical media
5a	454100	Electronic shopping and mail-order houses
5a	531100	Lessors of real estate
5a	531300	Activities related to real estate
5a	533100	Lessors of nonfinancial intangible assets (except copyrighted works)
5a	541900	Other professional, scientific, and technical services
5a	551100	Management of companies and enterprises
5a	561100	Office administrative services
5a	561300	Employment services
5a	561600	Investigation and security services
5a	561700	Services to buildings and dwellings
5a	561900	Other support services
5a	611400	Business schools and computer and management training
5a	813200	Grant-making and giving services
5b	325400	Pharmaceutical and medicine manufacturing
5b	523100	Securities and commodity contracts intermediation and brokerage
5b	531200	Offices of real estate agents and brokers
5b	541100	Legal services
5b	541600	Management, scientific, and technical consulting services
5b	541700	Scientific research and development services
5b	561400	Business support services
5b	561500	Travel arrangement and reservation services
5b	813900	Business, professional, labor, political, and similar organizations
5b	999100	Federal executive branch (OES designation)
5c	523200	Securities and commodity exchanges
5c	541200	Accounting, tax preparation, bookkeeping, and payroll services
5d	562900	Remediation and other waste management services
5e	541300	Architectural, engineering, and related services

management, and other professional/technical occupations. Cluster groups 5d and 5e are essentially "isolates", industries with very distinctive occupational classifications compared to others judged to be in their cluster. Cluster 5d is similar to cluster 12 and 13, while cluster 5e is similar to clusters 15b and 4c. Cluster 5d is dominated by employment in construction specialists and construction supervisors, while also having a concentration of managers, and professional and technical employees. Cluster 5e is dominated by employment in architectural and engineering occupations, but this cluster has lower levels of employment in management and other professional and technical occupations. The position of cluster 5d and 5e in the dendrogram is interesting, as it shows that not all professional service dominated industries are tightly clustered only with service industries.

A number of other clusters are identified in Fig. 5, and many of these are similar to industrial groupings used in previous analyses. In this paper these other groupings will not be discussed further with regard to the analysis of the 2006 BLS occupational $\times$ industry matrix. Table 4 provides a summary view of the occupational concentrations in the various clusters. In this table the occupational concentrations

Table 4 Summary occupational quotients by industry cluster

		Management	Business and finance	Computer and engineering	Architecture and engineering	Science	Community and social services	Legal	Education, training, library	Arts, design, entertainment, sport, media	Health care professional	Health care support	Protective services	Food preparation and serving	Building and grounds clearing and maintenance	Personal care and services	Sales	Office and administrative	Farming, fishing, forestry	Construction	Installation, maintenance, and repair	Production	Transportation and material moving
Food	1	0.50	0.03	0.00	0.00	0.00	0.00	0.00	0.00	0.10	0.00	0.00	0.12	10.58	0.21	0.03	0.28	0.05	0.00	0.00	0.04	0.06	0.30
Education	2	0.94	0.37	0.60	0.09	1.44	1.58	0.03	9.48	0.93	0.39	0.11	0.39	0.45	1.33	0.62	0.02	0.61	0.06	0.07	0.31	0.02	0.34
Government	3	1.09	1.57	0.85	1.24	2.11	4.57	3.17	0.39	0.40	0.68	0.59	10.07	0.14	0.96	1.42	0.07	1.16	0.74	1.13	0.98	0.24	0.30
Creative	4	1.42	1.07	7.18	0.90	0.87	0.02	0.18	0.34	9.30	0.02	0.00	0.14	0.11	0.16	0.71	1.43	1.11	0.20	0.05	1.78	0.90	0.46
Professional	5	1.46	2.35	1.61	3.06	3.08	0.21	4.73	0.10	1.04	0.51	0.39	1.74	0.11	2.89	0.29	0.57	1.43	0.54	0.39	0.77	0.64	0.76
Travel and recreation	6	0.76	0.34	0.10	0.03	0.14	0.22	0.02	0.20	1.37	0.03	0.11	1.31	2.42	5.27	8.01	0.43	0.94	0.18	0.04	0.13	0.12	0.46
Health	7	0.74	0.27	0.20	0.02	0.53	3.59	0.02	0.46	0.12	6.39	7.15	0.18	0.37	0.83	3.74	0.05	0.96	0.00	0.02	0.18	0.07	0.08
Construction	8	1.03	0.63	0.04	0.54	0.04	0.00	0.01	0.00	0.07	0.00	0.00	0.03	0.00	0.18	0.00	0.18	0.53	0.01	13.36	1.74	0.16	0.52
Social	9	0.79	0.44	0.21	0.18	0.17	0.69	0.04	0.05	0.46	0.32	0.10	0.08	0.11	0.25	0.24	4.04	1.17	1.01	0.10	1.46	0.35	2.09
Repair	10	0.56	0.40	0.21	0.10	0.04	0.00	0.01	0.00	0.02	0.00	0.00	0.04	0.00	0.18	0.00	0.94	0.68	0.00	0.04	11.57	0.52	3.19
Finance	11	1.73	5.42	2.20	0.02	0.45	0.04	1.40	0.00	0.25	0.13	0.02	0.10	0.00	0.09	0.00	0.58	3.08	0.01	0.00	0.06	0.00	0.00
Transportation	12	0.57	0.29	0.10	0.07	0.16	0.01	0.02	0.00	0.07	0.02	0.00	0.12	0.04	0.11	1.15	0.18	0.68	0.18	0.24	1.47	0.18	9.88
Misc.	13	1.36	1.19	0.99	4.34	4.16	0.00	0.23	0.00	0.23	0.05	0.00	0.22	0.00	0.16	0.00	0.19	0.94	0.05	0.80	5.50	3.25	0.61
Food mfg	14	0.52	0.20	0.05	0.10	0.30	0.00	0.00	0.00	0.24	0.17	0.05	0.06	0.94	0.31	0.02	2.41	0.92	1.77	0.02	0.65	3.54	2.00
Manufacturing	15	1.13	0.70	0.84	3.57	0.56	0.00	0.04	0.00	0.38	0.02	0.00	0.06	0.00	0.18	0.00	0.31	0.56	0.27	0.37	1.21	7.27	1.04
Petroleum and pipelines	16	1.09	0.70	0.27	2.76	3.02	0.00	0.00	0.00	0.03	0.04	0.00	0.05	0.00	0.07	0.00	0.16	0.82	0.55	1.50	2.78	5.40	1.07
Mining	17	1.24	0.89	0.41	2.31	3.16	0.00	0.60	0.00	0.04	0.08	0.00	0.04	0.00	0.06	0.00	0.12	0.56	0.10	7.53	2.08	1.13	2.16
Logging and support	18	0.35	0.08	0.02	0.01	0.62	0.00	0.00	0.01	0.04	0.01	0.00	0.05	0.00	0.41	0.62	0.05	0.33	211.95	0.03	0.53	0.16	1.80

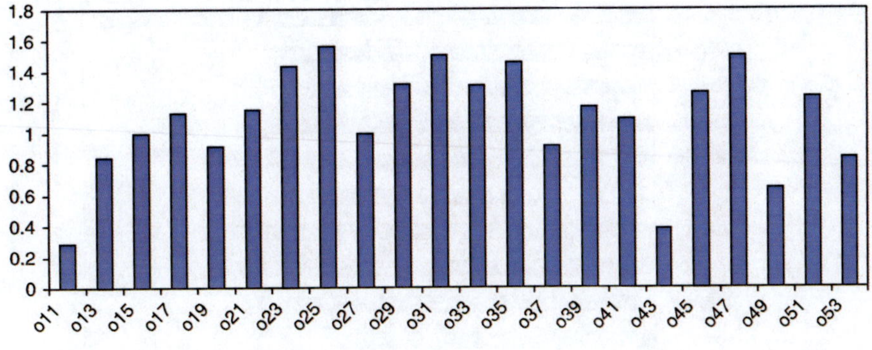

Fig. 6 Coefficient of occupational specialization

have been collapsed to the two-digit SOC level of detail. The clusters in this table with relatively high concentrations of human and creative capital – the focus of this conference – are education, creative, professional, health, and finance. Four of these sectors are also relatively strong sources of job growth over the 1998–2005 time period, and will be the examined later in this in terms of their role in patterns of regional growth (the exception is the education cluster – which includes only private education employment – while most education workers are employed in the public sector in the United States).

Two additional perspectives are presented with regard to the cluster analysis. These are also based on the two-digit occupational aggregations of the underlying occupational categories. Figure 6 presents estimates of coefficients of occupational specialization. The lower the index the more widely dispersed is the occupation. Thus, management and office and administrative support occupations are found to be relatively widely dispersed across all the industry clusters, while occupations with high values have concentrations only in a few industries. The inspection of Table 4 and Fig. 6 simultaneously reveals these patterns.

Another perspective is given in Fig. 7, that presents measures of cluster specialization, analogous to coefficients of regional specialization for regional statistics. In Fig. 7 we do not see nearly as much variation in the index as observed in Fig. 6, precisely because the cluster analysis algorithm has produced clusters that are differentiated from each other by their occupational structure.

3.1 *Development of Regional and Longitudinal Data*

The preceding section of this paper has developed a classification scheme for industrial employment, based on occupational structure. This section of the paper develops estimates of data for selected clusters by region, and over time.

The BLS data used for classification purposes in this paper were not available on a regional basis or for other years in the same format. To develop a database that

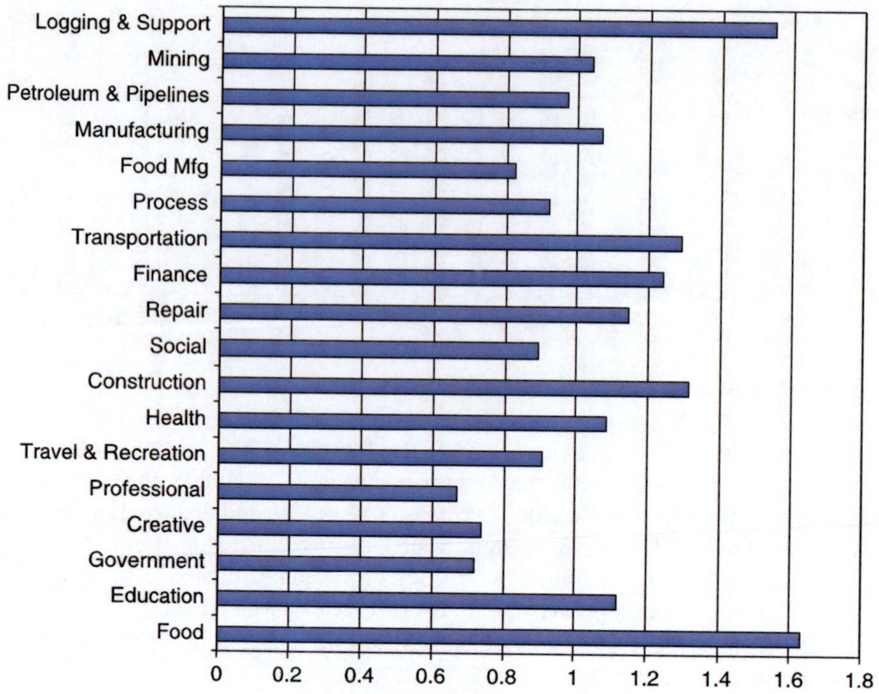

Fig. 7 Indices of cluster occupational specialization

would allow estimates for geographic regions, and over time, two other federal statistical series were used. They are US County Business Patterns (CBP), and the Nonemployer data series reported by the Census Bureau. Data were developed for the year 1998 and 2005 from these sources, using the classification of industries developed from the cluster analysis for US states. Table 5 reports results of this aggregation. The time frame used in this analysis is not long, but this is a byproduct of the change nationally to the NAICS classification system, whereby longitudinal data are not available for time periods before 1998 in this classification scheme at the level of industrial detail used in the paper. A critical consideration in the present paper was the inclusion of Nonemployer (proprietor) data, which are so important in many lines of services. The Nonemployer data are not entirely compatible with the CBP data, as in some cases only two-digit or three-digit levels of detail are provided, while the CBP data and clustering schema are based on two, three, and four-digit NAICS codes.

Table 5 reports on the composition and change in the cluster groups using the combination of CBP and Nonemployer data. The time period 1998–2005 is not long, and the US economy has been through a brief recession in this time period. Average growth was 11% over this relatively short time period. In the next section of this paper regional growth will be analyzed, and it will be of interest to see the contribution of the relatively rapidly growing food, creative, professional, health,

Table 5 Estimated employment by cluster

	1998	2005	Change	% Change
Food	7,908,842	9,394,854	1,486,012	18.8
Education	2,317,688	2,498,175	180,487	7.8
Government	129,052	171,686	42,634	33.0
Creative	8,224,712	9,438,000	1,213,288	14.8
Professional	22,388,085	27,052,198	4,664,113	20.8
Travel and recreation	4,049,768	4,403,528	353,761	8.7
Health	15,733,026	18,760,459	3,027,433	19.2
Construction	8,044,512	9,222,764	1,178,252	14.6
Social and retail/wholesale	21,684,646	23,842,154	2,157,508	9.9
Repair	1,467,051	1,524,078	57,027	3.9
Finance	5,986,479	7,164,584	1,178,105	19.7
Transportation	3,704,833	5,548,201	1,843,369	49.8
Process	1,057,668	918,831	−138,837	−13.1
Food mfg	4,896,453	4,371,522	−524,929	−10.7
Manufacturing	14,286,106	11,113,374	−3,172,732	−22.2
Petroleum and pipelines	154,475	160,880	6,405	4.1
Mining	579,258	597,010	17,752	3.1
Logging and support	323,641	311,934	−11,707	−3.6
Total	122,936,292	136,494,233	13,557,941	11.0

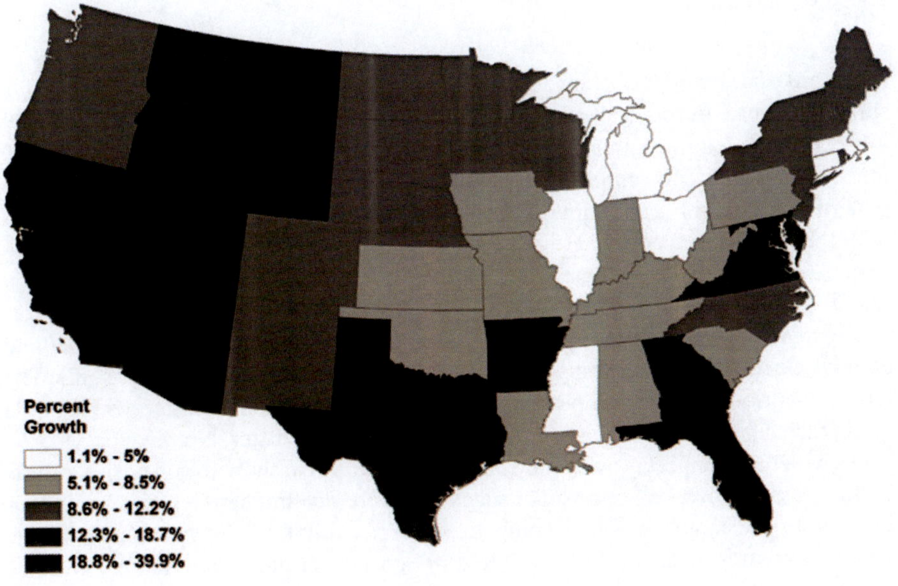

Fig. 8 Growth rate 1998–2005

finance, and transportation clusters to regional employment trends. Figure 8 reports the pattern of state growth rates, with strongest rates of growth in Florida, Arizona, Nevada, Idaho, and Wyoming. Slowest growth appears in sections of the old industrial belt, and in parts of the southern great plains and South.

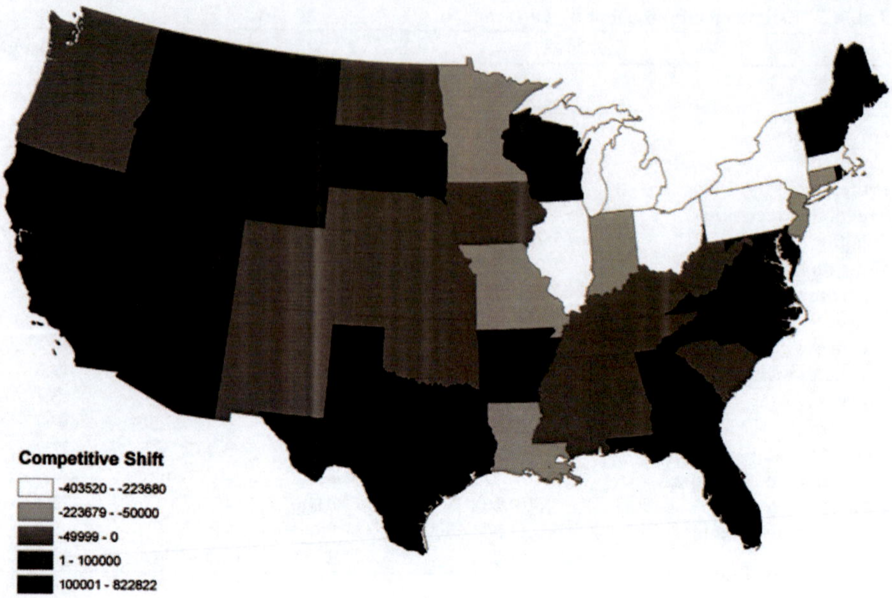

Fig. 9 Competitive shifts

A shift-share model was developed, and Fig. 9 reports the pattern of competitive shifts estimated in this analysis. The shift–share model indicates that ±2,644,219 jobs were created in regions outside where expected, given the share and industry mix components. This represents 19.5% of the change in national employment. The pattern reported in Fig. 9 is not dissimilar to that reported in Fig. 8. All of the states with the highest percentage growth rates also show up in the positive competitive shifts, while the pattern of negative competitive shifts is more clearly identified with the old industrial belt.

Figures 10 and 11 present patterns of competitive shifts for the professional and creative clusters. Space prevents inclusion of more maps of the competitive shifts. The largest contributor to the competitive shifts was the professional cluster, with Florida being the state with the largest competitive shift in this cluster. Not unexpected, the states with the strongest growth and competitive shifts also show strong positive shifts in this cluster. However, some other states that were not among the fastest growth or those with the largest positive competitive shift exhibit relatively strong positive competitive shifts in the professional cluster (Georgia, Virginia, Maryland). Figure 11 shows the pattern of competitive shifts in the creative cluster. Here a very different pattern is evident, with negative shifts in a band of states stretching from Colorado east to New Jersey and Massachusetts. New York, with its strong concentration of activity in the creative cluster stands out with its large positive shift. Washington State, with its strong concentration in software also stands out in this cluster. Further analysis is needed to determine the principle industrial contributors to the positive and negative competitive shifts for the professional and creative clusters.

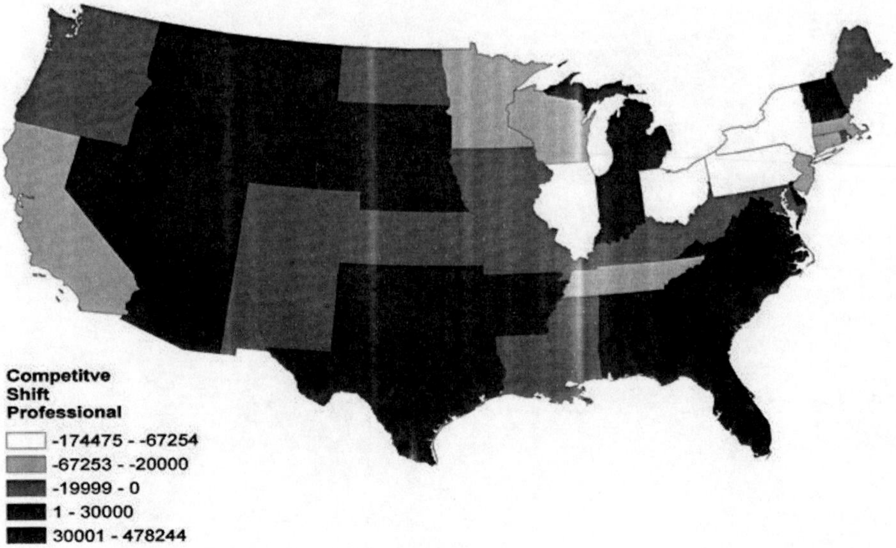

Fig. 10 Competitive shifts in the professional cluster

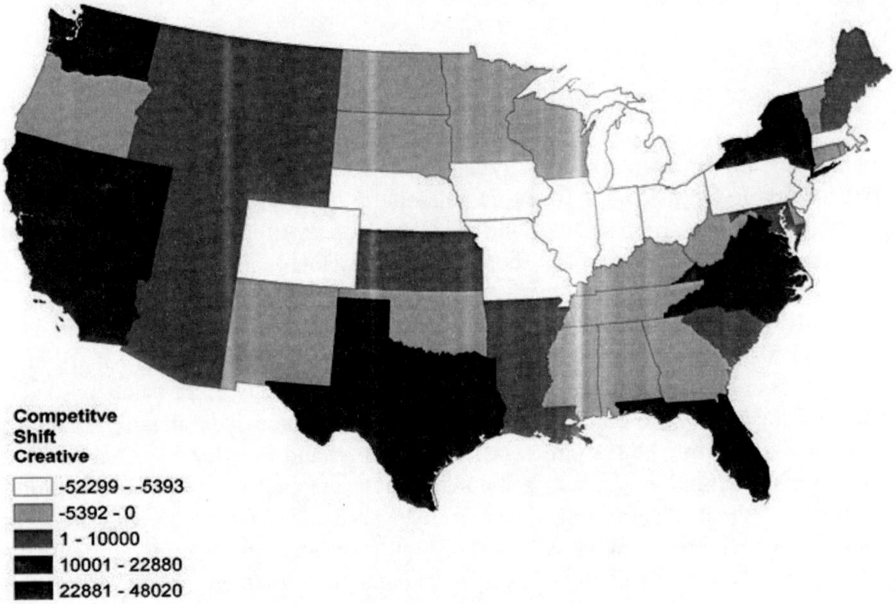

Fig. 11 Competitive shifts in the creative cluster

Two additional perspectives on the distribution of the clusters across the states is provided by estimates of coefficients of regional specialization, and coefficients of cluster concentration. Three states show very skewed cluster structures – West

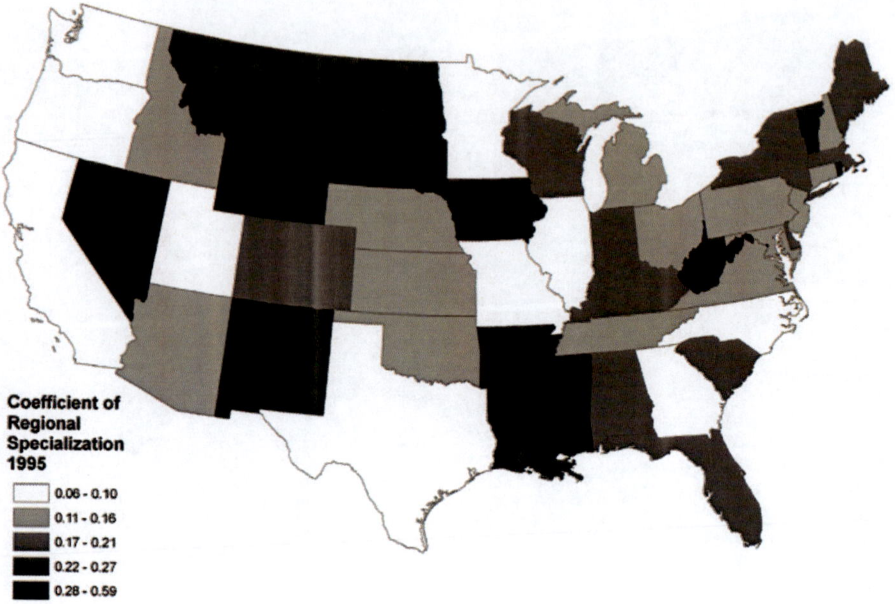

Fig. 12 Coefficients of regional specialization

Virginia and Wyoming due to mining, and Nevada due to high values for the travel and recreation cluster (gambling). Not all large states are in the lowest categories – a tendency for measures of this type. New York and Florida exhibit relative specialization; New York's index is related to its strong concentration in the creative, education, professional, and finance clusters. Florida's index is affected by strong values in the travel and recreation, professional, and government clusters (Fig. 12).

The distribution of the clusters across the states is quite variable, as indicated by coefficients of cluster concentration presented in Table 6. Some are distributed more or less in accord with the overall distribution of employment (such as social/retail-wholesale), while others exhibit much more regional specialization (such as travel and recreation, education, petroleum, mining, and forestry).

Patterns of employment among the states were mapped in the form of location quotients. Several of these maps are included here, for three of the clusters with high levels of creative and human capital. Figure 13 portrays location quotients for the creative cluster. High values in California and New York are no surprise, given the concentration of industries such as arts, film, and in the case of California computer related activities. The high value in Washington, Maryland, Virginia, and Colorado is related to the concentration of the information sector. The Professional cluster, as illustrated in Fig. 14, shows some similarity to the pattern in Fig. 13. However, Arizona and Florida enter the set of states with strong concentration in this cluster. Figure 15 presents a very different pattern. The strong concentrations in

Table 6 Coefficients of cluster concentration

	1998 CIC	2005 CIC	Change
Food	0.076	0.072	−0.004
Education	0.343	0.372	0.028
Government	0.463	0.437	−0.026
Creative	0.180	0.188	0.008
Professional	0.154	0.145	−0.009
Travel and recreation	0.243	0.231	−0.012
Health	0.101	0.101	0.000
Construction	0.131	0.143	0.012
Social and retail/wholesale	0.050	0.047	−0.003
Repair	0.103	0.097	−0.007
Finance	0.119	0.127	0.008
Transportation	0.107	0.156	0.048
Miscellaneous	0.314	0.305	−0.009
Food mfg	0.143	0.132	−0.010
Manufacturing	0.256	0.304	0.048
Petroleum and pipelines	0.580	0.594	0.013
Mining	1.002	1.037	0.036
Logging and support	0.505	0.488	−0.017

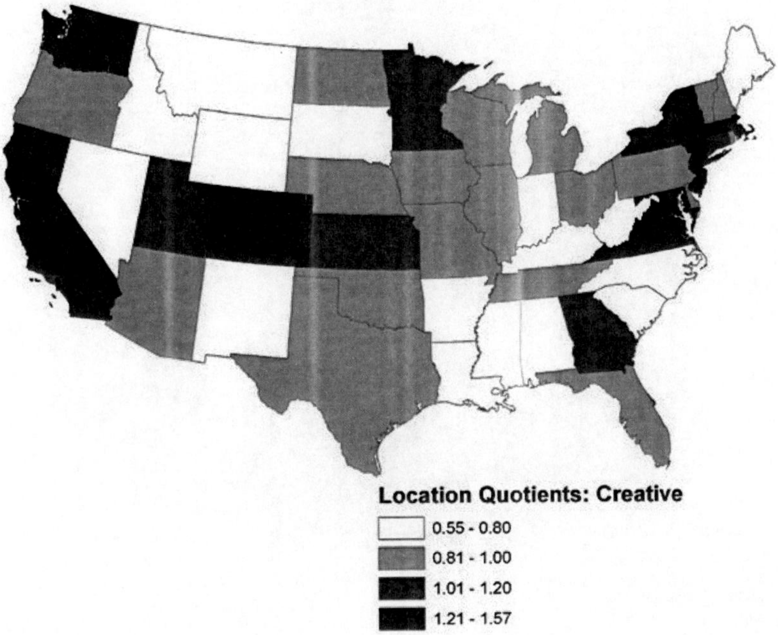

Fig. 13 Location quotients: creative cluster

the middle west may be related to demographics in rural farm states, with relatively large percentages of elderly people. This explanation may be improbable, as Florida, with its very large retired population does not show a strong concentration

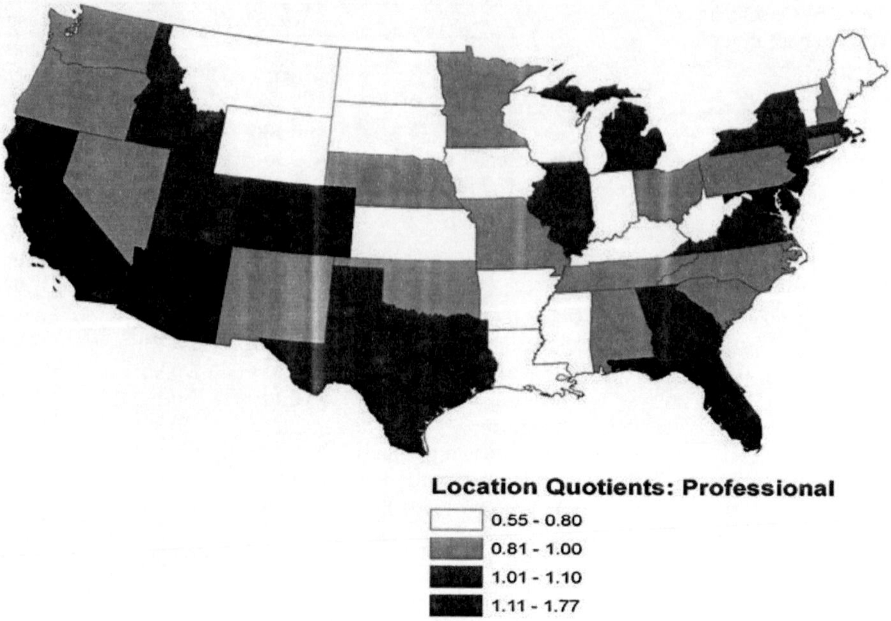

Fig. 14 Location quotients: professional cluster

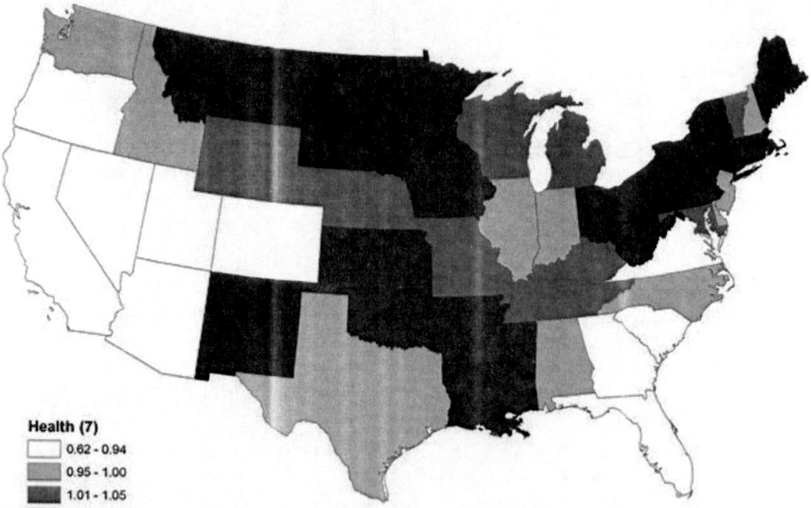

Fig. 15 Location quotients: health cluster

in health care. Further analysis is needed to better understand the patterns in these maps, and those found on other maps of clusters created but not included with this paper.

4 Correlates of State Growth, and Human and Creative Capital

The previous section reported results from a shift-share model that indicated 19.5% of growth was not where it was expected (was a part of the competitive shift). That section also reported that the professional cluster was a major contributor to the positive competitive shift. To further explain regional trends, various correlates of state growth rates over the relatively short 1998–2005 time period have been developed.

In economic development circles it has been fashionable to argue that the New Economy and High Tech are drivers of regional development. Using Atkinson's Correa's New Economy Index (Atkinson and Correa 2007), and a measure of high-tech developed by this author based on Hecker's occupationally based definition of high-tech, Table 9 reports results from this test. The variable Index is Atkinson's New Economy Index, while the variable HITECHLQ was the measured developed by this author. Atkinson also presents a high-tech index, essentially that of the American Electronics Association, which is a relatively limited definition of high-tech (HT_ATKIN). Table 7 reports no significant relationships between state employment growth rates and Atkinson's New Economy index. It reports a significant negative relationship between my measure of high-tech and regional growth rates. There is a strong positive correlation between my high-tech index and Atkinson's high-tech index, as well as Atkinson's New Economy Index Using Atkinson's and Correa's New Economy Index (Atkinson & Correa 2007), and a measure of high-tech developed by this author based on Hecker's occupationally based definition of high-tech.

Tables 8 and 9 present correlations between state growth rates and selected measures for 1998 and 2005. There is a significant positive correlation between state growth rates and the coefficient of regional specialization. This means that states with low coefficients of regional specialization have tended to have lower growth

Table 7 Correlations between state growth rates, the new economy index, and high tech measures

	GROWTH	INDEX	HITECHLQ	HT_ATKIN
GROWTH				
INDEX	0.062			
HITECHLQ	−0.386**	0.757**		
HT_ATKIN	−0.006	0.868**	0.779**	

**Correlation is significant at the 0.01 level (two-tailed)

Table 8 Correlations with 1998 location quotients

	GROWTH	CRS	COMPSHIFT	CREAT98	PROF98	HEALTH98	FINANC98
GROWTH							
CRS	0.360**						
COMPSHIFT	0.690**	0.090					
CREAT98	−0.065	−0.235	−0.025				
PROF98	0.152	0.098	0.156	0.701**			
HEALTH98	−0.441**	0.115	−0.312*	−0.142	−0.304**		
FINANC98	0.134	0.119	0.034	−0.402**	−0.460**	0.183	

*Correlation is significant at the 0.05 level (two tailed)
**Correlation is significant at the 0.01 level (two tailed)

Table 9 Correlations with 2005 location quotients

	GROWTH	CRS	COMPSHIFT	CREAT05	PROF05	HEALTH05	FINANC05
GROWTH							
CRS	0.360**						
COMPSHIFT	0.690**	0.090					
CREAT05	−0.063	−0.517**	−0.036				
PROF05	−0.326*	−0.376**	0.311*	0.559**			
HEALTH05	−0.545**	−0.010	−0.426*	−0.173	−0.563**		
FIANC05	−0.151	−0.310*	−0.137	0.342*	0.213	0.167	

*Correlation is significant at the 0.05 level (two tailed)
**Correlation is significant at the 0.01 level (two tailed)

rates than those that exhibit specialization. The significant correlation between state growth rates and the magnitude of the competitive shift is totally expected – states with slow growth rates are likely to have negative competitive shifts, and those with strong growth rates are likely to have positive competitive shifts. Interestingly, there is a significant negative relationship between state growth rates and 1998 location quotients for health care, and there is no meaningful relationship between state growth rates and the concentration in 1998 of the creative, professional, or financial clusters. Table 9 presents the same correlations, except using 2005 location quotients. In effect these models are indicating that the concentration or lack of it in two relatively rapidly growing clusters (creative and finance) are not good predictors of state growth rates, while there is a modest and significant correlation between concentration in the professional cluster and state growth rates. Moreover, there is a strong positive correlation between the location quotients for the creative, professional, and financial clusters. More work is needed to understand these results. Other correlates were explored, but not reported here. State growth rates and the share of the population over 65 was not a significant explanatory variable. Manufacturing cluster location quotients were negatively related to state growth rates, as could be expected.

5 Concluding Comments

This paper has presented a new classification of economic activities based on the occupational structure of industries in the United States. This classification scheme exposes industries with strong concentrations of workers in occupational categories that have high levels of human and creative capital. It also leads to the definition of alternative groupings of industries, groupings that are different across much of the service economy from the groupings based on industry definitions based on NAICS codes.

There is much more work that needs to be done to understand patterns of growth occurring in the United States through the lens of these industry clusters. This includes unpacking the contributions of particular industries to the growth trends in clusters. The development of this data base has been time consuming, but it promises to yield new insights into regional growth patterns in the United States.

One of the research needs is to develop measures of educational attainment by cluster group. While the occupational mix at the national level was used to define the clusters, it is likely that the industry mix at the regional level within these clusters is different than at the national level. Using national statistics on educational attainment at the level of the clusters would likely lead to fallacies of composition unless regional industry × occupation data were used, and related to regional measures of educational attainment by occupation. However, if such concordances can be made between industry, occupation, and educational data, then it would be possible to more precisely identify levels of human and creative capital in regions.

It is recognized that states are quite aggregate geographic units, and an alternative would be to develop data for the clusters identified in this paper at the level of the BEA economic areas, or for metropolitan areas. This is a large data assembly task, that would require grant support to develop the data base.

Another issue is whether the occupational structure of industries in the United States is similar to that found in other advanced economies. The author does not know if comparable industry × occupation data are available for European countries, or OECD countries. However, it would seem to be a research priority to raise the question as to whether there are occupational similarities across advanced economies, such that clusters similar to those developed in this paper could be defined for these regions.

Appendix

Table 10 Occupational categories used for cluster analysis

Occupation	Code	Employees
Executives	11_1	2,024,490
Managers	11_2	540,860
Service managers	11_3	1,374,590
Manufacturing managers	11_9	1,885,140
Agents	13_1	422,210
Adjusters	13_2	289,390
Cost estimators	13_3	444,820
Recruitment	13_4	284,000
Training	13_5	382,800
Logistics	13_6	585,040
Other business operations specialists	13_7	947,030
Accountants	13_8	1,213,180
Credit financial analysts	13_9	463,050
Loans	13_10	386,280
Tax specialists	13_11	249,760
Computer and information scientists, research	15_1	416,440
Computer programmers and software engineers	15_2	764,430
Computer support specialists	15_3	506,080
Computer systems analysts	15_4	436,780
Database and other computer system administrators	15_5	583,570
Other computer science occupations	15_6	170,960
Actuaries, mathematicians, operations research analysts, statisticians, other mathematical occupations	15_7	92,610

(continued)

Table 10 (continued)

Occupation	Code	Employees
Architects	17_1	187,630
Engineers	17_2	1,330,880
Drafters and technicians	17_3	737,100
Life, physical, and social science occupations	19-0000	1,220,470
Community and social services occupations	21-0000	1,738,410
Legal occupations	23-0000	970,690
Education, training, and library occupations	25-0000	8,197,750
Arts, design, entertainment, sports, and media occupations	27-0000	1,691,370
Healthcare practitioners and technical occupations	29-0000	6,642,630
Healthcare support occupations	31-0000	3,472,530
Protective service occupations	33-0000	2,970,040
Food preparation and serving related occupations	35-0000	11,019,380
Building and grounds cleaning and maintenance occupations	37-0000	4,344,830
Personal care and service occupations	39-0000	3,184,630
Supervisors retail	41_1	1,392,010
Cashiers and clerks retail	41_2	8,558,590
Service sales representatives	41_3	1,029,220
Wholesale sales representatives	41_4	1,868,640
Other sales representatives	41_9	880,600
Supervisors office workers	43_1	1,276,260
Switchboard operators	43_2	190,120
Billing and bookkeeping	43_3	3,622,050
Clerks	43_4	5,426,170
Cargo and freight agents, couriers, messengers, dispatchers	43_5	3,887,580
Secretaries	43_6	3,893,610
Data entry keyers, word processors and typists, desktop publishers	43_7	1,334,170
General office clerks, other clerks	43_8	3,026,720
Farming, fishing, and forestry occupations	45-0000	440,200
Construction supervisors	47_1	572,220
Construction tradesmen	47_2	5,012,950
Construction helpers	47_3	418,470
Construction specialists	47_4	346,970
Construction operators	47_5	230,380
Repair supervisors	49_1	445,150
Repairers	49_2	538,050
Repair technicians and mechanics	49_3	1,489,420
Other repairs	49_9	2,670,470
Supervisors-assembly	51_1	671,500
Assemblers	51_2	1,943,710
Bakers and butchers	51_3	649,010
Machine operators	51_4	2,079,130
Binders and printers	51_5	367,590
Laundry and dry cleaning, textiles, sewing	51_6	744,800
Cabinets, furniture	51_7	312,270
Plant operators	51_8	295,810
Miscellaneous operators and production workers	51_9	2,700,840
Transport supervisors, airline pilots and related, bus and ambulance drivers	53_1	1,378,470
Truck drivers	53_2	3,738,850
Truck tractor operators	53_3	612,790
Cleaners	53_4	321,150
Laborers	53_5	2,241,410
Packers	53_6	805,540

References

Atkinson RD, Correa DK (2007) The 2007 state new economy index. The Information Technology and Innovation Foundation, Washington, DC, p 87

Barbour E, Markusen A (2007) Regional occupational and industrial structure: does one imply the other. Int Reg Sci Rev 30(1):1–19

Beyers W (1989) The producer services and economic development in the United States: the last decade. US Department of Commerce, Economic Development Administration, Washington, DC

Beyers W (2007a) Occupational structure in industries in the United States: defining clusters and the regional distribution of selected clusters. Reser, Tampere, Finland

Beyers W (2007b) Services and regional development in the United States. In: Bryson J, Daniels PW (eds) The handbook of service industries. Cheltenham, Edward Elgar, pp 126–148

Browning HC, Singelmann J (1975) The emergence of a service society. National Technical Information Service, Springfield

Clark C (1957) The conditions of economic progress. Macmillan, London

Daniels PW (1985) Service industries, a geographical appraisal. Methuen, London

Daniels PW, Bryson J (2002) Manufacturing services and servicing manufacturing: changing forms of production in advanced capitalist economies. Urban Stud 39(5–6):977–991

Feser EJ (2003) What regions do rather than make: a proposed set of knowledge-based occupation clusters. Urban Stud 40(10):1937–1958

Fisher A (1939) Production, primary, secondary, and tertiary. Econ Rec 15:24–38

Florida R (2002) The rise of the creative class. Basic Books, New York

Gershuny JI, Miles ID (1983) The new service economy: the transformation of employment in industrial societies. Francis Pinter, London

Greenfield H (1966) Manpower and the growth of producer services. Columbia University Press, New York

Hutton TA (2007) Service industries, global city formation and new policy discourses within the Asia-Pacific. In: Bryson J, Daniels PW (eds) The handbook of service industries. Edward Elgar, Cheltenham, pp 149–167

Illeris S (1996) The service economy, a geographical approach. Wiley, Chichester, England

Markusen A (2006) Urban development and the politics of a creative class: evidence from the study of artists. Environ Plann A 38(10):1921–1940

Markusen A, Schrock G (2006) The artistic dividend: urban artistic specialization and economic development implications. Urban Stud 43(10):1661–1686

Markusen AR, Wassall G, DeNatale D, Cohen R (2008) Defining the cultural economy: industry and occupation approaches. Econ Dev Q 22(1):24–45

Noyelle T, Stanback T (1983) The economic transformation of American cities. Rowman and Allanheld, Totowa, NJ

Petit P (1986) Slow growth and the service economy. Francis Pineter, London

Scott AJ (2007) Capitalism and urbanization in a new key? The cognitive cultural dimension. Soc Forces 85(4):1465–1483

Thompson WR, Thompson PR (1985) From industries to occupations, rethinking local economic development. Econ Dev Comment 9:12–18

Thompson WR, Thompson PR (1987) National Industries and local occupational strengths: the cross-hairs of targeting. Urban Stud 24:547–560

Thompson WR, Thompson PR (1993) Cross-hairs targeting for industries and occupations. In: Barkley DL (ed) Economic adaptation, alternatives for nonmetropolitan areas. Westview, Boulder, pp 265–286

Wyatt ID, Hecker DE (2006) Occupational changes during the 20th century. Mon Labor Rev 129:35–57

Part IV
Outlook and Policy

Exploring Regional Disparities in Employment Growth

William Mitchell

1 Introduction

Australia has experienced a economic growth since the 1991 recession and convergence in labour market outcomes across space might reasonably have been expected given this 17 year span. The orthodox view of regional development suggests that "provided there are no major barriers to the operation of market forces, in an integrated national space economy there are strong pressures leading to the general *convergence* of regional incomes over time" (Martin and Sunley 1998: 201).

However, significant disparities in employment growth rates across Australia's metropolitan and regional areas have underpinned the persistence of unemployment rate differentials across the same spatial units and accompanying social disadvantage (Mitchell and Carlson 2005). Figure 4 shows the regions (SLAs) where employment grew (white) and where employment fell (black) for the period 1996–2006. This was a period when Australia enjoyed relatively strong overall employment growth as note previously and official unemployment fell to 4% (in February 2008 – the lowest level since December 1974). It is clear that the coastal areas (particularly along the populated east coast) and areas where mining industries are concentrated enjoyed the benefits of this employment growth. The rural decline outside of the mining areas is also very apparent. Overall, employment growth has not been sufficient overall to meet the preferences of the willing labour supply and has been spatially concentrated (Mitchell and Muysken 2008). It is hard to argue that an equilibrium convergence has resulted from the growth process (Fig. 1).

This paper examines the relative importance of regional-specific versus macroeconomic factors in determining regional employment outcomes which is an on-going debate among regional scientists and policy-makers (Rissman 1999). Keynesians typically argue that regional employment variations are caused by the

W. Mitchell
Centre of Full Employment and Equity, University of Newcastle, Callaghan, NSW 2308, Australia
e-mail: Bill.Mitchell@newcastle.edu.au

K. Kourtit et al. (eds.), *Drivers of Innovation, Entrepreneurship and Regional Dynamics*, Advances in Spatial Science, DOI 10.1007/978-3-642-17940-2_16,

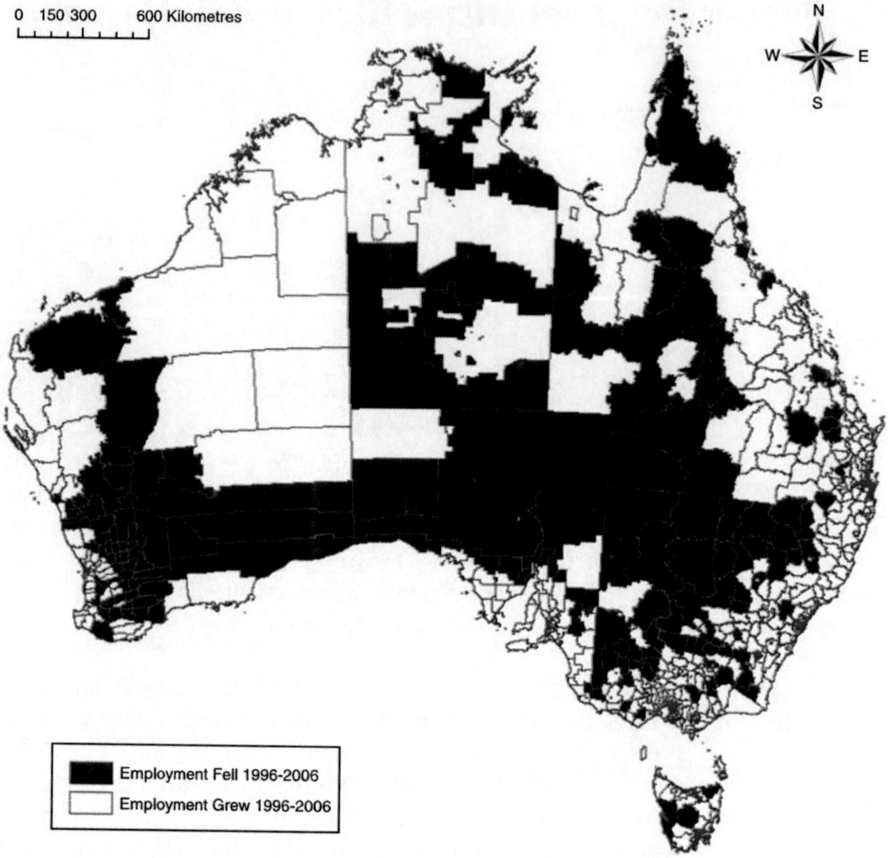

Fig. 1 Regional employment growth, 1996–2006
Source: ABS Population and Housing Census, Time Series Profiles, 2006

national business cycle impacting on growth rates across industries and reflect changes in aggregate factors, such as fiscal and monetary policy settings, business and consumer confidence and productivity trends. Thus, the cyclical sensitivity of regional outcomes reflects the impact of common aggregate shocks on a specific regional industry mix. Regions dominated by goods-production allegedly lose employment share in recessions relative to service-providing regions. The solution is for aggregate policy to maintain strong growth with industry policy attenuating structural shifts.

In this context, Australia's terms of trade up during the period studied were at their highest level in 30 years particularly benefitting regions involved in primary commodities (mining).

A rival regional development paradigm, termed "new regionalism", has become popular among regional development agencies since it emerged in the mid-1980s. It was inspired by case studies documenting economic successes in regions such as

Silicon Valley and Baden Württemberg (see, e.g., Sabel 1989; Saxenian 1994). Scott and Storper (1989) posited that regions had displaced nation states as sites of successful economic organisation and the emphasis should be on localised institutions and collaborations. Accordingly, the status of macroeconomic policy is considered peripheral to a particular region's growth potential (Castells and Hall 1994; Cooke and Morgan 1998). Despite the growing popularity of new regionalism, the claim that the region offers a convincing theoretical explanation of recent and future economic development is under-researched and has weak empirical underpinnings. There is little known about how the national economy and its regions interact. Further, no empirical evidence exists to verify assumptions of, first, the emergence of capitalism centred on spatialised, autonomous economies, and, second, a hollowed out, macro-weakened nation state (Lovering 1999; Markusen 1996).

Related to the new regionalist approach is endogenous (or new) growth (EGT) theory which was initially developed to challenge neo-classical growth theory and has been, more recently, extended to explain regional development (see Arrow 1962; Romer 1986; Aghion and Howitt 1997). The main attraction of EGT is that it recognises that technology is not static. It thus requires us to not only explain the growth process but also the reasons why technology (and the potential growth rate) varies over time. All the many variants of EGT stress the importance of knowledge in the growth process. Given that "ideas" are not limited and can be shared and recycled, EGT rejects the traditional concept of diminishing marginal returns that economists have considered limit the productive usage of other productive inputs. If knowledge, which is assumed to be subject to increasing returns, is generated and diffused locally regional development will occur. Stough and Nijkamp (2007: 749) argue that "the formulation of endogenous growth theory . . . brought back recognition that local resources and their deployment in a market fitted strategy make a significant contribution to growth. Local effort and resources are conditioned by a region's history, resources, market fit, institutions, leadership and orientation to entrepreneurship".

There is very little research on Australia that attempts to disentangle macroeconomic and regional growth factors as drivers of observed employment growth disparities across regions. This paper uses dynamic shift–share analysis (Arcelus 1984; Barff and Knight 1988) to decompose annual regional employment growth into three components (a) a national share (growth) effect, *NS* being "that part of the change in total employment in a region ascribed to the rate of growth of employment in the nation as a whole" (Barff and Knight 1988: 2). This component attempts to separate the regional growth that is attributable to nationwide economic progress independent of industry composition and specific-local factors; (b) an industry mix (proportional) effect, *IM* being "the change the region would have experienced had each of its industries grown at their national rates less the national growth effect" (Barff and Knight 1988: 2). So this effects arises because regions have different industry compositions and these industries grow (contract) at different rates over time; and a (c) regional share (differential or competitive effect), *RS* being "the difference between the actual change in employment and the employment change to be expected if each industrial sector grew at the national rate" (Barff and Knight 1988: 2). So this component focuses on local or regionally specific influences.

As an innovation, we modify the traditional shift–share framework to examine the regional impact of the increasing significance of part-time work in overall employment creation in Australia by decomposing employment dynamics into part-time and full-time components which helps us explore the spatial disparities more closely.

While we recognise the limitations of shift–share analysis raised long ago by Houston (1967), we consider that it provides a useful organising framework for more sophisticated hypothesis development and econometric modelling.

In relation to EGT, we might expect the regional shift component to be positive and dominant in regions where economic growth is strong. Stimson et al. (2005) clearly recognise that the shift–share components are constructed as accounting identities which exhaust total employment growth. Houston (1967: 578) says that "to be useful in explaining regional growth, there should be some *theoretical* basis for identifying the three separate components". In this context, Stimson et al. (2005) used the regional (competitive) share component from a shift–share analysis as the proxy for endogenous elements. The presumption is that there is a strong mapping between the regional (competitive) component and the elements that are conceptually identified as being central drivers in endogenous growth theory?

We do not challenge that presumption in this paper. Rather we seek to determine whether there is a *prima facie* case for pursuing EGT further by computing (in an accounting sense) the relative importance of the "unexplained" regional component.

We acknowledge the difficulty in disentangling the shift components (industry and regional). Houston (1967: 579) said that "supply changes, demand shifts, technological changes, locational shifts, any or all of these, may be behind either component". However, strong regional shifts provide indication that "local" factors are worth exploring further.

However, we emphasise that a major limitation of the shift–share framework is that it is unable to explain why the changes that are computed occur. Their correct interpretation requires a clear theoretical framework. While this paper is focused on measurement issues and is therefore largely descriptive the work by Holden et al. (1989) is a useful guide to the theoretical underpinning of the work which we develop in subsequent papers.

This issue is particularly relevant to the "two speed" economy of Australia where a major driving factor in the business cycle is the commodity price movements on international markets (Mitchell and Bill 2006). It doesn't make much sense to suggest that the mining regions of Western Australia or Queensland have enjoyed strong economic growth as a result of large competitive shifts.

The paper is organised as follows. Section 2 details the impact of terms of trade trends on the Australian economic in recent years. Section 3 outlines the data used and the definitions of the growth rates. Section 4 provides an overview of the disparate patterns of regional employment growth in Australia in recent years, while Sect. 5 decomposes these patterns by industry. Section 6 conducts the dynamic shift–share analysis which is then decomposed into full-time and part-time components in Sect. 7. Concluding remarks follow.

2 Terms of Trade Trends

To understand the pattern of regional employment growth in Australia one has to appreciate that economic activity has been dominated in recent years by primary commodity prices and the impact these have had on exchange rates and export incomes. Australia's terms of trade, which during the period studied were at their highest level in 30 years, clearly delivered uneven benefits to Australian regions – leading to commentators calling Australia the "two-speed" economy (Mitchell and Bill 2006).

Figure 2 shows that while rural commodities have faced fairly flat world prices, non-rural commodity prices, especially Base Metals have escalated substantially.

The accompanying industrial (and regional) shifts in demand for labour, and the impact of the boom on the price of intermediate goods, has exacerbated trends in other sectors of the economy. The booming terms of trade has led to significant exchange rate appreciation with a classic Dutch Disease situation emerging (see Corden 1984 for more explanation of this concept) Employment in mining has grown by a staggering 65% since 2002 (up to mid-2008). The strongest growth (2002–2008) has been in Queensland (97%) and Western Australia (59%).[1] Conversely, Manufacturing is generally declining with regionally concentrated costs (e.g., in NSW, Victoria and South Australia). The manufacturing heartlands

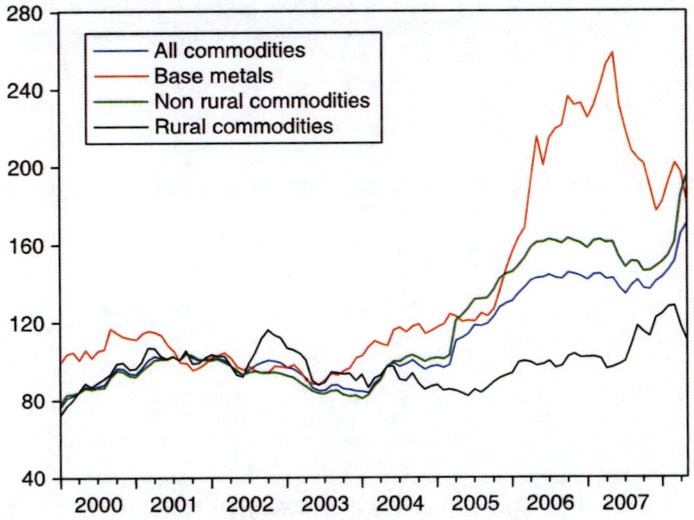

Fig. 2 Commodity prices in Australia, $A, 2001–2002 = 100, 2000–2008
Source: Reserve Bank of Australia, Index of commodity prices ($A)

[1]Australian Bureau of Statistics, Labour Force Survey, Employed Persons by Region, Sex, Industry division data cube.

of NSW, Victoria and South Australia have experienced negative growth in manufacturing employment between 2002 and 2008. However, in the mining concentrated regions, manufacturing is resisting the absolute decline in employment, but still losing employment share (see Sect. 5).

3 Data Issues

The typical unit of analysis for Australian regional studies, particularly in cross-national studies, has been the State/Territory (see Dixon and Shepherd 2001; Mitchell and Carlson 2005). While more detailed regional labour force data is available for 64 Major Statistical Regions collected through the Australian Labour Force Survey, the industry data is unreliable due to high sampling errors. To focus attention on industry employment movements the lowest reliable disaggregation available is at capital city (metropolitan)/rest of state level. Accordingly, we define the regions by capital city (metropolitan) (denoted _C) and rest of State (denoted _R) with the ACT and NT treated as complete regions. The quarterly data are available from 1978 for standard labour force categories and from 1985 for detailed employment data for the 17 ANZSIC industries (see Appendix for description of ANZSIC classification). The latter data are available for full-time, part-time and total employment by industry by region.

For the dynamic shift–share analysis in Sects. 6 and 7, annual industry employment data by region was used. Thus annual national employment growth is defined as:

$$g_n = (E_t - E_{t-1})/E_{t-1} \quad (1)$$

The annual growth in employment in industry i at national level is defined as:

$$g_{in} = (E_{it} - E_{it-1})/E_{it-1} \quad (2)$$

Finally, annual employment growth rate for industry i in region r is defined as:

$$g_{ir} = (E_{irt} - E_{irt-1})/E_{irt-1} \quad (3)$$

4 The Pattern of Regional Employment Growth 1987–2008

The employment levels for the regions indexed to 100 at November 1987 are shown in Fig. 3. The two-speed nature of the Australian economy that has emerged over the long growth cycle since the 1991 recession is clear (a) high growth regions (Queensland, Western Australia, and the Northern Territory); and (b) moderate to low growth regions (NSW and Victoria, South Australia and Tasmania).

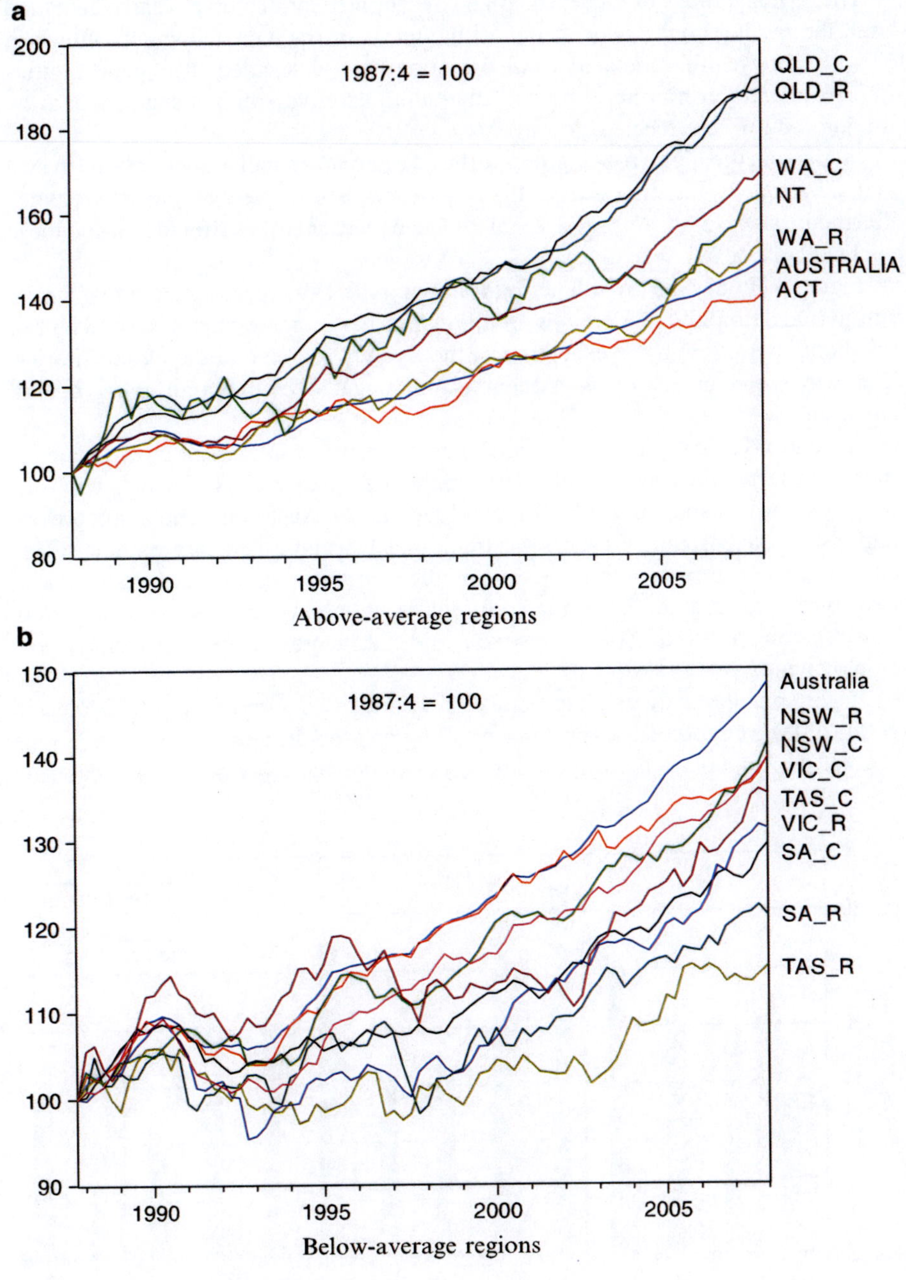

Fig. 3 Employment indexes, cities and regions, 1987:4 = 100
Source: ABS Labour Force

The regional areas of Queensland and Western Australia have clearly benefited from the mining boom associated with the terms of trade developments shown in Fig. 1. The wealth generated from the boom has also aided their capital cities (Brisbane and Perth) where many of the administrative structures supporting the mining industry are located.

In general, the high growth group suffered relatively smaller contractions in size and duration during the severe 1991 than the below-average growth regions. Tasmania and South Australia seem to have particularly suffered during these cyclical episodes.

Figure 4 divides the growth period following the 1991 recession into two 8-year sub-periods. Employment growth for all regions has increased in the second half of this boom period with the exception of the Sydney region and the two territories. The two speed economy assessment is not negated by the improving overall situation.

Appendix Table 6 presents more detailed summary employment growth data for the regions from the fourth quarter 1987 to the first quarter 2008. The data confirms the regionally disparate nature of employment in Australia. The high growth regions in Queensland, Western Australia and Northern Territory exhibit strong annual rates of growth with relatively smaller variance whereas the other regions are slower growing and have relatively higher variability over the sample period. The regional areas of South Australia and Tasmania exhibit particularly low employment growth rates.

If we split the growth period since 1991 into two samples (1992–2000 and 2000–2008) we note that variances are falling as growth rates improve among the poorest performers. It is not possible to conclude that convergence is occurring

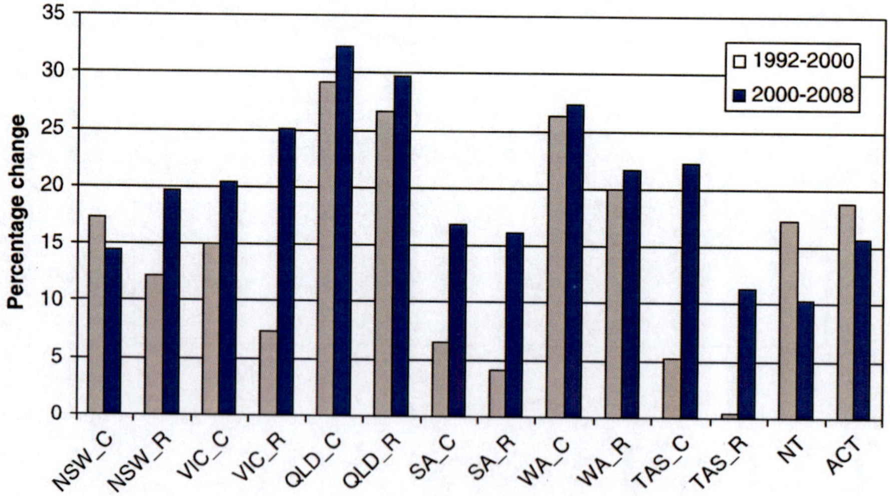

Fig. 4 Employment growth by region, 1992–2000 and 2000–2008, percentage change
Source: ABS Labour Force

because the most populated region (Sydney or NSW_C) has experienced deteriorating conditions in the latter period.

5 The Composition and Structure of Industry Employment by Region

The composition of industry employment and the pattern of change provide a clue to why regional employment growth in Australia is so disparate. However, given the number of regions in the analysis, it is difficult to depict the changing industry employment shares by region and nation over a long time-period in the space available. The highlights of such an analysis for the period 1985 and 2008 are:

- Agriculture has fallen sharply over the period but is still an important source of employment in regional NSW, Victoria, South Australia, Western Australia and Tasmania.
- Mining is not a large employer overall but is heavily concentrated in regional Queensland, Western Australia and Northern Territory. The combination of this concentration and the shift in commodity prices shown in Fig. 1 goes a long way to explaining why these regions have enjoyed strong employment growth.
- Manufacturing shares have fallen sharply over the period particularly in Sydney (NSW_C), Hobart (TAS_C) and Perth (WA_C). It remains an important employer in the Victoria, regional NSW, South Australia and regional Tasmania. Each of these labour markets has exhibited below-average employment growth.
- The growth in construction employment shares has been strong in the growth economies, particularly regional Queensland and Western Australia but has been muted or negative in the below-average regions.
- The large increase in shares in the Property and Business Services industry is striking and reflects the property boom that occurred in the last decade and which is still on-going in the high growth city centres of Perth and Brisbane.

6 Dynamic Shift–Share Analysis of Regional Employment

6.1 Industry-Region Decomposition

In this section we use dynamic shift–share analysis (Arcelus 1984; Barff and Knight 1988) to assess the extent to which the disparate regional employment growth patterns outlined in Sect. 3 reflect industry composition and regionally specific (locational) factors.

Where E_{ir}^t is employment in industry i in region r at time t (taken as the start of the period under scrutiny). The growth rates, g_n, g_{in} and g_{ir} are defined earlier. For each region, the individual industry components are summed to give NS_r, IM_r and RS_r.

As noted in the Introduction we decompose annual regional employment growth into three components (a) a national share (growth) effect, *NS*; (b) an industry mix (proportional) effect, *IM*; and a (c) regional share (differential or competitive effect), *RS*.

Total employment change for any region *r* and industry *i* is the sum of the three effects:

$$\Delta E_{ir} = NS_{ir} + IM_{ir} + RS_{ir} \tag{4}$$

The total shift (*TS*) measures the net variation in total employment that is not predicted by the national share and equals the actual change in employment minus the national share (or *IM* + *RS*).

There are 14 regions in the study (as defined above, six metropolitan areas, six rest of state areas, and two Territories) and 17 ANZSIC industries. The components for each industry *i* in region *r* are defined as:

$$\begin{aligned} NS_{ir} &= E_{ir}^t g_n \\ IM_{ir} &= E_{ir}^t (g_{in} - g_n) \\ RS_{ir} &= E_{ir}^t (g_{ir} - g_{in}) \end{aligned} \tag{5}$$

The results of the dynamic shift–share analysis are presented in Table 1 for three different periods: full sample – 1985–2008, 1992–2000, and 2000–2008. The shift–share components shown were derived as sums of the year-by-year components over the relevant time period.

A summary of the results is as follows:

1. The striking result is that non-metropolitan areas all suffered negative industry mix effects (for the overall sample and each of the sub-periods shown) which is in contradistinction to the good fortunes enjoyed by metropolitan areas. This means that over the period analysis the most dominant industries in these regions in terms of employment contribution have been declining relative to the national average. The large cities are thus gaining employment relative to regional areas as a consequence of their more favourable industry structure. We will examine more detailed industry breakdowns in Sect. 6.2 to identify the likely sources of these problems.
2. The employment growth in some non-metropolitan regions (QLD_R, WA_R) has been strongly supported by local or regionally specific factors that have more than offset the negative industry mix components identified previously. The strong employment growth in QLD and WA (both metropolitan and rest of state) has been driven by substantial regional share effects, which for the metropolitan areas of these states, has reinforced the positive industry mix components. While further research is required to identify what these local factors might be, these results provide some comfort for EGT proponents.

Table 1 Shift–share components for Australian regional employment, various periods, 1000s (% for last column)

	NS	IM	RS	Total change	Total shift	% Growth
1985–2008						
NSW_C	890.1	22.7	−187.6	725.2	−164.9	48.7
NSW_R	459.7	−44.2	−15.0	400.4	−59.3	51.0
VIC_C	756.2	12.5	−114.1	654.5	−101.7	50.6
VIC_R	271.2	−41.1	−2.1	228.0	−43.2	48.8
QLD_C	345.5	25.2	176.0	546.7	201.2	116.5
QLD_R	389.2	−11.4	215.5	593.3	204.1	109.5
SA_C	234.7	14.7	−98.4	151.0	−83.7	35.7
SA_R	61.9	−36.4	−4.9	20.6	−41.3	30.3
WA_C	296.4	42.8	69.3	408.5	112.1	94.6
WA_R	110.1	−16.4	25.5	119.3	9.2	69.5
TAS_C	41.0	4.1	−16.0	29.1	−12.0	40.6
TAS_R	54.4	−8.3	−20.6	25.5	−28.9	25.0
NT	40.7	5.9	−0.3	46.3	5.6	79.1
ACT	75.3	14.9	−16.4	73.8	−1.5	63.9
1992–2000						
NSW_C	282.9	3.4	2.7	289.1	6.1	17.3
NSW_R	145.4	−13.2	−22.5	109.7	−35.8	12.2
VIC_C	236.3	10.3	−34.9	211.8	−24.6	15.0
VIC_R	83.5	−10.1	−34.7	38.7	−44.8	7.4
QLD_C	111.1	6.2	62.8	180.2	69.1	29.2
QLD_R	124.2	−1.2	67.2	190.2	66.0	26.7
SA_C	74.3	5.1	−49.3	30.1	−44.2	6.6
SA_R	19.4	−10.4	−8.5	0.5	−18.9	4.1
WA_C	94.7	10.0	35.8	140.5	45.8	26.4
WA_R	35.3	−6.9	11.6	40.1	4.8	19.9
TAS_C	13.7	0.7	−10.3	4.0	−9.7	5.2
TAS_R	16.6	−1.8	−14.3	0.5	−16.1	0.4
NT	13.1	−0.7	1.2	13.6	0.5	17.3
ACT	24.4	2.3	0.6	27.3	2.9	18.9
2000–2008						
NSW_C	413.5	8.9	−134.4	288.0	−125.5	14.5
NSW_R	212.8	−16.0	4.2	201.0	−11.8	19.7
VIC_C	350.4	6.1	−19.3	337.3	−13.1	20.4
VIC_R	123.1	−20.0	41.7	144.8	21.7	25.1
QLD_C	173.5	10.5	68.2	252.2	78.7	32.4
QLD_R	195.9	−7.5	75.2	263.6	67.7	29.7
SA_C	104.7	4.5	−24.9	84.3	−20.4	16.9
SA_R	26.0	−13.2	9.2	22.1	−4.0	16.1
WA_C	144.6	19.4	19.6	183.6	39.0	27.4
WA_R	52.0	−4.1	5.0	52.9	0.9	21.7
TAS_C	17.8	2.2	−1.5	18.5	0.6	22.2
TAS_R	24.0	−4.5	−6.2	13.3	−10.7	11.3
NT	19.4	4.8	−14.9	9.3	−10.1	10.4
ACT	35.1	8.5	−16.8	26.9	−8.3	15.7

3. Conversely, the main city centres of Sydney, Melbourne, Adelaide and Hobart all experienced negative regional shifts which have exacerbated their poor industry mix performance.

4. The sub-period analysis shows however that in the recent growth phase (2000–2008) the regional areas of NSW, Victoria, and South Australia have enjoyed positive regional shares which has dampened the deteriorating industry mix.
5. Consistent with Fig. 3, only QLD, WA and the Northern Territory have experienced stronger employment growth than would be predicted if the regions had grown proportional to the national average. All other regions "underperformed" (total shift negative) with the sum of their industry mix and regional share effects being negative. However, there is considerable heterogeneity among these regions in terms of the balance between these effects.

6.2 Detailed Industry Shift–Share Analysis

The breakdown of the dynamic shift–share results into individual industries for each region provide a better understanding of which sectors have been responsible for the variations shown. While there is too much detail to present here (full tables are available on request from the author) some salient points include:

- National employment grew by 19% between 2000 and 2008. Three industries declined absolutely, Agriculture, Manufacturing and the Communication. The below-average growth industries included Wholesale Trade (5.1% growth) and Accommodation, Cafes, and Restaurants (11.8% growth). The latter had been one of the strongest growing industries of the 1990s as the services boom pushed the economy forward. The above-average growth industries were dominated by Mining (a staggering 74% growth), followed by Construction (44%), the Utilities (41%), Government and Defence (37%), Health and Community Services (33%), Property and Business Services (24.4%), Cultural and Recreational Services (24%), Education (22%), Transport and Storage (16%), Finance and Insurance (19%), Personal and Other Services (19%), and Retail Trade (19%).
- Mining employment has increased in all regions except ACT (where it has no presence) and TAS_R. The largest total changes are in WA_C (20.5 thousand), QLD_R (13.5 thousand), NSW_C (5.25 thousand) and WA_R (5.1 thousand). This reflects a combination of head office growth (in the cities) and an expansion of extractive employment (in the regions). Table 3 shows the shift share components and other summary measures for the important mining regions in Australia (either administrative or extraction sites). The interesting result is that despite favourable industry mix components in all regions, the total shift was negative in NSW_R, WA_R, TAS_R and the NT due to adverse regional shift components. This would appear to be a difficult result for those wanting to use the regional share component as a measure of endogenous growth. These three regions account for 28% of the total mining employment in Australia.
- The absolute national decline in manufacturing employment between 2000 and 2008 was resisted by NSW_R, VIC_R, QLD_C, QLD_R, SA_R, WA_C and WA_R, TAS_C, TAS_R, NT and ACT due to advantageous regional effects

offsetting the negative industry mix effects. So local or regionally specific factors (which are unidentified by the study) are supporting employment growth even though the specific industry composition in these regions is working against that. It is likely that some of the mining growth has spilled over into manufacturing in these regions (some of which specialise in metals industries). Conversely, the manufacturing strongholds of NSW_C, VIC_C and SA_C faced significant deterioration in their manufacturing with negative industry and regional shift components.

- While Accommodation, Cafes and Restaurants was the dominant service industry in the 1990s, in the 2000–2008 period its growth has been modest. Among the strong employment growth service industries, Property and Business Services (PBS) and Health and Community Services (HCS) have stood out. With the exception of WA_R and NT, the regional shares for PBS have been strong. The below-average growth regions have experienced negative regional shares in PBS. The pattern of regional shares for HCS is less clear and is not supportive of the view that the high growth regions enjoy favourable regional shares in the high growth industries.
- It remains true that the high growth regions generally had strong positive regional effects in above-average growth industries which reinforced the positive industry mixes.

The regional (competitive) shift component tells us whether a region is expanding or contracting its share of total industry employment. A positive component for a region's industry indicates that local employment in that industry is growing faster than the industry as a whole and hence the region is gaining industry share. The opposite holds for a negative regional component. Table 2 reports the regional shift components (RS) and the total employment change (Δ) for each region between 1985 and 2008 in thousands. The Δ row provides scale upon which the RS can be judged. Over the period covered, all industries bar Agriculture, Manufacturing, Electricity, Gas and Water and Wholesale Trade experienced positive growth.

It is apparent that regional shifts have been positive in most industries for the leading growth regions – QLD_C, QLD_R and WA_C. The same holds, albeit to a lesser extent for WA_R, The other top growth region, NT does not have the same. The scale of the regional shifts is fairly constant across these regions and their industries.

7 Full-Time and Part-Time Employment

7.1 Trends in Full-Time and Part-Time Employment in Australia

In addition to the vast sectoral changes noted in Sect. 5, there have also been substantial shifts in the employment mix between full-time and part-time across the regions since 1985 (see Tables 3 and 4). In 1985, 81.9% of total employment in

Table 2 Regional shift component (RS) and total employment change (Δ) by region and industry, 1985–2008, 1000s

		AGR	MIN	MAN	EGW	CON	WHO	RET	ACR	TAS	COM	FAI	PBS	GAD	EDU	HCS	CRS	POS
NSW_C	RS	1.8	0.9	−78.1	−9.0	−23.1	−23.3	−27.7	20.7	9.7	9.3	12.4	−25.9	−19.5	−3.0	−25.9	−6.5	−0.6
	Δ	−0.2	1.6	−86.5	−20.8	94.0	−9.1	103.6	77.7	47.9	14.9	56.1	216.4	6.0	55.6	96.5	32.4	39.1
NSW_R	RS	−2.1	−17.7	24.1	−2.9	−2.7	−0.4	23.1	−15.3	−18.0	−5.2	−5.1	−5.6	4.3	−9.4	15.0	−0.4	3.5
	Δ	−20.0	−10.6	21.0	−12.1	57.6	5.5	103.9	26.9	−2.4	−1.5	5.5	62.7	20.8	25.3	82.0	13.0	22.5
VIC_C	RS	−3.8	−1.4	−58.8	−0.1	4.9	10.9	−24.6	5.4	−4.1	3.5	1.9	6.6	−32.5	−6.1	−6.4	7.8	−17.4
	Δ	−5.3	0.2	−66.0	−7.5	98.9	24.5	95.9	46.8	21.8	9.9	33.7	199.4	−2.9	49.0	100.3	38.6	17.1
VIC_R	RS	−2.5	0.6	17.5	−2.4	−10.0	7.1	−4.6	−3.9	−0.6	−2.9	−2.3	0.2	−6.6	4.6	−0.6	0.4	3.9
	Δ	−14.1	2.1	16.8	−8.8	25.5	10.4	42.0	14.5	7.4	−1.1	2.9	32.9	6.9	26.0	41.3	8.6	14.7
QLD_C	RS	−0.7	1.0	39.8	4.5	6.8	−3.3	28.6	4.9	18.3	1.2	0.0	26.0	21.8	10.3	9.6	0.0	7.0
	Δ	−1.7	3.5	38.3	1.7	56.1	2.1	88.7	26.0	33.3	3.3	11.0	112.0	36.5	37.0	61.4	14.0	23.7
QLD_R	RS	−11.0	7.7	34.9	4.5	34.5	13.3	35.8	9.4	4.3	0.4	2.4	17.9	9.3	20.8	18.5	7.0	6.1
	Δ	−28.0	15.4	34.4	1.1	102.2	16.7	103.7	47.2	21.2	3.1	9.5	82.7	21.6	48.4	68.1	23.3	22.8
SA_C	RS	1.1	−0.4	−14.7	2.8	−15.2	−1.1	−16.6	−4.3	−8.0	−1.2	−3.7	−15.0	10.1	−12.9	−7.0	−7.5	−4.9
	Δ	0.3	0.7	−17.7	−0.3	12.4	3.6	24.0	11.2	−0.5	1.0	4.5	38.1	19.2	5.2	38.4	3.1	7.9
SA_R	RS	4.4	0.9	3.5	0.8	−4.8	−1.5	0.0	−4.8	−2.8	−2.8	−1.3	−2.7	1.4	−6.4	−1.2	−0.8	2.6
	Δ	−1.7	2.9	3.0	−0.6	4.5	−0.3	12.3	1.5	0.2	−1.9	−0.2	6.6	3.4	−0.7	11.0	1.1	6.1
WA_C	RS	0.5	17.6	17.3	3.5	8.9	2.9	−7.7	−4.3	4.3	2.1	−3.3	4.4	8.5	7.4	1.8	3.0	2.2
	Δ	−0.7	28.6	16.5	0.7	55.0	7.7	45.4	14.6	14.5	4.1	6.6	82.3	19.4	30.6	48.9	16.6	17.7
WA_R	RS	7.1	−4.4	13.2	0.7	8.5	0.8	6.8	−5.7	−3.1	−1.4	−1.2	0.9	−1.0	4.8	−1.7	−0.7	1.9
	Δ	0.7	2.7	13.3	−0.8	25.4	1.9	22.5	3.1	1.6	−0.5	0.1	14.8	3.4	12.5	10.3	1.9	6.4
TAS_C	RS	0.5	0.0	−0.1	−0.9	−4.2	−1.8	0.4	−0.4	−0.5	−1.8	−0.5	−2.2	−1.5	0.3	−1.3	−1.1	−1.0
	Δ	0.2	0.1	−0.4	−1.7	0.9	−1.2	7.3	3.3	0.7	−1.3	0.9	5.5	1.5	3.9	6.6	1.2	1.5
TAS_R	RS	4.8	−3.0	0.5	−1.1	−3.9	0.0	−6.4	−2.1	−1.1	−1.0	−0.7	−2.0	0.5	−1.3	−2.4	0.4	−1.7
	Δ	2.0	−2.1	0.0	−1.9	2.2	1.1	2.7	2.4	1.2	−0.8	0.3	4.6	2.2	2.8	6.3	2.0	0.4
NT	RS	0.3	−1.6	1.4	−0.9	−0.1	−3.1	−0.7	1.9	3.4	−0.6	0.2	−2.4	0.6	−1.1	3.2	0.2	−0.8
	Δ	−0.7	−1.2	1.4	−0.9	5.8	−2.6	5.3	5.7	5.1	−0.3	1.1	4.7	8.0	2.4	8.2	2.6	2.1
ACT	RS	−0.4	−0.1	−0.6	0.6	0.4	−0.5	−6.4	−1.5	−1.9	0.6	1.2	−0.3	4.6	−8.1	−1.5	−1.8	−0.7
	Δ	−0.1	−0.1	−0.6	0.1	8.9	0.1	5.2	4.2	−0.3	1.3	2.7	19.4	19.6	−1.1	8.4	3.2	3.2

Agriculture; manufacturing; electricity, gas and water; and wholesale trade all contracted nationally
Industry acronyms are provided in Appendix

Table 3 Full-time and part-time employment in Australia, 1985 and 2008

	1985		2008	
	000s	% of total	000s	% of total
Full-time	5,488.8	81.9	7,677.0	72.3
Part-time	1,214.2	18.1	2,948.7	27.7

Source: ABS Labour Force Survey, Australia. 2008 is the average to May

Table 4 Trends in employment generation, 1985 and 2008

	Full-time as % total employment		1985–2008			
	1985 %	2008 %	Total Δ 000s	Full-time Δ 000s	Part-time Δ 000s	Δ PT % of total
NSW_C	83.8	75.0	702.8	394.6	308.3	43.9
NSW_R	81.7	68.3	405.6	173.1	232.5	57.3
VIC_C	82.7	72.1	644.8	327.5	317.3	49.2
VIC_R	80.2	69.0	212.7	92.6	120.1	56.5
QLD_C	81.2	73.4	523.3	345.7	177.6	33.9
QLD_R	81.4	72.0	577.1	363.2	213.9	37.1
SA_C	79.4	69.1	147.3	57.6	89.7	60.9
SA_R	79.6	70.6	47.6	19.6	27.9	58.7
WA_C	79.4	72.5	394.9	255.6	139.3	35.3
WA_R	80.7	73.7	110.8	68.9	41.9	37.8
TAS_C	82.4	71.1	26.6	10.5	16.1	60.6
TAS_R	81.8	72.2	21.3	5.2	16.1	75.6
NTE	88.3	80.8	38.7	26.3	12.4	32.0
ACT	80.7	76.5	69.4	48.0	21.4	30.8
AUST	81.9	72.3	3,922.7	2,188.2	1,734.5	44.2

The operator Δ refers to the absolute change
Source: ABS Labour Force Survey

Australia was full-time (5,489,000). By 2008 (average January to May), this share had fallen to 72.3% (7,677,000). Of the 3,923,000 jobs created since 1985 in Australia 44% (1,734,000) have been part-time.

Two questions are relevant for this paper (a) Is the spatial superiority of the cities in employment generation overstated once we allow for part-time employment trend? and (b) Where the regional shares are strongly positive, what percentage of the net job creation is part-time? If the regional employment growth is being driven by a proliferation of precarious low-pay part-time jobs then we have to question the meaning of the regional shares.

In the period between 1985 and 2008, net part-time job creation constituted 44% of the total employment change, reflecting the increasing fractionalisation of employment in Australia. This trend was not consistent across regions.

Table 4 indicates that overall the full-time share in total employment has fallen by 9.6% points since 1985. No region has resisted that trend. However, with the exceptions of Sydney (NSW_C) and regional South Australia (SA_R), the loss of full-time share has been lowest in the above-average employment growth regions (Queensland, Western Australia). The ACT is an outlier largely due to the concentration of public service jobs which tend to be full-time.

7.2 *Modified Shift–Share Analysis to Account for Full-Time and Part-Time Trends*

To consider whether the regional shares are weighted towards full-time or part-time work across the regional space, we modify the shift–share model outlined in Sect. 6 to account for separate movements in full-time and part-time employment. The modified shift–share identity now explains total employment change for any region r and industry i and employment category s (where s is either full-time or part-time) as the sum of four effects:

$$\Delta E_{irs} = NS_{irs} + IM_{irs} + RS_{irs} + EC_{irs} \tag{6}$$

The previously defined components *NS*, *IM* and *RS* have the same meaning as before except now they can be computed for the two employment categories. The employment category shift, EC_{irs} indicates the shift in employment category s in industry i in region r due to faster or slower employment growth in that category relative to average employment growth in that industry and region.

The components for each industry i in region r and category s are now defined as:

$$\begin{aligned} NS_{irs} &= E^t_{irs} g_n \\ IM_{irs} &= E^t_{irs}(g_{in} - g_n) \\ RS_{irs} &= E^t_{irs}(g_{ir} - g_{in}) \\ EC_{irs} &= E^t_{irs}(g_{irs} - g_{ir}) \end{aligned} \tag{7}$$

where E^t_{irs} is employment in industry i in region r in category s at time t (the start of the period). The growth rates, g_n, g_{in} and g_{ir} are defined earlier. The category s employment growth in industry i and region r is defined as $g_{irs} = (E_{irst} - E_{irst-1})/E_{irst-1}$. For example, if an industry in a region is experiencing faster growth in full-time employment relative to total industry employment in that region, the EC_{irs} component will be positive and measures the shift away from part-time employment. Obviously $EC_{irf} + EC_{irp} = 0$ (where f is full-time and p is part-time). However, this decomposition allows us to examine the impact of the shifting full-time/part-time ratio within a region on the other components *NS*, *IM* and *RS*. For an industry as a whole in any region the total change in employment is the sum of the change in the two s categories (f and p). For the region as a whole, these individual industry components are then summed to give NS_r, IM_r and RS_r.

The results of the dynamic shift–share applied to the four-shift model are presented for the period 1985–2008 (with dynamic sums being shown) in Table 5. We are now able to appreciate the impact of the shifting full-time/part-time ratio in a spatial sense. The various totals correspond to the total shares displayed in Table 3. The national shares are simply the employment change in full-time and part-time if they had both grown at the annual national employment growth rate without any changes in the industry mix or regionally specific factors.

Table 5 Shift-share components for Australian regions, by full-time and part-time, 1985–2008, 1,000s

	NSW_C	NSW_R	VIC_C	VIC_R	QLD_C	QLD_R	SA_C	SA_R	WA_C	WA_R	TAS_C	TAS_R	NTE	ACT	AUST
NS															
Full-time	696.4	338.1	575.6	197.3	259.5	290.1	171.4	61.9	218.0	82.7	29.9	40.1	32.5	57.3	3,050.9
Part-time	193.7	121.6	180.6	73.9	86.0	99.1	63.3	23.5	78.3	27.4	11.1	14.3	8.2	18.0	998.8
Total	890.1	459.7	756.2	271.2	345.5	389.2	234.7	85.4	296.4	110.1	41.0	54.4	40.7	75.3	4,049.7
IM															
Full-time	−133.8	−129.3	−134.8	−90.7	−38.6	−78.2	−40.8	−36.4	−21.0	−33.2	−5.8	−18.6	−2.4	−1.5	−765.1
Part-time	156.5	85.1	147.3	49.7	63.8	66.9	55.4	13.8	63.8	16.9	9.9	10.2	8.3	16.4	764.0
Total	22.7	−44.2	12.5	−41.1	25.2	−11.4	14.7	−22.6	42.8	−16.4	4.1	−8.3	5.9	14.9	−1.1
RS															
Full-time	−142.5	−40.2	−107.1	−2.3	140.4	159.9	−71.1	−4.9	64.5	25.8	−10.2	−11.4	4.1	−5.0	0.0
Part-time	−45.2	25.2	−7.0	0.1	35.6	55.5	−27.3	−10.7	4.8	−0.3	−5.8	−9.2	−4.4	−11.5	0.0
Total	−187.6	−15.0	−114.1	−2.1	176.0	215.5	−98.4	−15.6	69.3	25.5	−16.0	−20.6	−0.3	−16.4	0.0
FTPT															
Full-time	−103.6	−92.4	−118.2	−51.0	−42.7	−60.7	−37.5	−10.2	−40.3	−13.3	−4.4	−6.7	−8.0	−6.4	−595.1
Part-time	103.6	92.4	118.2	51.0	42.7	60.7	37.5	10.2	40.3	13.3	4.4	6.7	8.0	6.4	595.1
Total	0.0	0.0	0.0	0.0	0.0	0.0	0.0	0.0	0.0	0.0	0.0	0.0	0.0	0.0	0.0
Total shift															
Full-time	−276.3	−169.5	−242.0	−93.0	101.8	81.7	−111.8	−41.3	43.5	−7.4	−16.1	−29.9	1.7	−6.4	−765.1
Part-time	111.4	110.2	140.3	49.8	99.4	122.4	28.1	3.2	68.7	16.6	4.1	1.0	3.9	4.9	764.0
Total	−164.9	−59.3	−101.7	−43.2	201.2	204.1	−83.7	−38.2	112.1	9.2	−12.0	−28.9	5.6	−1.5	−1.1
Total change															
Full-time	420.1	168.6	333.6	104.3	361.3	371.8	59.6	20.6	261.5	75.3	13.8	10.2	34.2	50.9	2,285.8
Part-time	305.1	231.8	320.9	123.7	185.4	221.5	91.4	26.6	147.0	44.0	15.3	15.3	12.1	22.9	1,762.9
Total	725.2	400.4	654.5	228.0	546.7	593.3	151.0	47.2	408.5	119.3	29.1	25.5	46.3	73.8	4,048.7

The negative industry mix effects noted earlier are now more transparent. Overall, the nation has seen a loss of employment due to industry shifts with the loss of full-time employment being marginally above the gains in part-time employment. All regions lost full-time opportunities as a consequence of industry shifts. However, this decline was more than offset by gains in part-time work arising from industry shifts in all the urban areas (NSW_C, VIC_C, QLD_C, SA_C, WA_C, TAS_C, ACT and NT). In stark contrast, the substantial loss of full-time work in regional areas (NSW_R, VIC_R, QLD_R, SA_R, WA_R, TAS_R) was not offset by positive part-time industry mix effects. This suggests that the job generating potential of the growth industries in cities is superior to regional areas although the gains are in part-time work at the expense of full-time employment.

In terms of the breakdown of regional share effects, the growth in QLD and WA (both urban and regional) employment arising from local factors is heavily weighted towards full-time employment. This stands in contrast to the other Australian regions which suffered negative regional effects overall and in their full-time job generation. In some cases, the regional effects of part-time employment were positive and attenuated the regional shift losses (NSW_R and VIC_R).

In total, the shifting full-time/part-time landscape has seen 765,000 less full-time jobs than would have been the case if the full-time ratio had have remained at its 1985 level. This is in the context of a change in total employment of four million over the 1985–2008 period. These are substantial shifts and the loss of full-time work has hurt regional areas more than metropolitan centres.

8 Conclusion

In this paper, dynamic shift–share analysis has been used to explore the disparate patterns of regional employment growth in Australia. We were motivated by an aim to generate more detailed breakdowns of the regional shift to inform the debate surrounding EGT.

It is clear that changes in industry composition have favoured the metropolitan labour markets. This result applies to high, moderate and low growth regions alike. Overall, the nation has seen a loss of employment due to industry shifts with the loss of full-time employment being marginally above the gains in part-time employment. However, without exception, the industry mix gains in metropolitan areas have manifested in part-time employment. Further, the regional areas have all failed to take advantage of the shifting industry mix because they have not been able to offset substantial full-time employment losses with commensurate part-time employment growth. It appears that the job generating potential of the growth industries in cities is superior to regional areas although the gains are in part-time work at the expense of full-time employment. Given that the incidence of precarious work has increased over the period studied, the positive industry shifts are not unambiguously favourable.

It is also clear that the ability to benefit from positive regional factors goes some of the way to explaining the separation of high from low growth regions. The high growth regions all have strong positive regional (local-specific) factors operating in their favour. With the exception of WA_R and NT, these regions have regional gains in both full-time and part-time employment. The growth in QLD and WA (both urban and regional) employment arising from local factors is heavily weighted towards full-time employment. In sharp contrast, the below-average growth regions have negative regional shifts overall. Several experienced unfavourable regional shifts in both full-time and part-time job generation whereas, for NSW_R and VIC_R, positive part-time regional effects attenuated the regional shift losses. More research is needed to determine the sources of these local advantages.

We repeat the difficulty that exists in disentangling the industry and regional shift components especially in the face of the commodity price boom identified in Sect. 2. In the latter part of the study period, economic growth was driven by this boom which, in turn, was spatially concentrated. The high growth regions identified with the strong regional shifts were the primary beneficiaries. In that context, it would be hard for an EGT to argue that the beneficial economic growth in recent years was the result of large competitive shifts. The spatial specificity of primary commodity deposits is the primary reason for the favourable outcomes enjoyed by these regions. Clearly, no policy tool can influence these advantages.

In terms of the issues raised in the introduction, the results support the previous conclusions of Mitchell and Carlson (2005) who argued that neither traditional Keynesian nor new regionalist strategies were likely to provide a sound basis for sustained regional development. It is clear that national factors remain dominant in determining a region's labour market outcomes. However, changing industry structure and unspecified local factors also play a significant role in employment growth across the regions.

The results generally suggest that any regional development policy aimed at stimulating local factors (e.g., knowledge accumulation) should be buttressed with strong aggregate demand policies to ensure there is no spending gap at the aggregate level. This will generally require positive national government deficits in the face of a desire to net save by the private sector (Mitchell and Muysken 2008). However, indiscriminate Keynesian expansion without regard to its spatial distribution is unlikely to reverse the trends identified in this paper. To ensure that this spending is spatially distributed to regions that have declining industry and negative regional factors operating targetted regional development policies incorporating infrastructure and industry development are required.

The results of the paper thus demonstrate that a mix of Keynesian aggregate demand policies and region-specific policies (favoured by EGT) should comprise the regional development strategy aimed at sustaining strong employment growth without significant disparities across space.

Appendix

Table 6 Summary average annual rates of employment growth and dispersion, various periods

Region	1987:4 to 2008:1			1992:1 to 2000:1			2000:1 to 2008:1		
	Average annual growth (% p.a.)	Standard deviation	CoV (%)	Average annual growth (% p.a.)	Standard deviation	CoV (%)	Average annual growth (% p.a.)	Standard deviation	CoV (%)
QLD_C	3.1	2.0	63.5	3.00	1.79	59.7	3.32	1.70	51.1
QLD_R	3.1	2.2	72.0	2.68	1.78	66.5	3.22	1.52	47.0
WA_C	2.5	2.0	80.4	2.70	2.07	76.5	2.81	1.36	48.5
NT	2.5	5.4	218.2	2.40	5.05	210.9	1.53	4.00	261.6
WA_R	2.0	2.6	130.5	2.06	2.17	105.0	2.26	2.78	122.9
Australia average	1.9	1.6	85.2	1.68	1.42	84.4	2.24	0.70	31.4
ACT	1.6	2.1	132.6	1.60	2.75	171.4	1.78	1.84	103.1
VIC_C	1.6	2.3	143.3	1.47	1.98	134.5	2.07	1.07	51.6
NSW_R	1.6	2.3	142.1	1.16	2.91	250.4	2.11	1.70	80.4
NSW_C	1.5	1.7	112.5	1.71	1.79	104.6	1.51	0.89	59.2
VIC_R	1.4	3.0	223.9	0.41	2.87	707.4	2.70	1.75	64.8
TAS_C	1.3	3.2	236.7	0.28	2.93	1,041.6	2.17	2.43	112.2
SA_C	1.2	1.9	157.6	0.64	1.50	233.6	1.68	1.29	76.7
SA_R	0.9	3.1	345.5	0.85	3.97	466.3	1.68	1.52	90.9
TAS_R	0.6	2.9	475.3	−0.03	2.92	n/a	1.43	2.53	176.9

Source: ABS Labour Force data. Regions ranked by growth rate over full sample (1987:4 to 2008:1). CoV is the coefficient of variation in percent

Table 7 The Australian and New Zealand standard industrial classification at one-digit level

Industry	Mnemonic used in paper
Agriculture, forestry and fishing	AGR
Mining	MIN
Manufacturing	MAN
Electricity, gas and water supply	EGW
Construction	CON
Wholesale trade	WHO
Retail trade	RET
Accommodation, cafes and restaurants	ACR
Transport and storage	TAS
Communication services	COM
Finance and insurance	FAI
Property and business services	PBS
Government administration and defence	GAD
Education	EDU
Health and community services	HCS
Cultural and recreational services	CRS
Personal and other services	POS

Table 8 Regional mnemonics

Mnemonic	State/Territory/Region
NSW	New South Wales
NSW_C	Sydney main statistical region
NSW_R	NSW balance of state
VIC	Victoria
VIC_C	Melbourne main statistical region

(*continued*)

Table 8 (continued)

Mnemonic	State/Territory/Region
VIC_R	VIC balance of state
QLD	Queensland
QLD_C	Brisbane main statistical region
QLD_R	QLD balance of state
SA	South Australia
SA_C	Adelaide main statistical region
SA_R	SA balance of state
WA	Western Australia
WA_C	Perth main statistical region
WA_R	WA balance of state
TAS	Tasmania
TAS_C	Hobart main statistical region
TAS_R	TAS balance of state
NT	Northern Territory
ACT	Australian Capital Territory

References

Aghion P, Howitt P (1997) Endogenous growth theory. MIT, Cambridge

Arcelus FJ (1984) An extension of shift-share analysis. Growth Change 15:3–8

Arrow K (1962) The economic implications of learning by doing. Rev Econ Stud 29:155–173

Barff RA, Knight PL (1988) Dynamic shift-share analysis. Growth Change 19:1–10

Castells M, Hall P (1994) Technopoles of the world: the making of the twenty first century industrial complexes. Routledge, London

Cooke P, Morgan K (1998) The associational economy: firms, regions and innovation. OUP, Oxford

Corden WM (1984) Boom sector and Dutch disease economics: survey and consolidation. Oxf Econ Pap 36:362

Dixon R, Shepherd D (2001) Trends and cycles in Australian state and territory unemployment rates. Econ Rec 77(238):252–269

Holden DR, Nairn AGM, Swales JK (1989) Shift share analysis of regional growth and policy: a critique. Oxf Bull Econ Stat 51(1):15–34

Houston DB (1967) The shift and share analysis of regional growth: a critique. South Econ J 33 (4):577–581

Lovering J (1999) Theory led by policy: the inadequacies of the new regionalism. Int J Urban Reg Res 23:379–395

Markusen A (1996) Sticky places in slippery space: a typology of industrial districts. Econ Geogr 72:293–313

Martin R, Sunley P (1998) Slow convergence? The new endogenous growth theory and regional development. Econ Geogr 74(3):201–227

Mitchell WF, Bill A (2006) The two-speed economy: the deterioration of the Sydney labour market. People Place 14(4):14–24

Mitchell WF, Carlson E (2005) Exploring employment growth disparities across metropolitan and regional Australia. Australas J Reg Stud 11(1):25–40

Mitchell WF, Muysken J (2008) Full employment abandoned: shifting sands and policy failures. Edward Elgar, Aldershot

Rissman E (1999) Regional employment growth and the business cycle. Econ Perspect 4:21–39, Federal Reserve Bank of Chicago

Romer P (1986) Increasing returns and long-run growth. J Polit Econ 94(5):1002–1037
Sabel C (1989) Flexible specialisation and the re-emergence of regional economies. In: Hirst P, Zeitlin J (eds) Reversing industrial decline? Berg, Oxford
Saxenian A-L (1994) Regional advantage: culture and competition in Silicon Valley and Route 128. Harvard University Press, Boston
Scott AJ, Storper M (1989) The geographical foundations and social regulation of flexible production systems. In: Dear M, Wolch J (eds) The power of geography: how territory shapes social life. Unwin, Boston, pp 21–40
Stimson RR, Robson APL, Shyy T (2005) Modeling determinants of spatial variation in regional endogenous growth: non-metropolitan regions in the mainland states of Australia. In: Proceedings of 29th annual conference of the Australian and New Zealand regional science association international, Sydney, Australia
Stough RR, Nijkamp P (2007) Entrepreneurship in regional economic development: some methodological applications. Ann Reg Sci 41:749–752

Regional Branching and Regional Innovation Policy

Ron Boschma

1 Introduction

Regional diversification is high on the scientific and political agenda. As many regions are currently facing economic decline due to the economic crisis, there is increasing awareness that there is a need to develop new economic activities, in order to compensate for losses in other parts of their regional economies. Economic geographers have raised the question how to develop new growth paths in regions over and over again, but this question has largely remained unanswered until recently (Scott 1988; Storper and Walker 1989; Martin and Sunley 2006; Simmie and Carpenter 2007). For instance, there is still little understanding of how old industrial regions may overcome structural problems, such as congestion, overspecialization, a bad image, and inflexible institutions, which, according to many, make them unlikely places for new industries to emerge. However, some do quite well, while others do not, but there is still little known what are the reasons behind that (Hassink 2005).

More recently, researchers have taken up this crucial question (Feldman et al. 2005; Cooke 2010; Fornahl et al. 2010). This paper focuses attention on two closely related concepts that might impact on regional diversification, that is, technological relatedness and related variety. More in particular, we explore the extent to which the entry and growth of a new industry depends on (a variety of) industries to which it is technologically related. Due to a lack of space, we leave out other dimensions that might be considered crucial in the process of regional diversification, such as institutional reforms, among others (Hassink 2010). Section 2 briefly presents some of the latest empirical insights concerning the importance of technological relatedness and related variety for regional development and regional branching (Boschma and Frenken 2011). Section 3 takes up some implications for regional innovation policy, and explains how technological relatedness may provide an input for effective policy making. Among other things, we claim that public policy should

R. Boschma
Department of Economic Geography, University of Utrecht, PO Box 80 115, 3508 TC Utrecht, The Netherlands
e-mail: r.boschma@geo.uu.nl

K. Kourtit et al. (eds.), *Drivers of Innovation, Entrepreneurship and Regional Dynamics*, Advances in Spatial Science, DOI 10.1007/978-3-642-17940-2_17,

avoid picking winners that do not fit into the regional industrial space (see also others, like Todtling and Trippl 2005), and should prevent supporting declining industries that take a quite peripheral position in industry space in the respective region. In Sect. 4, we sketch some policy options that use the concepts of technological relatedness and related variety as basic inputs to move regional economies into new directions. More in particular, we direct attention to various mechanisms through which new industries may be stimulated to connect to technologically related industries at the regional level. Section 5 draws some conclusions.

2 Technological Relatedness, Related Variety and Regional Development

Knowledge has a number of interesting features. Knowledge is not reduced when it is used, but accumulates through learning-by-doing (Arrow 1962). This cumulative nature of knowledge is embodied in individuals and organizations. Knowledge is also often tacit, that is, difficult to articulate and codify. For these reasons, knowledge is actor-specific and far from easy to imitate. Therefore, variety in an economy is the rule, and knowledge accumulation is its prime mover (Saviotti 1996). Knowledge is also a non-rival good: its use by one firm does not preclude its use by others. This implies that other firms may also benefit from the accumulation of knowledge.

However, space seems to be a constraining factor: knowledge spillover effects are often geographically localized. For instance, these effects tend to become weaker the higher the distance from the source of knowledge (Audretsch and Feldman 1996). This suggests that geographical proximity is a prerequisite for knowledge diffusion. There is, however, a need to rethink and reconsider this position. Boschma (2005) has argued that geographical proximity is neither a necessary nor a sufficient condition for interactive learning. There is increasing awareness that other barriers of knowledge diffusion need to be overcome first, such as cognitive and social distance (see e.g., Torre and Rallet 2005; Balland 2009). Some, if not all of these forms of proximity between actors need to be secured in order to make them connected, and to enable knowledge transfers (Lagendijk and Oinas 2005). Cognitive proximity has attracted most attention in this respect. Besides absorptive capacity (Cohen and Levinthal 1990), knowledge transfer requires cognitive proximity, that is, actors need to share similar knowledge and expertise to enable effective communication (Nooteboom 2000).

Having said that, there is increasing awareness that too much cognitive proximity between actors may also harm real learning. When two actors know exactly the same, they cannot mean much to each other in terms of learning, because they have not much to add to what the other already knows (Nooteboom 2000). And when these two actors with almost identical knowledge would establish a knowledge network link (which is most likely because they enjoy cognitive proximity), they run the risk of not being exposed anymore to external knowledge that is new to the both of them. But even when they get access to that knowledge, they might lack the capacity to understand and absorb it (Broekel and Boschma 2011).

Here comes in the relevance of cognitive or technological relatedness for knowledge creation and learning. That is, two actors need to have some degree of cognitive proximity to enable communication and mutual understanding, but not too much cognitive proximity, to avoid cognitive lock-in (Nooteboom 2000). In other words, when two actors are technologically related, they share different but related competences that might facilitate and enhance effective knowledge transfer. This shows strong resemblance with the concept of technology system developed in the 1990s in particular, which emphasized technological interdependencies that exist across industries (e.g. Carlsson and Stankiewicz 1991).

From a geographical perspective, it is relevant to investigate whether geographical proximity between two actors with related knowledge enhances interactive learning. This is a topic that is relatively unexplored, but which opens a whole new research agenda (Boschma and Frenken 2011). For example, one can expect that the higher the number of actors with related knowledge in a region (i.e., related variety), the more learning opportunities there are for local firms, and the more it will contribute to regional growth. So it is not so much regional variety per se (Jacobs' externalities) that matters, as stressed by the spatial externalities literature, because too much cognitive distance between sectors might be involved. Accordingly, a high number of technologically related industries in a region should be more beneficial than a diversified but unrelated set of industries, because related industries combine cognitive distance with cognitive proximity, in which the positive effects of variety across and relatedness among industries can be exploited. The notion of related variety has been measured at the regional level on the basis of the degree of variety in technologically related sectors, and shows a positive impact on regional growth (Frenken et al. 2007). This effect of technological relatedness may also cross regional boundaries. A study of Boschma and Iammarino (2009) suggests that a region can benefit from extra-regional knowledge when it originates from sectors that are related or close, but not quite similar to the sectors present in the region. In those circumstances, cognitive proximity between the extra-regional knowledge and the knowledge base of the region is not too small but also not too large.

The concept of relatedness is also useful to explain regional diversification. Apart from the fact that relatedness among industries in a region drives the growth of existing industries through agglomeration externalities, it may also drive the entrance of new industries in a region. Many case studies have shown that new local industries are often being deeply rooted in related activities in the region (Glaeser 2005; Klepper 2007). The post-war economic expansion of the Italian region of Emilia Romagna is a prime example. Historically, this region has a diffuse knowledge base in engineering. After the Second World War, many important new sectors like packaging, ceramic tiles and robotics emerged out of this pervasive and generic knowledge base, diversifying the industrial base of Emilia Romagna.

There is also more systematic evidence that shows that territories are more likely to expand and diversify into industries that are closely related to their existing activities (Hausmann and Klinger 2007; Hidalgo et al. 2007; Neffke 2009). Focusing on shifts in export portfolios over time, Hausmann and Klinger (2007) have demonstrated that countries predominantly expand their export portfolio by moving into products that are related to their current export mix. In addition, rich countries that have a wide range of related export products have more opportunities to diversify into new related export products, and thus more opportunities to sustain economic growth, in comparison to poorer countries. This process by which new variety (products or industries) arises from technologically related industries in regions has been termed "regional branching" (Boschma and Frenken 2011).

Neffke et al. (2009) have looked at the probability of new industries entering a region and the probability of existing industries disappearing from a region, and how that is affected by the degree of technological relatedness with other industries in the region. They have followed the evolution of 174 manufacturing industries in 70 Swedish regions during the period 1969–2002. What they found is that a new industry is more likely to enter a region when it is technologically related to other industries in that region. Another interesting finding was that an existing industry had a higher probability to exit a region when that industry was not or weakly technologically related to other industries in the region. Consequently, the rise and fall of industries seem to be subject to a path-dependent process at the regional level.

Which mechanisms are responsible for this process of regional branching is largely unclear. There is some evidence from studies on the life cycle of industries that entrepreneurship might be one of the driving forces. These studies tend to show that old sectors give birth to new sectors, and that new firms in new industries have a higher survival rate when the entrepreneur originates from related industries (Klepper 2007). Boschma and Wenting (2007) have shown that in the early development stage of the UK automobile industry, firms had a higher survival rate when their entrepreneurs had previously worked in related industries, like bicycle making, coach making, or mechanical engineering, and when their regions featured a strong presence of these related industries. So, when diversifying into the new automobile sector, these types of entrants could exploit related competences, which improved their life chances, as compared to start-ups lacking those related competences.

3 Possibilities to Intervene Publicly in the Process of Regional Diversification

The previous studies show that a new industry tends to emerge and develop in a region when that industry can connect locally to other industries to which it is technologically related. As explained above, the Neffke et al. (2009) study examined the process of regional diversification in all Swedish regions together, in which the data of every region were pooled. This has the advantage of providing systematic evidence on the process of regional diversification. However, one can also use the same idea and methodology to draw a picture of the industrial evolution of one individual region. This is also quite useful when it comes to discussing possible policy implications, to which topic we turn now.

In Fig. 1, we show the long-term evolution of the industrial structure of the Linkoping region during the period 1969–2002. This has been taken from Neffke et al. (2009). First of all, this figure represents the position of each of the 174 Swedish industries based on the degree of technological relatedness with other industries in the whole Swedish economy. Each node stands for one industry. Each link between two nodes means that the two industries are technologically related above a certain threshold. The degree of technological relatedness between industries has been assessed by co-occurrence analysis of products that are produced in combination at the plant level. It is not the purpose of this paper to go into details how this has been measured (for details, see Neffke and Svensson Henning 2008; Neffke et al. 2009). In this figure, only the 2,500 most important links are displayed. More closely related industries are positioned in the more dense parts of this industry space.

With respect to the Linkoping region, the white dots represent industries that have never been present in this region during the period 1969–2002, the blue dots stand for industries that were constantly present in the region, while the green and red dots denote industries that have entered and exited the region respectively during that period. What can be observed first is that the majority of new industries

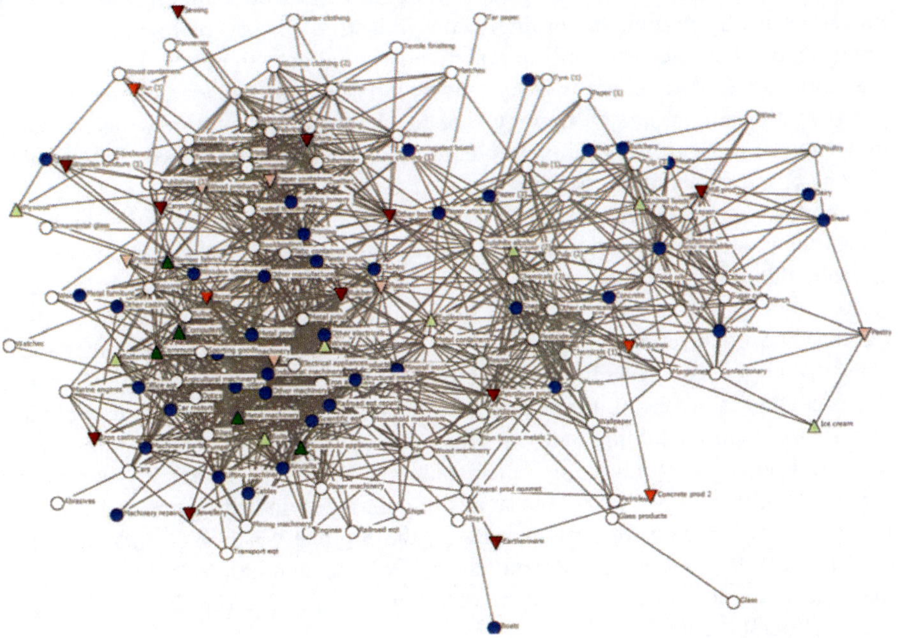

Fig. 1 Long-term evolution of the industrial structure of the 160 Linkoping region during the period 1969–2002
Source: Neffke et al. (2009), p. 47

that entered the region of Linkoping are positioned quite close to other existing industries to which they are technologically related confirmed. In addition, industries that disappear from the region tend to take a more peripheral position with respect to other industries in the region, which is in line with expectations.

What kinds of policy implications for the Linkoping region can be drawn from this figure? Of course a full picture is lacking. For instance, we have no information whatsoever on whether the industries that emerged in the Linkoping region got whatever support from public policy. Therefore, we have to be cautious. Nevertheless, this picture shows that one should be careful to pick out new industries that are considered most promising nation or world wide (like nanotech or biotech), but which take a rather peripheral position in the industry space of the Linkoping region. This means that these new industries will not connect easily to other industries in the region, because there are no other regional industries to which they are technologically related. Because public policy is about spending public money, it may be wiser to follow a risk-averse policy strategy in this case. This means that regional policy should better bet on those new industries that can more easily connect to and be embedded in the existing industrial structure, because that might increase the probability of policy success. This is in line with literature stating that it is wrong to follow a "one-size-fits-all" policy, which is often found practice in regional policy (Todtling and Trippl 2005; Raspe and Van Oort 2006; Asheim et al. 2011). There is a need for tailor-made policy strategies to capitalise on region-specific assets that are grounded in the variety of technologically related industries in the region. Accordingly, the industrial history of regions provides opportunities but also sets limits to effective regional growth policy (Lambooy and Boschma 2001). This is not to deny that regional policy focused on creating new growth paths remains a risky strategy. Nobody can predict the future winners, and it may even be difficult, if not impossible, to predict what new technological combinations of industries might occur in the near future.

Another policy implication to be drawn from Fig. 1 is that backing declining industries in a region is not necessarily a bad thing. Given the perspective of this chapter, this may be useful when it concerns industries that are technologically related to other industries in the region. This is not to deny that there might be other reasons like the size and high number of input–output linkages with other local firms that might provide additional underpinnings of regional innovation policy, but this is beyond the scope of this chapter. From a relatedness perspective, it might not be wise to support declining industries that already take a peripheral position in the industrial portfolio of the region because they have a high probability to exit the region at some point of time. This stands in contrast to industries that have strong technological ties with other industries in the region. When such industries are confronted with a temporary demand fall, the loss of such industries would jeopardize the existence and development of other regional industries to which it is technologically related.

When targeting region-specific assets with the aim of broadening and renewing the economic base of regions, policy should avoid selecting only a very limited number of regions. In principal, every region can be made part of such a policy approach. Nor do specific sectors (low or high tech, creative or not) have to be excluded from such a policy approach. The only selection criterion is to target industries based on their degree of technological relatedness with other industries in the region. However, the objective is not to make strong sectors even stronger, but the objective is to enhance interaction and exchange between different but complementary activities, in order to support new variety in the region.

4 More Concrete Policy Actions for Regional Diversification

Recently, related variety has been taken up as a major input for regional innovation policy (Asheim et al. 2006, 2010; Cooke 2010; Harmaakorpi et al. 2011). Some of these authors have referred to platform policies, which aim to connect different industries and explore the possibilities of recombinations to enhance regional development. In the remaining part of this chapter, we will briefly direct our attention to three knowledge transfer mechanisms through which industries may be connected at the regional level. These three mechanisms are: spinoffs, labor mobility and collaborative networks.

Spinoffs are considered to be key drivers of knowledge diffusion, entrepreneurship and regional development (Helfat and Lieberman 2002). Spinoffs are new ventures that are established by entrepreneurs that have acquired experience in the same or a related industry. Spinoff companies often perform better than other types of start-ups because they can build on relevant knowledge and experience acquired in parent organizations in the same or related industries (Klepper 2007). Since spinoffs tend to locate near their parents, they may be considered a local knowledge transfer mechanism that may provide a basis for regional innovation policy. Spinoffs have also played a crucial role in the emergence of some sectors. As a consequence, spinoffs also may be seen as a mechanism through which regional economies diversify into new sectors while building on knowledge acquired by entrepreneurs in existing sectors (Boschma and Wenting 2007). A policy option is to target potential entrepreneurs (instead of supporting any new start-up), by taking into consideration whether the founder brings related knowledge into the new firm.

Regional innovation policy based on relatedness could also focus on encouraging labor mobility. Labor mobility is a crucial mechanism through which related skills and experience are transferred from one company to the other (Boschma et al. 2009). Since most labor mobility takes place at the regional level, policies promoting labor mobility may enhance knowledge transfer and innovation at the regional level. Since labor mobility may take away the incentive of firms to invest in their personnel, public policy should invest heavily in education and life-long learning at the same time. Aghion et al. (2006) argue that labor markets need to be more flexible in order to smooth the process of creative destruction and lower the costs of such adjustments. This needs to be complemented by a policy of life-long-learning that increases the capability of individuals to confront changes and to move from one job to the other.

Another policy measure based on relatedness is to encourage the immigration of skilled labor because it may bring new ideas and knowledge into the region. One way to achieve this is through international exchange programmes for students. Incoming students bring in new talents and skills from abroad, and combine these with new skills that are acquired in high education institutes in the host country. If the host country is capable of maintaining this group of high-skilled students after graduation (policy can most certainly play a role here), they will contribute to the economy as skilled employees or as founders of new firms. Outgoing students will acquire new skills in research and education institutes abroad, and may return to their home region after a while, where they will exploit their newly acquired skills in an environment they are familiar with (Saxenian 2006). Policy could target those outgoing groups and provide incentives to return to their home region.

Another policy option is to stimulate networks as effective settings through which knowledge circulates and interactive learning takes place. Policy makers may act as intermediairs or knowledge brokers, or establish policy platforms that facilitate knowledge to spill over and diffuse from sectors to related ones. In doing so, policy should avoid that vested interests of established firms take over and dominate these networks, and deny access to small firms and newcomers. In a similar vein, competition policy could aim at stimulating the establishment of network alliances between firms in related industries as a way of diversifying regional economies into new but complementary fields of activity.

This type of network policy should acknowledge that knowledge networks frequently cross boundaries of regions. Regional innovation policy should stimulate extra-regional networks, because it may brings new and related knowledge into the region. Besides new infrastructure development and international exchange programs, a way to accomplish this is to encourage foreign investments. Universities may also play a crucial role in exploiting inter-regional linkages, because they are extremely well connected to international networks. After their graduation, students will exploit and diffuse this knowledge in the regional economy. Academic spinoff policy and other policy measures may be implemented to ensure that the knowledge of universities will be further exploited economically at the regional level.

5 Conclusions

Related variety is crucial for regional development. Knowledge spills over more intensively when regions are endowed with related industries that share a common knowledge base. This makes regional economies to diversify into new directions and start up new growth paths, which is crucial for their long-term development. We have explored how related variety may serve as an underpinning for effective regional innovation policy. Instead of copying best practices, policy should take the history of the region as a basic starting point, and identify regional potentials and bottlenecks accordingly. Policy actions focusing on spinoff activity, labor mobility and networks may enhance the effect of related variety on regional development, because these mechanisms of knowledge transfer are basically taking place at the regional level, and because they can make regions move into new growth paths while building on their existing assets.

References

Aghion P, David PA, Foray D (2006) Linking policy research and practice in 'STIG systems'. Many obstacles, but some ways forward. Paper presented at SPRU conference, 11–13 Sep 2006, pp 26

Arrow KJ (1962) The economic implications of learning by doing. Rev Econ Stud 29:155–173

Asheim B et al. (2006) Constructing regional advantage. Principles, perspectives, policies. Final report, DG Research, European Commission, Brussels

Asheim B, Boschma R, Cooke P (2011) Constructing regional advantage: Platform policies based on related variety and differentiated knowledge bases, *Regional Studies*, doi: 10.1080/00343404.2010.543126 (forthcoming)

Audretsch DB, Feldman M (1996) Spillovers and the geography of innovation and production. Am Econ Rev 86:630–640

Balland PA (2009) Proximity and the evolution of collaboration networks. Evidence from R&D projects within the GNSS industry, Papers in Evolutionary Economic Geography, no. 09.14. Department of Economic Geography, Utrecht University

Boschma RA (2005) Proximity and innovation: a critical assessment. Reg Stud 39(1):61–74

Boschma R, Frenken K (2011) Technological relatedness and regional branching, In H. Bathelt, M.P. Feldman and D.F. Kogler (eds.), *Beyond Territory. Dynamic Geographies of Knowledge Creation, Diffusion and Innovation*, Routledge, Taylor and Francis, 64–81

Boschma RA, Iammarino S (2009) Related variety, trade linkages and regional growth. Econ Geogr 85(3):289–311

Boschma RA, Wenting R (2007) The spatial evolution of the British automobile industry. Ind Corp Change 16(2):213–238

Boschma R, Eriksson R, Lindgren U (2009) How does labour mobility affect the performance of plants? The importance of relatedness and geographical proximity. J Econ Geogr 9(2):169–190

Broekel T, Boschma R (2011) Knowledge Networks in the Dutch Aviation Industry - The Proximity Paradox, *Journal of Economic Geography* (forthcoming)

Carlsson B, Stankiewicz R (1991) On the nature, function and composition of technological systems. J Evol Econ 1:93–118

Cohen WM, Levinthal DA (1990) Absorptive capacity: a new perspective on learning and innovation. Adm Sci Q 35:128–153

Cooke P (2010) Transversality and transition. Branching to new regional path dependence. Paper prepared for workshop Trinity College, Oxford, pp 5–7 September

Feldman MP, Francis J, Bercovitz J (2005) Creating a cluster while building a firm. Entrepreneurs and the formation of industrial clusters. Reg Stud 39:129–141

Fornahl D, Henn S, Menzel M (eds) (2010) Emerging clusters. Edward Elgar, Cheltenham

Frenken KV, Oort FG, Verburg T (2007) Related variety, unrelated variety and regional economic growth. Reg Stud 41(5):685–697

Glaeser EL (2005) Reinventing Boston: 1630–2003. J Econ Geogr 5:119–153

Harmaakorpi V, Tura T, Melkas H (2011) Regional Innovation Platforms, In: P. Cooke, B. Asheim, R. Boschma, R. Martin, D. Schwartz and F. Todtling (eds.), *Handbook on Regional Innovation and Growth*, Cheltenham, Edward Elgar, 556–572

Hassink R (2005) How to unlock regional economies from path dependency? From learning region to learning cluster. Eur Plan Stud 13(4):521–535

Hassink R (2010) Locked in decline? On the role of regional lock-ins in old industrial areas. In: Boschma R, Martin R (eds) The handbook of evolutionary economic geography. Edward Elgar, Cheltenham, pp 450–467

Hausmann R, Klinger B (2007) The structure of the product space and the evolution of comparative advantage, CID working paper no. 146. Center for International Development, Harvard University, Cambridge

Helfat CE, Lieberman MB (2002) The birth of capabilities. Market entry and the importance of pre-history. Ind Corp Change 11(4):725–760

Hidalgo CA, Klinger B, Barabási A-L, Hausmann R (2007) The product space conditions the development of nations. Science 317:482–487

Klepper S (2007) Disagreements, spinoffs, and the evolution of Detroit as the capital of the US automobile industry. Manag Sci 53:616–631

Lagendijk A, Oinas P (eds) (2005) Proximity, distance and diversity. Issues on economic interaction and local development. Ashgate, Aldershot

Lambooy JG, Boschma RA (2001) Evolutionary economics and regional policy. Ann Reg Sci 35 (1):113–133

Martin R, Sunley P (2006) Path dependence and regional economic evolution. J Econ Geogr 6 (4):395–437

Neffke F (2009) Productive places. The influence of technological change and relatedness on agglomeration externalities. PhD thesis, Utrecht University, Utrecht

Neffke F, Svensson Henning M (2008) Revealed relatedness: mapping industry space. Papers in Evolutionary Economic Geography, no. 8.19. Department of Economic Geography, Utrecht University

Neffke F, Henning M, Boschma R (2009) How do regions diversify over time? Industry relatedness and the development of new growth paths in regions. Papers in Evolutionary Economic Geography, no. 9.16. Department of Economic Geography, Utrecht University

Nooteboom B (2000) Learning and innovation in organizations and economies. Oxford University Press, Oxford

Raspe O, van Oort FG (2006) The knowledge economy and urban economic growth. Eur Plan Stud 14:1209–1234

Saviotti PP (1996) Technological evolution, variety and the economy. Edwrad Elgar, Cheltenham

Saxenian AL (2006) The new argonauts. Regional advantage in a global economy. Harvard University Press, Cambridge

Scott AJ (1988) New industrial spaces. Flexible production, organization and regional development in North America and Western Europe. Pion, London

Simmie J, Carpenter J (eds) (2007) Path dependence and the evolution of city regional development, Working Paper Series No. 197. Oxford Brookes University, Oxford

Storper M, Walker R (1989) The capitalist imperative. Territory, technology and industrial growth. Basil Blackwell, New York

Todtling F, Trippl M (2005) One size fits all? Towards a differentiated regional innovation policy approach. Res Policy 34:1203–1219

Torre A, Rallet A (2005) Proximity and localization. Reg Stud 39(1):47–60

Beyond the Creative Quick Fix Conceptualising Creativity's Role in a Regional Economy

Jane Andrew and John Spoehr

1 Introduction

Over the last decade creativity has joined innovation as one of the strategic hallmarks of growth oriented regional economies. In an attempt to emulate the growth trajectories of regions as diverse as Emilia Romagna and Silicon Valley, policymakers have sought to identify universal principles of regional economic success and prosperity. Creativity has emerged as the focus of considerable interest throughout the world (Leadbetter 2000; Florida 2003; Reich 2001; Landry 2000). In the state of South Australia in Australia creativity has been elevated to one of six key objectives in *South Australia's Strategic* Plan released in 2004. This policy stance has been influenced by a range of thinkers who have been engaged by a number of State Government departments.

This chapter examines the conceptualisation of creativity in South Australia, arguing that a tendency to confine policy debates on creativity to the arts and cultural sectors obscures the wider challenge of thinking more holistically about creativity. In this context we ask whether the approach adopted in regions like South Australia is an adequate foundation for the development of creative capacity on an economy wide basis or whether the dominant tendency is for a creative quick fix. To answer this we identify and analyse recent developments in South Australia in the context of popular policy debates on the role of creativity in regional innovation systems. The chapter provides insights into the early experiences of one region, recognising that there is likely to be a great deal of analytical and practical value in building on this foundation through future research comparing this experience with that of other regions. It is hoped that this preliminary work

J. Andrew (✉)
PhD Candidate, Australian Institute for Social Research, University of Adelaide Lecturer, Director matchstudio, Art, Architecture and Design School, University of South Australia
e-mail: jane.andrew@unisa.edu.au

J. Spoehr
Australian Institute for Social Research, The University of Adelaide
e-mail: john.spoehr@adelaide.edu.au

K. Kourtit et al. (eds.), *Drivers of Innovation, Entrepreneurship and Regional Dynamics*, Advances in Spatial Science, DOI 10.1007/978-3-642-17940-2_18,

might stimulate this more ambitious project. We concur with Tepper (2002, p. 159) that "... rather than spend time calculating the impact or size of the creative economy, we should direct our analytical and policy energies toward better understanding how creative work and institutions are changing and what might be done to foster a more robust, more creative and more diverse [economy] and cultural life".

2 Conceptualising Creativity: Broad and Narrow

Creativity can be conceptualised in broad and narrow terms, informing the focus of creativity policies and strategies. Situating particular policies is the starting point for an examination of case studies like that in South Australia.

Mihály Csikszentmihalyi's broad conception of creativity focuses on the creative process, the potential of its application and the filters and agents through which its value is defined. Rather than seeking to understand the mechanics and sequence or "the science" of the creative process, Csikszentmihalyi seeks to gain a more nuanced understanding of the role creativity plays in our economic and social life by examining not *what* creativity is but *where* it is. Csikszentmihalyi (1996) observed that it is the interrelationship of what he conceptualises as a system made up of the three main parts; the domain; the field; the person[1] that stimulate the creative process and the translation of ideas into economic goods.

Jason Potts argues that creativity and the creative industries contribute not just to value-added products, services and jobs, but more importantly, to the evolutionary process by which economic systems grow. He states:

> Creativity is perhaps the generic name for the set of forces that supply new ideas as new solutions to problems to connect new technologies with new human lifestyles. The creative industries do not just facilitate the origination of novelty, but also work to facilitate the adoption and adaptation of new technologies – through design and advertising, for example – along with the embedding of new technologies and their ongoing maintenance. The creative industries are involved in all stages of the innovation process. (Potts 2007, p. 8)

These broader conceptions of creativity were echoed at the 2005 Creative Capital conference in Amsterdam which defined creative capital "as the combined assets of society that enable and stimulate its people and organisations to be innovative and creative" (van den Steenhoven et al. 2005, p. 11). Following this view about the embedded character of creativity, Chris Bilton asserts that "individual creativity needs to be integrated with organisational resources, capacities and systems if new ideas are ever to bear fruit" (Bilton 2007, p. xv).

---

[1]• Csikszentmihalyi describes the domain, the field and the creative person thus:

- The domain – consists of a set of symbolic rules and procedures, i.e. symbolic knowledge a or culture shared by a group of people for example architects or a particular society
- The field – includes all the individuals who act as gate keepers to the domain
- The person – someone whose thoughts or actions transforms and existing domain (n with the explicit support from the field responsible for it) or establishes a new domain

In order to understand the role creativity plays in stimulating innovation and regional growth more broadly across the economy, it is necessary to utilize a more flexible means of identifying the actors, institutions and instruments that contribute to a creative economy. This has not been the starting point for many as the following examination of definitional models of the creative industries/creative economy highlights. This suggests a tendency in creative economy policy debates to reduce the debate about creativity to instrumentalist discussions about what constitutes a creative product, industry or occupation.

3 Reconceptualising the Creative Economy's Definitional Framework

Policy debates surrounding creativity commonly draw from relatively narrow conceptualizations of creativity, bounded by instrumental imperatives of developing and sustaining a creative economy or more specifically export oriented creative industries. One of the key contributors to this debate has been cultural economist David Throsby (2000) who argues that the creative industries might be best understood as a set of concentric circles (see Fig. 1). He argues that individual

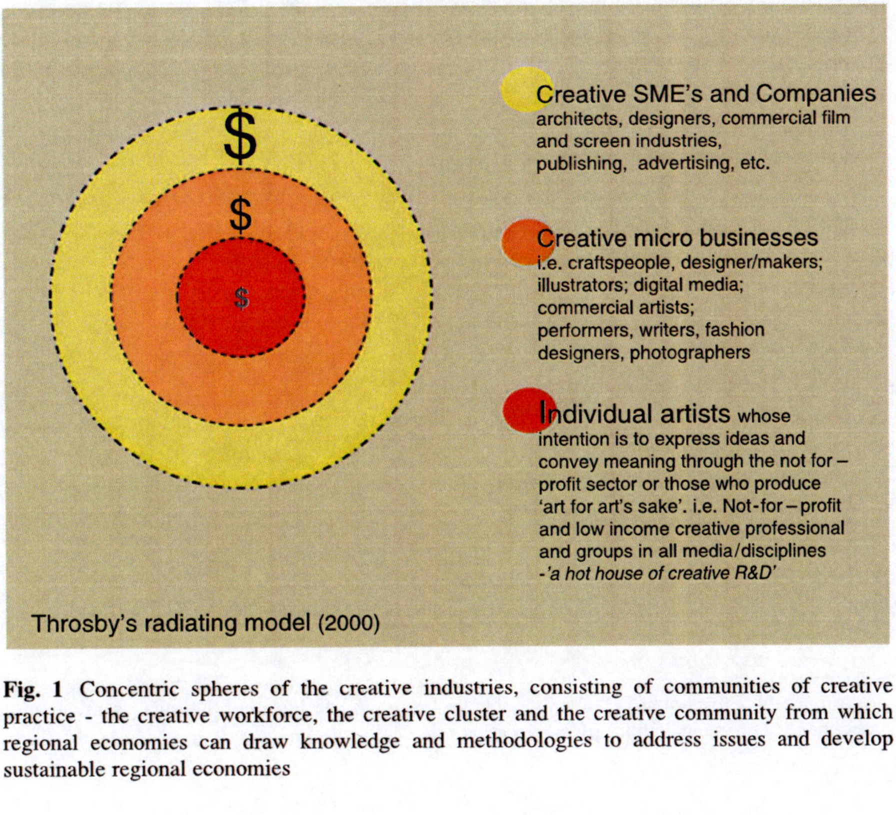

Fig. 1 Concentric spheres of the creative industries, consisting of communities of creative practice - the creative workforce, the creative cluster and the creative community from which regional economies can draw knowledge and methodologies to address issues and develop sustainable regional economies

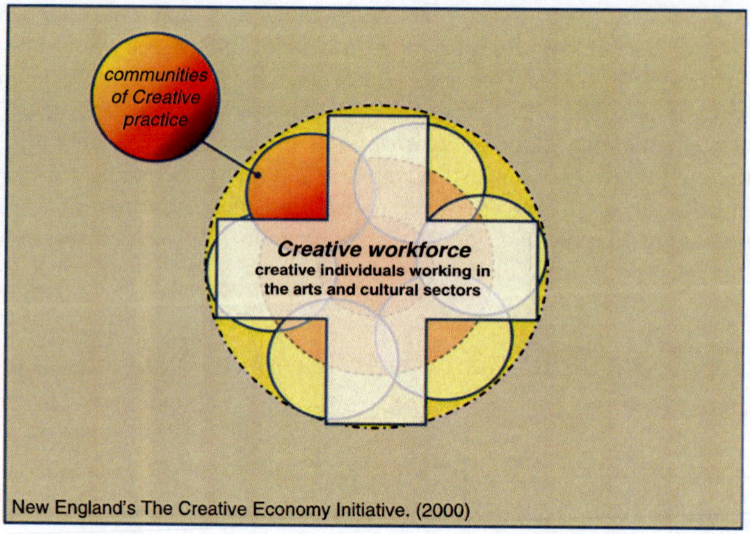

Fig. 2 The Creative Cluster comprises enterprises and industries populated by the creative workforce as well as sectors that contribute to creative and cultural production

artists whose intention is to express ideas and convey meaning through the not for – profit sector or those who produce "art for art's sake" lie at the core of the creative/ cultural economy. This sector is what Landry suggests is a hot house of creative R&D.[2] As the circles radiate outward the creative intention and resulting product or service becomes increasingly commercially orientated.

New England's Creative Economy Council offers a conceptualisation of a creative economy that is built upon creative individuals that constitute a *creative workforce*. The diversity and application of creative capacity within the creative workforce is illustrated by the multiple communities of practice that work within and across the creative economy's spheres. In addition to the web of individuals and businesses within the creative workforce, are businesses and individuals who act as interacting agents working with the creative workforce and are critical in the realisation, development and delivery of cultural goods. The Creative Economy Council terms this *creative cluster*. The sustained interaction and acknowledgement of the value of creative ideas, services and products mediated by the local market is referred to by the Creative Economy Council as the *creative community*. The conceptualisation of a creative community can be considered within any scale of geographic region such as a precinct, a city, a state or a larger region. The following three diagrams (Figs. 2–4) represent Throsby's concentric circles model overlaid with the New England Creative Economy Council; model.

---

[2]We propose that this hot house of R&D includes the art and design educators such as universities and colleges of technical and further education (TAFE'S).

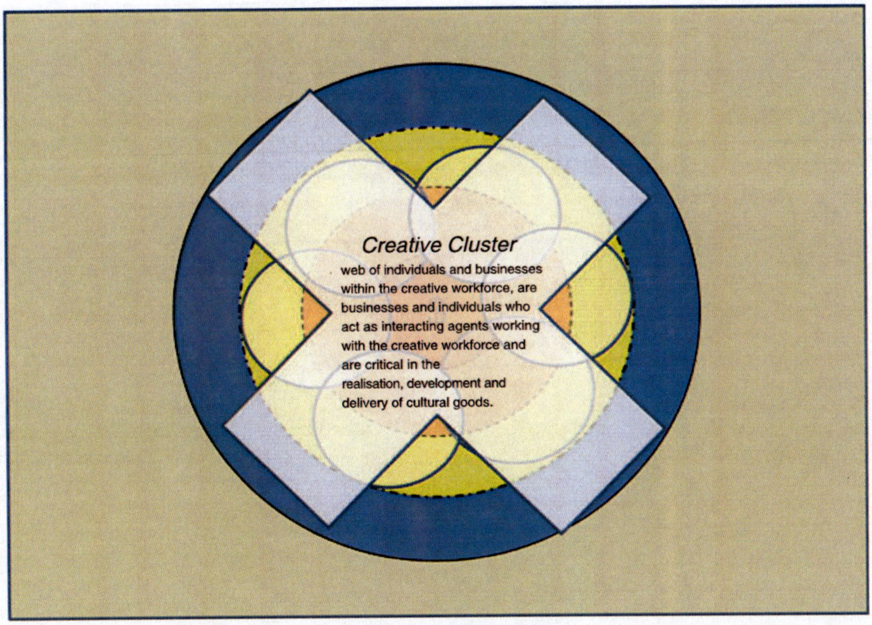

Fig. 3 A Creative Community is defined as a geographic area with a concentration of creative workers, creative businesses, cultural organisations and supporting agencies

Jason Potts broadens the conceptual landscape of creativity arguing that we should view the "creative industries" as part of a "creative system". He suggests that the creative system is not just focused on the origination, innovation and initial adoption of physical technology; but rather on "*all* novel ideas" (Potts 2007). Like Csikszentmihalyi, Potts argues for the exploration of a Systems Theory approach to understanding the creative economy and the potential for it to contribute more significantly to the broader economy. To this end drawing from debates surrounding regional innovation systems is informative when attempting to reconceptualise the creative industries, how they fit into and contribute to the mainstream economy.

Discourse informing regional innovation systems theory has a long history stemming back to Schumpeter in 1911 with his work on innovation. Emerging out of this tradition in the late 1990s Porter, Henton and Walesh were influential in South Australia in the mid 1990s through their research on industry clusters and networks. Porter (2002) proposed that the enduring competitive advantage in a global economy is often heavily local, arising from a concentration of highly specialised skills and knowledge, institutions, related businesses and customers in a particular region. In South Australia, this led to the establishment of a number of precincts populated by business within a specific industry sector or supply chain. Although much attention and investment on the development of on clusters focuses on particular industries and their supply chains, Lagendijk suggests that "regional economic success is increasingly attributed to the performance of particular networks and institutional configurations, with an emphasis on idiosyncratic forms of

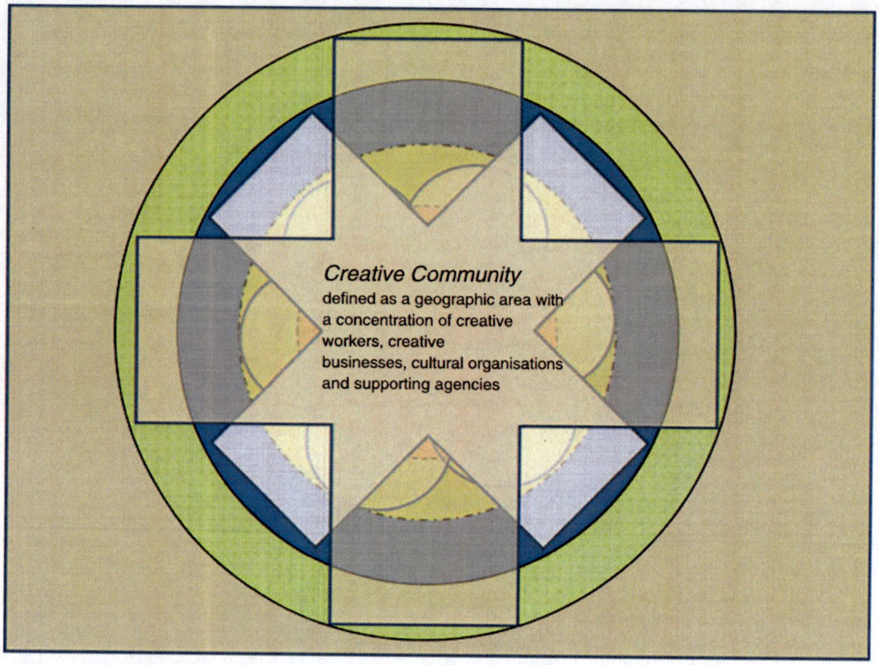

Fig. 4 The Creative Community as a catalyst for policy innovation. A more integrative approach to enabling the creative industries to contribute more broadly to regional economies would be to consider how the creative sectors do or could in the future contribute to addressing issues within policy domains such as health and wellbeing, the environment, industry and commerce, and society and culture

knowledge, interaction and strategic power". To this end Lagendijik's work has informed this papers proposition that reconceptualising the creative economy as an integral part of the broader economy's innovation system can inform policy makers of the creative industries potential to contribute to regional economic growth across industry sectors and policy domains.

Like Lagendijik, Cooke and Memedovic (2006) argue that successful regional economies knit together a mix of regional innovation policies and institutions with knowledge flows, and the systems on which they rely. They observe successful systems tend to display a number of common characteristics such as:

- Intensive co-operation among firms
- High quality workforces
- Flexible work structures
- Dense infrastructures of supporting institutions and organisations
- Innovative regional cultures
- Activist regional governments

It is common for innovative activity to be recognised within the realms of science and engineering. Jason Potts suggests that this heavy focus on innovation as a technical search and discovery process by universities and businesses "largely ignores the more complex interactions between producers and consumers, as well as subsequent phases beyond technology innovation, such as adoption and adaptation of a novel product or service to human lifestyles, along with its retention and normalisation by a population of carriers" (Potts 2007, p. 7).

This brief discussion of broad and narrow conceptions of creativity provides a foundation for examining the evolution of creativity as a concept and policy objective in South Australia. How might engagement with the concept be characterised in South Australia? Has a particular conception of creativity prevailed and what is the likely trajectory for creativity as an object of government policy given our view about this? To begin to answer these questions we need to provide some context to the emergence of creativity as a core objective of South Australia's Strategic Plan.

4 Creativity, Art and Culture: Changing Models of Production, Definition and Government Support

Creativity has been commonly viewed in Australia through the lens of the arts and cultural sectors. Inspired by the Keynesian model of support for the arts in Great Britain, a view that culture is a national good, from which locally specific cultural expression builds national, and regional identity has prevailed. Following the British experience the establishment of the Australia Council for the Arts in the 1970s enshrined government support for the arts and cultural sectors, manifesting in a wide range of Federal and state based arts funding programs that support organisations and individual artists to develop and exhibit new work in a wide range of forms including Dance, Music, Theatre, and Literature, Visual arts, craft and design.

More recently, traditional forms of cultural expression have been influenced by the introduction of new technologies, which has led to increasingly fuzzy boundaries between what were once specialised artistic genres. With an increasing emphasis in art colleges on conceptual development as opposed to developing exceptional production skills, many emerging artists are choosing to side step the purist gatekeepers in their chosen field or domain such as gallery curators, publishers or music producers and are turning into cultural producers in their own right. Using new technologies and the internet has enabled self publishing of texts and "blogs", producing and distributing music, and the production and distribution of animations, films and images. In order to disseminate their ideas images and sounds to a global audience, this group of cultural producers develop and utilise their large social networks to support and promote their work rather than more traditional market intermediaries. Mapping and measurement studies of regional creative economies are consistently identifying these increasingly hybrid forms of artistic and cultural production which include the use of new technologies in the production and distribution of their creative content as having the potential to become increasingly economically significant.

As part of their strategies to promote a region to potential business investors, local and state governments are increasingly emphasising the role of arts and culture in promoting vibrant, creative and innovative communities as part of a broader economic growth strategy. However, the utilisation and promotion of the cultural assets of South Australia does not necessarily equate to significantly increased investment in the individuals, agencies and programs that enable the creation, development and presentation of our of our home grown arts and cultural assets. Rather, government expenditures on the arts and cultural sectors are increasingly absorbed by long standing cultural institutions in the maintenance of their collections and buildings, whilst arts development and support organisations that foster the development of creative capital and knowledge sharing find it increasingly difficult to sustain their programs with increasing industry and funders expectations despite their limited funding and capacity to generate earned income.

Within this uncertain environment of support for the arts sectors, there is tension in debates surrounding how best to evaluate the contribution of creativity to the economy and how broadly or narrowly conceived the concept of the creative industries should be. As public expenditures on the not for profit arts and culture sector struggle to find justification in the face of commercialisation pressures, there has been a growing tendency to focus policy on those parts of the cultural sector that are seen to have potential to generate commercial returns. These are commonly areas of the arts and cultural sectors that intersect with science and technology and have the greatest export potential. These are commonly described as the creative industries. Defining what constitutes the creative industries and quantifying the significance of these has been the focus of great deal of recent research.

There has been a considerable debate about what the "creative economy" or "creative industries" are. Definitions range in scope and detail, from conceptually based descriptors to lists of products and services or occupations as the following table illustrates. Richard Caves' definition of the creative industries is bounded by the commonly held view that the cultural, artistic and entertainment sectors represent a creative core. David Throsby focuses on a conceptual description of cultural capital and the areas of the economy in which he argues it resides. Others such as the United Kingdom's Department of Culture, Media and Sport (DCMS), Charles Landry, and PMSEC[3] offer an expanded list of creative industries and include those that are on the cusp of arts, technology and entertainment.

[3]This report was prepared by an independent working group for the Prime Minister's Science, Engineering and Innovation Council (PMSEIC) and was presented at the 14th meeting of PMSEIC on 2 December 2005. The report provides recommendations for leveraging the intellectual and creative wealth of the nation.

(Prime Minister's Science Engineering and Innovation Council Working Group and (PMSEIC) (2005). Imagine Australia: The Role of Creativity in the Innovation Economy, PMSEIC).

Richard Caves (2000) Creative industries: contracts between art and commerce	David Throsby (2000) Economics and culture	Department of Culture Media and Sport (DCMS, UK) (2001) Creative industries taskforce
Creative industries Supplying goods and services that we broadly associate cultural artistic, or simply entertainment value. Book and magazine publishing Visual arts (painting, sculpture) Performing arts (theatre, opera, concerts, dance) Sound recordings Cinema and TV films Fashion Toys and games	Cultural capital Core creative arts Location of the primary artistic producers at the centre producing text, sound, image in both old and new art forms Wider cultural industries Film, television, publishing, video games, etc	Creative industries Advertising Architecture Music Art and antiques markets Performing arts Computer and video games Crafts Publishing Design Software Designer fashion Film and video Television and radio

More prescriptive definitions based on production outputs, are offered by the Australian Department of Communication, Information Technology and the Arts (DOCITA) who base their definitional parameters around ANZSIC classifications. Richard Florida bases his theoretical argument on a human capital model and describes the main resource of a creative economy as its stock of creative capital which he refers to as the creative class (Florida 2003, pp. 8, 249).

Richard Florida (2003) The rise of the creative class Creative industries: contracts between art and commerce	Charles Landry (2003) Rethinking Adelaide capturing imagination	DOCITA (2007)
Creative class Computer and mathematical occupations Architecture and engineering occupations Life, physical and social science occupations Arts, design, entertainment, sports, and media occupations Management occupations Business and Financial operations and	Creative industries The creative industries are those industries that are based on individual creativity, skill and talent. They are also those that have the potential to create wealth and jobs through developing intellectual property Architecture Music Art and antiques markets	Culture related industries Based on the Australian and New Zealand Standard Industrial Classification (ANZSIC) Printing Newspaper printing and publishing Book and other publishing Architectural services Advertising services Commercial art and display services Film and video production and distribution Motion picture exhibition Radio and TV services (including broadcasting) Libraries and museums Parks and gardens Music and theatre production Creative arts Musical composition, the literary arts, and visual arts Such as painting, drawing, sculpture, pottery etc.

(*continued*)

Richard Florida (2003) The rise of the creative class Creative industries: contracts between art and commerce	Charles Landry (2003) Rethinking Adelaide capturing imagination	DOCITA (2007)
occupations Legal occupations Healthcare practitioners and technical occupations High end sales and sales management	Performing arts Computer and video games Crafts Publishing Design software Designer fashion Film and video Television and radio	Services to the arts Operating sound recording studios; operating performing arts venues (e.g., entertainment centres, concert halls, play houses and opera houses) Providing services to the arts. Such as casting agency operations, costume design services, set design services and theatre ticket agencies Video hire outlets and photographic studios

5 Innovation and Creativity Must Be at the Centre of Everything We Do: A South Australian Perspective

South Australian politicians and policymakers have been particularly influenced by the work of Richard Florida and Charles Landry with both being invited on separate occasions to advise the South Australian government on measures to foster creativity and innovation in South Australia.

Since the early 1990s Landry has focused his research and consultancy work on the idea of the "Creative City". He argues that cities possess, to varying degrees, a collection of broadly defined cultural resources which he describes as:

> the skills and creativity of local people, the concrete manifestations of people's work (buildings, manufactured products, artifacts) and more intangible, yet significant qualities such as social milieu, people's memory and the reputation of the place. These three types of cultural resources can be exploited in different ways and require different kinds of intervention. (Landry and Bianchini 1994, p. 16)

The publication of *The Creative City: A toolkit for urban innovators* by Landry in 2000 caught the attention of policy makers and politicians in South Australia with Landry being invited by the South Australian Government to be one of it's first Thinkers in Residence. The focus of his residency was on:

- Helping Adelaide unlock its creative potential
- Enhancing understanding of why creativity is so important in achieving social and economic progress
- Developing new connections between the city and the northern suburbs
- The role of culture in building stronger and more cohesive communities

In his *Rethinking Adelaide* report to the South Australian Government, Landry (2003) argues that Adelaide should consider its creative and cultural assets as an industry. He observed that the government funded arts and cultural sector often acts as an important hub for research and development that fuels innovation in the wider economy. He argued that consideration should be given to the development of links between the subsidised and commercial sectors to better enable this. In many regions this is characterised by "arts" led approaches to economic regeneration which assert that a lively arts and cultural scene is symbolic of a creative and innovative region – a creative place that attracts tourists with discretionary income, and most importantly for sustained economic activity businesses and their employees.

In policy terms, this commonly manifests in major events and arts and cultural festivals, and in the development of cultural precincts. In South Australia the support for major events and festivals is most notably illustrated by the Adelaide Festival of Arts established in 1960, and WOMADelaide the world music event established in Adelaide in 1992. Using evidence from economic impact statements, many argue that these events contribute to regional development. McGregor Tan Research calculated that South Australian businesses experienced a 15% increase in turnover during the Adelaide Bank 2006 Festival of Arts, generating around 190 FTEs equivalents of employment (Adelaide Festival of Arts 2008, http://festival.fusion.com.au). In 2008, 855 artists and performers participated in the Festival, including 277 from overseas and 232 from interstate (ibid). It is important to bear in mind that the Adelaide Festival of Arts lasts for two weeks; however the development and preparation leading up to presenting at the festival can take years. In addition the presenters form interstate or overseas often develop their productions and exhibitions outside of the city in which it is to be presented, thus the economic benefit of increased artist activity and expenditure on goods and services during the development phase is not providing benefit to the city in which the festival is hosted. Economic impact studies are commonly undertaken to justify public expenditure in festivals and events, however no research has been undertaken to demonstrate the role that investments in the development of creative capital across the creative industries make to a more broadly conceptualised "creative economy" embraced within the day to day mainstream economy rather than an annual demonstration of creative capacity.

6 Conceptualising a Creative Class: Useful or Not in Informing Regional Economic Developments Strategies?

In 2004, Richard Florida presented his "Creative Class" theory to a cross section of politicians' and policy makers in Adelaide. Florida argued that "regional economic growth is powered by creative people who prefer places that are diverse, tolerant and open to new ideas – a so called creative class". He asserts that the most successful places are those that combine the three "T′s" – tolerance, talent and technology (Florida 2003, p. 266).

Florida argues that the creative class comprises two elements: the Super Creative Core including scientists and engineers, university professors, poets and novelists, artists, entertainers, actors, designers and architects as well as the thought leadership of modern society: nonfiction writes, editors, cultural figures, think tank researchers, analysts and other opinion makers' (Florida 2003, p. 69). Beyond the core group

Florida defines the "creative professionals" who work in areas such as accounting, high tech sectors, legal and health care professionals and business management who all he argues utilise creative problem solving in their work. Attempting to break down the rigid typologies of academic disciplines and industries Florida's Creative Class theory argues that "creativity" rather than any single industry or policy intervention per se, is the fundamental source of economic growth. To this end he argues that:

> The best route to continued prosperity is by investing in our stock of creativity in all its forms, across the board. This entails more than just pumping R&D spending or improving education .... It requires increasing investments in the multidimensional and varied forms of creativity- arts, music, culture, design and related field – because all are linked and flourish together. (Florida 2003, p. 320)

As a result of Florida's presentations in Adelaide in 2004, there was heightened interest in understanding how his indices of creativity might be applied to enable South Australia to benchmark itself against other regions and set targets to improve its national rating. Gibson and Klocker observe this increased attention to creativity is dominated by a perception of its "power to transform the images and identities of places. This has constituted a 'cultural turn', of sorts, away from an emphasis on macro-scale projects and employment schemes, towards an interest in the creative industries, entrepreneurial culture and innovation" (Gibson and Klocker 2005, p. 93).

Since the initial interest in Florida's work, there is growing recognition that Florida's indices do not adequately reflect and measure the multidimensionality of the creative sector, nor do they take account of regional scale, politics, policies, assets and economic environment as key variables associated with the development of creative economies. Whilst Florida's indices relating to technology, talent and tolerance, such as the Bohemian Index, the Gay Index, the High Tech Index, the Melting Pot Index and the Creative Index enable regions to benchmark themselves against other regions, they do not provide enough detail for the development of endogenous policy that might assist in attracting or developing a regions "creative class". Nor does he explicitly identify the frameworks and processes that enable the translation of investments in human capital to create "the creative class" which in turn must be cultivated to produce the raw commodity "creative capital" to be utilised on an economy wide basis. In Australia, Gibson and Klocker (2005, p. 99) observe that these constructions of class have been used to label both individuals and geographical regions, and the resulting discourse contributes to what Howlett (2003, cited Gibson and Klocker 2005, p. 99) identifies as "a damaging conflation of socio-economic inequality and cultural attributes". Regardless of the debate surrounding the robustness of Florida's theory, his work has attracted the attention of policy makers and focused attention on a broader notion of creativity that transcends art and culture. This and narrower conceptualisations of creativity appear to co-exist in the South Australian creativity policy landscape. Much of the practical attention however appears to have been upon defining and quantifying creative industries.

7 The Creative Industries in South Australia

In 2003, the South Australian Government began to contemplate the changing nature of the local arts and cultural sectors, in particular how new technologies intersected with art forms, and the increasingly fuzzy boundaries between craft and design (designer-makers). A cross-government steering committee representing Arts SA, Department of the Premier and Cabinet (DPC), Department of Trade and Economic Development (DTED), and the Department of Further Education, Employment, Science and Technology (DFEEST) commissioned a study to map and measure the creative industries in South Australia. The objective of the study was to provide an assessment of the economic significance of the creative industries to South Australia (SA). The final report *The Creative Industries in South Australia* discusses the complexity of defining the creative industries, and states the approach to definition and measurement they adopted was the most pragmatic approach to take. The report authors qualify their remarks by suggesting the chosen grouping of creative sectors is "linked by the use of similar creative and artistic inputs and produce products and services that fall into the categories of entertainment, education and art" (Arts S.A, Department of Premier and Cabinet et al. 2005, p. 21). The Creative Industries in South Australia report, like similar studies in Australia identified the following sectors as constituting the creative industries:

- Audio-visual, media and digital media
- Advertising
- Craft, visual arts and indigenous arts
- Design (including architecture, fashion, and graphic, urban, industrial and interior design)
- Film and television
- Music
- Publishing
- Performing arts
- Cultural heritage/institutions

The report argues that the sectors with the greatest potential for growth are those that are based on digital technologies. Typically, these are companies that are near the definitional borderline between creative industries and information and communications technology (ICT). The report defines these as the "digital creatives" – or those companies exploiting creativity and technology to drive growth, high skill employment and exports. The diversity of business products and services in this sector ranges from software for mobile phone producers to clay animation. *The Creative Industries in South Australia* report concluded that at the time of writing "there are probably fewer than ten significant players" in the sector (Arts S.A, Department of Premier and Cabinet et al. 2005, p. 83) Together with approximately the same number of emerging businesses the report estimates that total employment in the sector as of 2004 was 480 people (ibid).

This more narrowly focused and pragmatic conception of the value of creativity has been followed by a more ambitious attempt to elevate creativity to a central objective of state policy in South Australia's Strategic Plan.

8 Creativity, Innovation and Regional Growth: The Agenda Broadens

In April 2003 the Economic Development Board (EDB), was formed by South Australia's Rann Labor government. The EDB was charged with the responsibility of guiding long-term economic development in the state. The EDB's report *A Framework for Economic Development in South Australia* (2003) contained 72 major recommendations for action by government, business and the community. Section 3 of the document outlined a broad framework and strategy, identifying key sectoral strengths as automotive, wine, water technology, food, defence, electronics tourism and the creative industries. Importantly the EDB recommended that South Australia develop a whole of government Strategic Plan based on the State of Oregon, US, Comprehensive Plan (1999).

In 2004, the first iteration of South Australia's Strategic Plan (SASP) was released and "Fostering Creativity" was included as a central policy objective. The SASP stated that:

> innovation and creativity provide South Australia's future capital for growth and expansion. The Government recognises its role in providing the right environment for these attributes to flourish in sectors ranging from the arts to manufacturing, and its ability to provide a lead for the rest of the community. Our capacity to do things differently will be one of the keys to achieving all of our objectives. (Department of the Premier and Cabinet 2004, p. 3)

The "Fostering Creativity" objective identified creativity as a key to future prosperity, arguing that South Australia had a long history of creativity, which was exemplified by South Australia's Nobel Prize winners, award winning South Australian filmmakers and innovative manufacturing. In introducing the Fostering Creativity Objective the SASP proclaimed that:

> Our priority is to reinforce South Australia as a place that thrives on creativity and innovation. This capacity to do things differently will be one of the keys to achieving all of our objectives. The focus will be on fostering a culture of creativity, on developing creative, innovative and enterprising people, on investing in science and research, and in innovation infrastructure, and on converting ideas into practice. (Department of the Premier and Cabinet 2004, p. 39)

The Key Targets of the 2004 *Fostering Creativity* objective were:

- Achieve a ranking in the top three of Richard Florida's Creativity index within 10 years
- Increase patent applications to exceed our population share of all Australian applications within 5 years

- Significantly grow and expand South Australia's share of the national feature film industry to match our population share.

Investment in Science, Research and Innovation

- Exceed the national average of business expenditure on research and development (SA a percentage of GSP) and approach the OECD within 10 years

Providing Support Infrastructure

- Increase the level of internet use in metropolitan and regional South Australia by 20% within 10 years

Developing Creative and Innovative People

- Improve learning outcomes in the arts and other curriculum areas that utilise enterprise education
- Improve the connections between educational institutions and industry to enhance creativity and innovation
- Increase the number of families participating in the Learning Together and school-community arts and recreation programs

The 2007 edition of the SASP involved substantial changes to key targets. Notably these included the omission of a target to achieve a ranking in the top three nationally in Richard Florida's "Creativity index" within 10 years. More broadly the "Fostering Creativity" objective became "Fostering Creativity and Innovation", reflecting an important shift in thinking about the relationship between the two concepts. This did not appear to translate into the presentation of targets in the plan which separated the innovation targets from the creativity targets with the latter focusing heavily on science and technology R&D.

The targets for Creativity in the 2007 SASP included:

- TARGET 4.1: Creative industries: increase the number of South Australians undertaking work in the creative industries by 20% by 2014.
- TARGET 4.2: Film industry: double the number of feature films produced in South Australia by 2014.
- TARGET 4.3: Cultural engagement – institutions: increase the number of attendances at South Australia's cultural institutions by 20% by 2014.
- TARGET 4.4: Cultural engagement – arts activities: increase the number of attendances at selected arts activities by 40% by 2014.
- TARGET 4.5: Understanding of Aboriginal culture: Aboriginal cultural studies included in school curriculum by 2014 with involvement of Aboriginal people in design and delivery.

Valuable though these targets may be they reflect the narrow but historically dominant conceptualisation of creativity centring on the arts and cultural sector. The wider objective of "Fostering Creativity" in the SASP appears to have been reduced through these targets to the more pragmatic goal of fostering business development at the interface of design and technology. This less ambitious project is not the end of the story however. More recent developments suggest that a broader conception of creativity may be emerging. The establishment in late 2009 of an "Integrated Design Commission" by the State Government represented a significant broadening of the creativity agenda, engaging a range of professions and agencies in the future development of South Australia's urban environment. This is perhaps the most tangible manifestation of the more ambitious underlying purpose of the "Fostering Creativity" objective in South Australia's Strategic Plan.

9 Conclusion

Concepts like creativity are more rather than less likely to preoccupy researchers and policymakers over decades to come as the search for answers to multi-dimensional challenges like climate change and population ageing demands that we become more creative. This in itself is a challenge. Brewer (1999, p. 328) observed that "the world has problems, universities have departments". This might also be true of governments where agencies compete with each other and often struggle to interpret and operationalise complicated concepts like creativity. The narrow focus on creative industries that emerged in South Australia reflects the dominant player in the creativity agenda within government – the Department of Trade and Economic Development which approached the challenge through a commercialisation and export orientation lens. Until recently other government agencies have been relatively peripheral to the "Fostering Creativity" objective.

The challenge ahead in South Australia is to build on attempts to operationalise a broad conception of creativity, one that moves beyond creative quick fixes and the digital creatives to one that is holistic, multi-sectoral, multi-disciplinary and multi-dimensional. It is likely that attempts in South Australia to engage with the concept of creativity will resonate with other regions seeking to acknowledge the significance of creativity in our social and economic lives. Translating a broad conception of creativity into an integrated strategic whole of government response is about as difficult as policy challenges can get. Moving beyond creative quick fixes will require both the elevation of creativity as an overarching objective of strategic policy within government and processes that enable it to be operationalised in a systematic and integrated way. Some of the foundations appear to have been laid in South Australia.

References

Adelaide Festival of Arts (2008) "Festival facts." Retrieved 28 June 28 2008, from http://www.adelaidefestival.com.au/Content.aspx?p=62

Arts S.A, Department of the Premier and Cabinet et al (2005) The creative industries in South Australia, Cross-Government Steering Committee, Adelaide

Bilton C (2007) Management and creativity - from creative industries to creative management, Blackwell
Brewer G (1999) "The challenges of interdisciplinarity." Policy Sciences (32):327–337
Caves RE (2000) "Contracts between art and commerce" The Journal of Economic Perspectives 17(2):73–83(11)
Cooke P, Memedovic O (2006) Regional innovation systems as public goods. United Nations Industrial Development Organization, Vienna
Csikszentmihalyi M (1996) Creativity: flow and the psychology of discovery and invention. Harper Collins, New York
Department of Culture Media and Sport (DCMS) (2001). Creative industries mapping document. London, Department for Culture Media and Sport
Department of the Premier and Cabinet (2004). South australian strategic plan: Creating opportunity. Department of the Premier and Cabinet. Adelaide, Government of South Australia
Department of Communications Information Technology and the Arts (DOCITA) and The National Office for the Information Economy (NOIE) (2002). Creative industries cluster study- stage one report, DOCITA
Economic Development Board (2003) A framework for economic development in South Australia. Government of South Australia, Adelaide
Florida R (2003) The rise of the creative class: And how it's transforming work, leisure, community and every day life. Melbourne, Pluto Press
Gibson C, Klocker N (2005) The 'cultural turn' in Australian regional economic development discourse: neoliberalising creativity? Geogr Res 43(1):93–102
Howlett M and Ramesh M (2003) Studying public policy: Policy cycles and policy subsystems, Oxford University Press, Canada
Landry C (2003) Rethinking Adelaide. 'Capturing Imagination'. Department of Premier and Cabinet, South Australian Government, Adelaide, 51
Landry C, Bianchini F (1994) The creative city: Indicators of a creative city a methodology for assessing urban viability and vitality
Landry C (2000) The creative city: A tool kit for urban innovators. London, Earthscan
Leadbetter C (2000) Living on thin air: The new economy; with a new blueprint for the 21st century. London, The Penguin Group
Porter M (2002) Clusters and the new economics of competition. Harv Bus Rev 76:77–90
Potts J (2007) Innovation: an evolutionary economic view of the creative industries, UNESCO observatory, Faculty of Architecture, Building and Planning, The University of Melbourne Refereed E-Journal
Prime Minister's Science Engineering and Innovation Council Working Group and (PMSEIC) (2005). Imagine Australia: The role of creativity in the innovation economy, PMSEIC
Reich R (2001) The future of success: Working and living in the new economy. New York, Vintage
Tepper (2002) "Creative assets and the changing economy" The Journal of Arts Management, Law and Society 32(2):159–168
Throsby D (2000) Economics and culture. Cambridge University Press, New York
van den Steenhoven J, Stikker M et al (2005) Final report creative capital conference. Creative Capital conference, Amsterdam, www.creativecommons.nl